Expertise

ENGAGING PHILOSOPHY

This series is a new forum for collective philosophical engagement
with controversial issues in contemporary society.

Disability in Practice
Attitudes, Policies, and Relationships
Edited by Adam Cureton and Thomas E. Hill, Jr.

Taxation
Philosophical Perspectives
Edited by Martin O'Neill and Shepley Orr

Bad Words
Philosophical Perspectives on Slurs
Edited by David Sosa

Academic Freedom
Edited by Jennifer Lackey

Lying
Language, Knowledge, Ethics, and Politics
Edited by Eliot Michaelson and Andreas Stokke

Treatment for Crime
Philosophical Essays on Neurointerventions in Criminal Justice
Edited by David Birks and Thomas Douglas

Games, Sports, and Play
Philosophical Essays
Edited by Thomas Hurka

Effective Altruism
Philosophical Issues
Edited by Hilary Greaves and Theron Pummer

Philosophy and Climate Change
Edited by Mark Budolfson, Tristram McPherson, and David Plunkett

Applied Epistemology
Edited by Jennifer Lackey

The Epistemology of Fake News
Edited by Sven Bernecker, Amy K. Flowerree, and Thomas Grundmann

The Ethics of Social Roles
Edited by Alex Barber and Sean Cordell

The Ethics of Surveillance in Times of Emergency
Edited by Kevin Macnish and Adam Henschke

Expertise

Philosophical Perspectives

Edited by

MIRKO FARINA, ANDREA LAVAZZA,
AND DUNCAN PRITCHARD

OXFORD
UNIVERSITY PRESS

Great Clarendon Street, Oxford, OX2 6DP,
United Kingdom

Oxford University Press is a department of the University of Oxford.
It furthers the University's objective of excellence in research, scholarship,
and education by publishing worldwide. Oxford is a registered trade mark of
Oxford University Press in the UK and in certain other countries

© the several contributors 2024

The moral rights of the authors have been asserted

All rights reserved. No part of this publication may be reproduced, stored in
a retrieval system, or transmitted, in any form or by any means, without the
prior permission in writing of Oxford University Press, or as expressly permitted
by law, by licence or under terms agreed with the appropriate reprographics
rights organization. Enquiries concerning reproduction outside the scope of the
above should be sent to the Rights Department, Oxford University Press, at the
address above

You must not circulate this work in any other form
and you must impose this same condition on any acquirer

Published in the United States of America by Oxford University Press
198 Madison Avenue, New York, NY 10016, United States of America

British Library Cataloguing in Publication Data
Data available

Library of Congress Control Number: 2023951626

ISBN 978–0–19–887730–1

DOI: 10.1093/oso/9780198877301.001.0001

Printed and bound by
CPI Group (UK) Ltd, Croydon, CR0 4YY

Links to third party websites are provided by Oxford in good faith and
for information only. Oxford disclaims any responsibility for the materials
contained in any third party website referenced in this work.

Contents

Contributors   vii

PART 1 INTRODUCTION

1. Expertise   3
   Mirko Farina, Andrea Lavazza, and Duncan Pritchard

PART 2 EXPERTISE AND TRUST

2. The Problem of Pseudoscience: Who Should We Trust When It Comes to Fringe Beliefs?   13
   Massimo Pigliucci

3. Authority, Legitimacy, and the Expert-Layman Problem   31
   Gloria Origgi

4. Institutionalized Expertise: Trust, Rejection, and Ignorance   44
   Daniel R. DeNicola

PART 3 SITUATED AND GROUP EXPERTISE

5. Expertise as Perspectives in Dialogue   65
   Michael Larkin, Lisa Bortolotti, and Michele Lim

6. Affective, Cognitive, and Ecological Components of Joint Expertise in Collaborative Embodied Skills   85
   John Sutton

PART 4 EXPERTISE AND PUBLIC POLICY

7. Expert Judgement without Values: Credences not Inductive Risks   107
   Rivkah Hatchwell and David Papineau

8. From the Right to Science as an Epistemic-Cultural Human Right to the Right to Expertise   126
   Michela Massimi

9. Studies of Expertise and Experience: Demarcating and Defending the Role of Science in Democracy   147
   Harry Collins and Robert Evans

PART 5 EXPERTISE AND VIRTUE

10. Humility for Experts 171
    Linda Zagzebski

11. Expertise-in-Action: The Importance of Intellectual and Moral Virtue(s) to Experts' Epistemic Authority 189
    Andrea Lavazza, James Giordano, and Mirko Farina

PART 6 EXPERTISE ABOUT VALUE

12. Experts in Aesthetic Value Practices 213
    Dominic McIver Lopes

13. Moral Expertise and Socratic AI 232
    Emma C. Gordon

PART 7 NEW DIRECTIONS

14. Decolonising Experts 253
    Veli Mitova

15. Public Expertise and Ignorance 274
    Duncan Pritchard

Index 287

Contributors

Lisa Bortolotti, University of Birmingham, UK

Harry Collins, Cardiff University, UK

Daniel R. DeNicola, Gettysburg College, USA

Robert Evans, Cardiff University, UK

Mirko Farina, Institute for Digital Economy and Artificial Systems, Xiamen University, China and Lomonosov Moscow State University, Russian Federation

James Giordano, Georgetown University, Washington DC, USA

Emma C. Gordon, University of Glasgow, UK

Rivkah Hatchwell, King's College London, UK

Michael Larkin, Aston University, UK

Andrea Lavazza, University of Milan; Centro Universitario Internazionale, Arezzo, Italy

Michele Lim, University College London, UK

Michela Massimi, University of Edinburgh, UK

Dominic McIver Lopes, University of British Columbia, Canada

Veli Mitova, University of Johannesburg, South Africa

Gloria Origgi, Institut Jean Nicod, Paris, France

David Papineau, King's College London, UK; City University of New York, USA

Massimo Pigliucci, City College of New York, USA

Duncan Pritchard, University of California, Irvine, USA

John Sutton, Macquarie University, Australia; University of Stirling, UK

Linda Zagzebski, University of Oklahoma, USA

PART 1
INTRODUCTION

1
Expertise

Mirko Farina
Institute for Digital Economy and Artificial Systems,
Xiamen University and Lomonosov Moscow State University

Andrea Lavazza
University of Milan; Centro Universitario Internazionale, Arezzo, Italy

Duncan Pritchard
University of California, Irvine, USA

1. Expertise in Contemporary Debate

Experts and expertise are the natural outcome of the division of intellectual (and other) labour. As the complexity of tasks has increased, along with the level of skill and experience required to perform them, it is inevitable that specialization, and thus the development of expertise, has been the result. Accordingly, we increasingly defer to experts in our existence both for knowledge as such and to navigate the intricacies of life in complex, high-tech societies (Holst and Molander 2019; Quast 2018; Singer 2006; Selinger and Crease 2006; Williamson 2011).

The task of philosophy has been to clarify (and this is not a paradox) with its expertise what the criteria are for demarcating the expert in a primarily epistemic sense. This challenge has a special contemporary resonance due to the recent COVID pandemic, which has brought the public role of experts into an especially sharp relief, as experts have urged various strategies to deal with the health crisis (including offering sometimes inconsistent advice, Farina and Lavazza 2021a; Farina and Lavazza 2021b; Lavazza and Farina 2020; Pietrini et al. 2022). This high-profile manifestation of expertise in action has also led to prominent forms of pushback against expertise (or, at least, the use of this expertise within public policy) and to related phenomena such as science denialism (Oliver and Rahn 2016; Pappas 2019; Lukacs 2005). While this is the most conspicuous example of this pattern of public expertise and popular pushback, the pattern continues for a range of issues of public importance, from climate change to tax policy, from the courtroom to sport (think about the use of Video Assistant Referee (VAR) in football matches). In all such contexts, the extent to which experts can be trusted is a subject of persistent and contentious debate (Collins and Evans 2019; Watson 2021).

The reasons why experts are challenged in some areas of public life often has to do with political or ideological orientations. This is exacerbated by the new modes of public communication that are based around social media. These favour short, assertive, and emotional messages, where there is less of a role for objective data and analytical arguments. The result is that all kinds of expertise have come under attack, even scientific expertise, which risks tilting towards forms of irrationalism.

Expertise is therefore a topic that is of tremendous contemporary significance, raising questions that stretch right across the theoretical spectrum (Farina and Lavazza, 2022; Farina and Lavazza, 2024). What is an expert? Who decides who the experts are? Should we always defer to experts? How should expertise inform public policy? What happens when the experts disagree? Must experts be unbiased? Should all experts be treated the same, or does it matter what the source of the expertise is? How should the testimony of experts be reported by the media?

One distinctive feature of the contemporary philosophical debate about expertise is its strong interdisciplinary flavour, with major contributions by sociologists (see e.g., Collins and Evans 2002) and psychologists (see e.g., Ericsson 2006). In addition, thanks to the development of social epistemology (Goldman 2001; Haddock, Millar, and Pritchard 2010; Fuller 2012), and to the integration of research between sociology, political theory, and philosophy of science, there has been the development of new frameworks for such questions as how to define the role of experts in society, how laypeople can critique experts, as well as the social and ideological character of expert advice and its incorporation into governmental bureaucracy and policy making (Nicols 2018). As often occurs where there are different disciplinary perspectives in play, however, one ends up with divergent answers to some of the fundamental questions that concern us. It is part of the task of philosophy to find ways to adjudicate between these divergent proposals. This volume, in building and expanding on previous work in the field, sets out to achieve this by bringing together a range of contemporary figures working on expertise, from multiple perspectives, to explore the central debates about expertise.

2. Overview of the Chapters

The volume is arranged in six parts, each corresponding to a fundamental topic in the contemporary philosophy of expertise. Part 1 is devoted to the role of trust in evaluations of expertise. Part 2 explores the specifically social and situated nature of expertise. Part 3 is devoted to the question of how expertise informs public policy. Part 4 examines the role of the virtues in the nature of expertise. Part 5 looks at the specific question of expertise in the context of value, such as aesthetic and ethical expertise. Finally, Part 6 considers some new directions in the philosophy of expertise.

Part 1 (Expertise and Trust) consists of three chapters. In 'The Problem of Pseudoscience: Who Should We Trust When it Comes to Fringe Beliefs?',

Massimo Pigliucci explores the epistemology of expertise in the context of the demarcation problem from philosophy of science—essentially, how are we to demarcate genuine science from non-science? The analogue challenge in the context of expertise arises when we consider 'fringe' beliefs that run counter to the status quo. Should we make space for genuine expertise that advocates for such fringe beliefs, and if so, how do we go about doing that? Pigliucci is particularly concerned to show how this issue is not a merely theoretical challenge, as it concerns a number of highly important social questions, including some that have existential significance.

In 'Authority, Legitimacy, and the Expert-Layman Problem', **Gloria Origgi** starts from the fact that although expertise is ubiquitous in our techno-scientific societies, we rarely are in direct contact with the experts. It follows that we need to assess the legitimacy of experts from indirect cues. The author explores how the public navigates the social information that surrounds an expert to come up with an evaluation of her reputation and her role in society. She notes that what signals most the reputation of an expert is the reputation of those who recommend the expert. Accordingly, one needs to consider a series of biases when one comes to trust those who recommend an expert.

In 'Institutionalized Expertise: Trust, Rejection, and Ignorance', **Dan DeNicola** navigates both the significant alterations in contemporary professional expertise and the important sources of current populist suspicion regarding expertise. Key to his discussion is a focus on 'open-loop' professions—those in which experts have a practice that involves advising clients. The author argues that systemic changes in institutionalized expertise may fuel the anti-expertise sentiments. He sketches ways in which democratic equalitarianism and expertise may both be respected and concludes with a discussion of the responsibilities that both experts and clients have for beneficial interactions.

Part 2 (Situated and Group Expertise) consists of two chapters. In 'Expertise as Perspectives in Dialogue', **Lisa Bortolotti, Michael Larkin,** and **Michele Lim** draw on mental health research to examine the notion of gaining expertise through experience and to consider some of the objections commonly raised against its legitimacy. They argue that the best way to characterize the integration of different forms of expertise is to describe the process as a case of perspectives in dialogue. In this sense, a perspective is a way of referring to how something appears from a particular standpoint, which acknowledges the relevance of that standpoint to what is foregrounded.

In 'Affective, Cognitive, and Ecological Components of Joint Expertise in Collaborative Embodied Skills', **John Sutton** analyses the ecological and situational components underlying expert joint action in practice in domains such as art and sports. Sutton distinguishes and summarizes some key cognitive affective resources that he claims are crucial to shaping the cognitive and affective components of joint expertise. He further considers how these resources may ground successful performance.

Part 3 (Expertise and Public Policy) consists of three charters. In 'Expert Judgement Without Values: Credences not Inductive Risks', **Rivkah Hatchwell** and **David Papineau** analyze the ideal of a value-free science and the consequent claim that it is inevitable and proper that scientific findings should dictate evaluative responses. They maintain that it does not follow from this ideal that scientists are well-placed to take responsibility for policy decisions. In the real world, political judgement requires not only identifying the good and bad upshots of potential actions, but also weighing these consequences against each other. They argue that this is a responsibility for democratically answerable elected politicians, not for members of a scientific elite.

In 'From the Right to Science as an Epistemic-Cultural Human Right to the Right to Expertise', **Michela Massimi** discusses the 'right to science' originally recognized by the United Nations (UN) Declaration of Human Rights. She aims to give the philosophical arguments for construing this right to science as an epistemic-cultural human right (a right concerning scientific knowledge), and to articulate a notion of expertise that can do the philosophical 'heavy-lifting' for interpreting the right to science along these lines. She concludes by claiming that a right to expertise exists that is a natural supplement to the right to science.

In 'Studies of Expertise and Experience: Demarcating and Defending the Role of Science in Democracy', **Harry Collins** and **Robert Evans** take expertise to be a collective achievement that individuals acquire via their socialization into their groups. They argue that science is endangered by basing its use in political decision-making on its infallibility since the public is bound to be disappointed by failures when science is confronted with the complex demands of real-world problems. But giving too little credit to science leads to a dissolution of the boundary between expert and non-expert. In response, the authors argue for the importance of keeping science distinguishable from politics.

Part 4 (Expertise and Virtue) consists of two chapters. In 'Humility for Experts', **Linda Zagzebski** explores what the virtue of humility looks like when applied to someone who is an expert in their field. Acknowledgements of one's limits, she claims, is a part of humility even when one's limits are not as great as those of the non-expert. The interaction between experts and their peers differs from the interaction between an expert and the public. Taking responsibility for mistakes and miscommunication is an aspect of humility. Zagzebski thus analyses how humble experts relates to themselves, to their peers, and to the public, and the vices that impede their humility.

In 'Expertise-in-Action: The Importance of Intellectual and Moral Virtue(s) to Experts' Epistemic Authority', **Andrea Lavazza** and **Mirko Farina** draw on Zagzebski's account of expertise to advocate for the development of a new notion of expertise, which they call 'expertise in action'. Expertise in action mandates: (1)

explicit and specific epistemic and responsibilist expertise in informing public decisions, as well as (2) relative transparency about the extent, role, and contribution of any such expertise as well as the adherence to a basic set of virtues or character traits (such as benevolence, fidelity to trust, veracity, intellectual humility, pathos, fortitude, and practical wisdom).

Part 5 (Expertise About Value) consists of two chapters. In 'Experts in Aesthetic Value Practices', **Dominic McIver Lopes** describes three different ways of appealing to expertise regarding aesthetic value. Against the backdrop of traditional approaches that suggest that aesthetic experts either provide authoritative access to aesthetic value or they constitute it, the author offers a new framework for understanding aesthetic values. In this new framework aesthetic values are relative to social practices and expertise determines which practices are the ones that we belong to. Aesthetic experts, therefore, are neither guides to aesthetic value nor those whose responses constitute it.

In 'Moral Expertise and Socratic AI', **Emma Gordon** connects traditional work in social epistemology with contemporary research in bioethics (on moral enhancement) and artificial intelligence (regarding 'Socratic AI,' an Artificial Intelligence (AI) assistant that engages in Socratic dialogue with users to non-prescriptively assist in ethical reasoning). The aim of this interdisciplinary chapter is to investigate whether Socratic-AI assisted moral enhancement might be compatible with manifesting genuine moral expertise, and how such a capacity to improve moral reasoning might influence our criteria for identifying moral experts.

The final part of the volume, Part 6 (New Directions), consists of two chapters. In 'Decolonising Experts', **Veli Mitova** argues that the experts from marginalized groups—for instance, the traditional healers of Southern Africa— are genuine experts, on epistemic and moral grounds. Since existing accounts of expertise cannot accommodate the claim that marginalized experts are bona fide experts, she develops an alternative view of expertise that she terms 'communitarian functionalism'. On this proposal, a person counts as an expert in a domain in virtue of her role in her epistemic community, the needs of this community regarding that domain, and whether she responsibly lives up to this role.

In 'Public Expertise and Ignorance', **Duncan Pritchard** argues from a normative account of ignorance to the claim that one important role of public expertise is not just to combat ignorance but also in a sense to manufacture it. What he terms 'informative public expertise' functions to create the social epistemic conditions under which a body of information is such that the public ought to be aware of it (and hence they count as ignorant of this information if they continue to be unaware of it). Pritchard also discusses a second form of public expertise—'critical public expertise'—that serves an explicitly critical social function by revealing hidden ignorance.

References

Collins, H. and Evans, R. (2019) *Rethinking Expertise*. Chicago: University of Chicago Press.

Collins, H. M. and Evans, R. (2002) "The third wave of science studies: Studies of expertise and experience." *Social Studies of Science* 32(2), 235–96.

Ericsson, K. A. (2006) "The influence of experience and deliberate practice on the development of superior expert performance," in *The Cambridge Handbook of Expertise and Expert Performance*, (eds.), K. A. Ericsson, N. Charness, P. J. Feltovich, and R. R. Hoffman, 683–703. Cambridge: Cambridge University Press.

Farina, M. and Lavazza, A. (2021a) "Advocating for greater inclusion of marginalized and forgotten populations in COVID19 vaccine rollouts." *International Journal of Public Health* 66, 1,604,036.

Farina, M., Lavazza, A. (2021b) 'The Meaning of Freedom after Covid-19'. History and Philosophy of the Life Sciences, 43, 3, doi: https://doi.org/10.1007/s40656-020-00354-7.

Farina, M., Lavazza, A. (2024) Philosophy, Expertise and The Myth of Neutrality. London, UK: Routledge.

Fuller, S. (2012) "Social epistemology: A quarter-century itinerary." *Social Epistemology* 26(3–4), 267–83.

Goldman, A. I. (2001) "Experts: Which ones should you trust?." *Philosophy and Phenomenological Research* 63(1), 85–110.

Haddock, A., Millar, A., and Pritchard, D. (2010) *Social Epistemology*. Oxford, UK: Oxford University Press.

Holst, C. and Molander, A. (2019) "Epistemic democracy and the role of experts." *Contemporary Political Theory* 18, 541–561.

Lavazza, A. and Farina, M. (2020) "The role of experts in the Covid-19 pandemic and the limits of their epistemic authority in democracy." *Frontiers in Public Health* 8, 356.

Lavazza, A. and Farina, M. (2022) "Experts, naturalism, and democracy." *Journal for the Theory of Social Behaviour* 52(2), 279–97.

Lukacs, J. (2005) *Democracy and Populism: Fear & Hatred*. New Haven, CT: Yale University Press.

Nichols, T. (2018) *The Death of Expertise: The Campaign against Established Knowledge and Why It Matters*. Oxford: Oxford University Press.

Oliver, J. E. and Rahn, W. M. (2016) "Rise of the Trumpenvolk: Populism in the 2016 election." *The ANNALS of the American Academy of Political and Social Science* 667(1), 189–206.

Pappas, T. S. (2019) *Populism and Liberal Democracy: A Comparative and Theoretical Analysis*. Oxford: Oxford University Press.

Pietrini, P., Lavazza, A., and Farina, M. (2022) "COVID-19 and biomedical experts: When epistemic authority is (probably) not enough." *Journal of Bioethical Inquiry* 19(1), 135–142.

Quast, C. (2018) "Expertise: A practical explication." *Topoi* 37, 11–27.

Selinger, E. and Crease, R. P. (eds.) (2006) *The Philosophy of Expertise*. New York: Columbia University Press.

Singer, P. (2006) "Moral experts." *The Philosophy of Expertise, New York* 187, 189.

Watson, J. C. (2021) *A History and Philosophy of Expertise: The Nature and Limits of Authority*. London: Bloomsbury.

Williamson, T. (2011) "Philosophical expertise and the burden of proof." *Metaphilosophy* 42(3), 215–29.

PART 2
EXPERTISE AND TRUST

2
The Problem of Pseudoscience
Who Should We Trust When It Comes to Fringe Beliefs?

Massimo Pigliucci
Department of Philosophy, the City College of New York

1. A Problem as Old as Socrates

Why should you trust anything you read in this book? Certainly not because it is printed black on white, as there are plenty of books out there that are not worth the paper they are printed on (or the pixels they use, in the case of e-books, see Plato 1985).[1] Nor because you will find the text peppered with scholarly-looking notes, since those are just as easy to come by in books and articles that promote pseudoscience. It isn't enough that the contributors to this volume have a PhD after their names, since a number of charlatans do as well, unfortunately. Moreover, plenty of reasonable, informative, and insightful books are written by people with no academic degrees at all. And whatever our credentials here, you will likely not have time to fact-check every assertion we make in the book, or to read it carefully enough to analyze the fine structure of the arguments being presented, searching for logical fallacies or weak links.

This problem, arguably a major issue underlying the debates about science and pseudoscience in the public arena, boils down to questions of expertise and trust. What qualifies someone as an expert? And what degree of trust should we put in what that person says about a given topic? We are by now used to the notion prevalent in the media that whenever a controversial issue is discussed, we should abstain from taking sides until we have had time to consider a "balanced" treatment, ideally giving equal time to "both sides." But often there are more than two sides, and not infrequently one side is downright silly or plainly wrong, despite the fact that it has marshaled "experts" to defend its positions. What is the average intelligent citizen to do?

---

[1] Parts of this essay are adapted and updated from chapter 12 of my *Nonsense on Stilts*, University of Chicago Press, 2nd edition, 2018.

Massimo Pigliucci, *The Problem of Pseudoscience: Who Should We Trust When It Comes to Fringe Beliefs?*
In: *Expertise: Philosophical Perspectives*. Edited by: Mirko Farina, Andrea Lavazza, and Duncan Pritchard, Oxford University Press. © Massimo Pigliucci 2024. DOI: 10.1093/oso/9780198877301.003.0002

The problem of expertise on technical and scientific matters is not a new one. Plato tackles it in one of his Socratic dialogues, the *Charmides*, in which Socrates is engaged in a discussion of what constitutes temperance. The *Charmides* is a strange early Platonic dialogue, and scholars are not even sure what the subject matter really is. Nonetheless, there is a small section of it that is pertinent to our inquiry, in which Socrates takes a comment by one of his interlocutors as a reason to explore the relation between wisdom and medicine. This in turn leads him to ask how "a wise man" might distinguish good from bad medicine, a doctor from a quack. As it is often the case, this is not really a dialogue so much as Socrates—who does most of the talking—pursuing a point with the assistance of a willing interlocutor who simply nods along the way:

SOCRATES: Let us consider the matter in this way. If the wise man or any other man wants to distinguish the true physician from the false, how will he proceed?... He who would inquire into the nature of medicine must test it in health and disease, which are the sphere of medicine, and not in what is extraneous and is not its sphere?
OTHER GUY: True.
SOCRATES: And he who wishes to make a fair test of the physician as a physician will test him in what relates to these?
OTHER GUY: He will.
SOCRATES: He will consider whether what he says is true, and whether what he does is right, in relation to health and disease?
OTHER GUY: He will.
SOCRATES: But can anyone pursue the inquiry into either unless he has a knowledge of medicine?
OTHER GUY: He cannot.
SOCRATES: No one at all, it would seem, except the physician can have this knowledge—and therefore not the wise man. He would have to be a physician as well as a wise man.
OTHER GUY: Very true.

That, in a nutshell, is the problem of expertise (to which Socrates, by the way, did not have a general solution): it would seem that the only way to assess whether an expert is in fact an expert is to be an expert oneself. On the positive side, this is what justifies peer reviews in science: if I wish to write a paper, say, on quantum mechanics or evolutionary theory, I need to submit it to the (usually anonymous) review of several scientists whose specialty is quantum mechanics or evolutionary theory. The idea is not that experts are infallible, but rather that they are the best (often the only) people qualified to make a judgment on technical matters.

On the negative side, it would seem that if I think (and I do) that Deepak Chopra talks nonsense when he tells people about quantum mechanical elixirs of youth, I would first have to become an expert in quantum mysticism. But the problem is that quantum mysticism is (I think) quackery, and that therefore there is no such thing as "expertise" in quantum mysticism. This despite the fact that quantum mechanics is one of the best-established scientific theories of all time (then again, I'm not an expert in quantum mechanics either). So, once more, how is the average intelligent person (or Socrates's "wise man," for that matter) supposed to distinguish between science and pseudoscience without becoming an expert in both? Moreover, how is it even possible to become an expert in nonsense?

In what follows I cannot possibly articulate a full treatment of these questions, nor do I have the space to review the extensive pertinent literature in epistemology (see e.g., Hikins and Cherwitz 2011; Origgi 2015; Quast 2018; Kilov 2020) and the philosophy of pseudoscience (see Pigliucci and Boudry 2013). What I will do instead is to sketch an approach based on three interrelated topics. First, I will discuss what makes an expert. I will tackle the issue from an empirically informed perspective in lieu of an always tricky and often not very useful formal definition of expertise, which is the more common approach in the philosophical literature. Second, I will summarize the classic treatment of trust and expertise due to Alvin Goldman, which I and many others consider the most practically useful available to date, despite Goldman's own further evolution in his thinking (Goldman 2018) and a number of criticisms the approach has received (Quast 2018; Croce 2019; Grundmann 2021). Thirdly, I will raise the special issue of bullshit, *sensu* (Frankfurt 2005), as recently articulated by Victor Moberger (2020) and affecting the topic at hand. My hope, again, is not to be exhaustive, but rather to be useful, not just to colleagues interested in the technical aspects of the issues being discussed but especially to the person in the street, so to speak. After all, it is the latter who is often confronted with dubious claims coming from self-professed experts, the acceptance or rejection of which may have dire consequences for one's health, finances, and general wellbeing.

2. How Does One Become an Expert?

As it turns out, much scientific research has been done on expertise, so in some sense we can say that there are experts on expertise (Boshuizen, Gruber, and Strasser 2020; Tynjälä, Kallio, Heikkinen 2020). It may be instructive to take a look at what they have found. For instance, Anders Ericcson and Jacqui Smith begin their discussion of expertise with two conclusions from their work: first, a single success, however spectacular, does not make one an expert. Second, being an expert in a given field has little to do with one's general abilities to retain information (Ericsson and Smith 1991).

The first point is not exactly surprising: someone can pull off a once-in-a-lifetime success, like penning a best-selling novel or writing a song that makes it to the top of the charts, without necessarily being an expert in how to write best-selling books or songs. That said, few people win the Nobel Prize more than once, yet it is hard to conceive of an amateur achieving such an honor in today's sophisticated and specialized world, regardless of whether the discipline in which the Nobel is bestowed is a scientific one (where the expectation of expertise is almost a given) or not.

The point about memory, however, is a bit more subtle: while a good general memory certainly will not hamper anyone, studies have shown that experts in a field do not have better general memory than novices; rather, the difference is in the way the expert structures her memories of the pertinent knowledge. In this sense, experts seem to work like the appropriately named "expert systems" in computer science (Liebowitz 2019). Expert systems are not just databases of information that can be retrieved by brute force. Rather, the software focuses only on a particular domain (of expertise) and gains its power from the fact that the programmers organize that special knowledge in ways that reflect our best understanding of whatever the specific field happens to be, whether medicine, law, engineering, or something else. It will be interesting, of course, to see how expert systems will be improved or even replaced by artificial intelligence (AI), but that's a discussion for another occasion.

According to the research summarized by Ericcson and Smith, it takes time, not surprisingly, to become an expert. Studies of chess have become particularly useful in understanding expertise, since the game is characterized by a small number of rules with an open-ended universe of possible outcomes, making it a good model system for social scientists. Most studies on chess distinguish novices from expert players and the latter from masters. The results show that it takes about three thousand hours of playing experience to become an expert, but about ten times as much to qualify as a master.

Interestingly, the rule of thumb that a decade or so is the time frame that allows one to become an expert in a field applies not only to chess playing, but also to science and the arts. A typical doctoral degree in the sciences takes four to six years in the United States, to which one has to add four years of college, and increasingly a number of years as a postdoctoral associate. It is not unusual to begin a tenure-track academic job after one has already spent a decade learning about specialized aspects of one's field. But expertise is about seeing patterns, not sheer amount of knowledge. In the case of chess, according to Ericcson and Smith, and contrary to popular belief, masters don't actually calculate a much larger number of moves ahead when compared to novices; instead, their experience allows them to identify patterns on the chessboard that suggest which move(s) will be most profitable, given the current state of the system.

Based on research conducted over the span of decades, Ericcson and Smith propose a three-stage process for the acquisition of expertise: at the initial, *cognitive stage*, a person focuses on understanding what the task is, trying to discriminate between relevant and irrelevant information. This is the situation for an amateur chess player or a beginning undergraduate student. The second, *associative stage* is characterized by quantitative improvements in the ability to retrieve information, mostly because such information is filed in a structured way, with associations being noticed and built among disparate pieces of information. A chess expert (but not yet a master) or a graduate student may be at this level. Finally, one reaches the *autonomous stage*, where performance becomes largely automatic, with little actual conscious thought going into most of the tasks at hand. A chess master or a professional academic is at this level. Speaking from personal experience, one develops a "feeling" for what happens, an intuitive sense of what is relevant information and what constitutes background noise, or for what will likely turn out to be a good approach to a problem, as opposed to a waste of time. Of course, the expert can reflect on her intuitions and articulate reasons for what she is doing, but the answer comes "in a flash," as they say, not as the result of hours of explicit logical pondering (Segalowitz 2007).

As we saw previously, questioning what constitutes expertise began (as far as we know) with Plato's inquiry into the difference between a doctor and a quack. It is appropriate, therefore, to look at what Barbara Daley has found out about expertise in the medical profession, with particular respect to nurses (Moberger 2020). The results paint a picture that parallels the one emerging from the study of chess playing remarkably closely, despite the obvious dissimilarities between the two fields. Again, generalized memory has little to do with it; even skilled professionals' ability to recall raw information is pretty limited. What the experts do instead is to use their accumulated structural knowledge of what works and doesn't work as they have encountered it in their experience. Expert nurses, unlike novices, see things more systemically, being aware of a variety of issues including organizational structure, availability of resources, and even politics. The experts augment their learning through informal encounters, and they rely much more on their own sense of what they need to learn. Beginners, by contrast, need more structured, formal education; in the words of one of the interviewed nurses, they "did not even know what they did not know," and "what I didn't know as a novice, and what I learned in my career, was how to learn."

Similar results were obtained by Gaea Leinhardt (1989), who studied teaching techniques as implemented by novice and expert math teachers. These are people who all understand math very well; what they differ in is their experience in *teaching* math. The two skills are not at all the same, as Murray Sperber (2000) points out when he denounces the "good researcher—good teacher" myth that he says is propagated by both faculty and administrators at universities worldwide.

Sperber contends that there is not a shred of empirical evidence for the somewhat intuitive idea that because someone understands a subject very well, even excellently well, one is capable of explaining (teaching) it well. Leinhardt's results lend empirical foundation to Sperber's contention. Novices who had a similar background in math to that of experts, but had much less teaching experience structured their lesson plans in more superficial and ineffective ways, demonstrating lack of depth in their ability to connect concepts, to come up with innovative examples, and to build on previously taught ideas. As Leinhardt puts it (74), "Teachers seem to grow [with experience] from being structured and somewhat rigid to being introspective, flexible, and responsive." Expertise means familiarity with a complex territory that is a confusing terra incognita for non-experts.

Another author who has written on the nature of expertise, drawing from a variety of empirical studies, is James Shanteau (1992). His conclusion is not only that novices and experts use about the same number of "clues" to arrive at their decisions, but also that novices have a tendency to use too much information, getting stuck because of their inability to discriminate between what constitutes a useful clue and what doesn't. The clues used by experts to arrive at a particular decision are more informative than those used by novices, because experts have a conceptual model of the field that makes use of natural correlations among pieces of information: when information is structured, a small number of items may embody a significant wealth of detail that can be read by someone who knows what to do with it. Unfortunately, concludes Shanteau, it may take a lifetime to figure out which variables to pay attention to and which to ignore.

As Philip Ross (2005) points out, moreover, expertise does not translate from one field to another, not even across closely related fields. Ross cites research conducted by psychologist Edward Thorndike, who a century ago discovered that studying Latin did not improve one's understanding of another language, such as English; nor did proficiency in geometry improve one's performance in logic. Despite both of these being persistent myths in pedagogy.

Time is a necessary but not sufficient condition to become an expert: Ross refers us to research by Anders Ericcson showing that to become an expert, one needs to engage in "effortful study." An amateur musician, say, may spend thousands of hours practicing an instrument without ever getting past a relatively early stage of development in his musicianship. In order to get to the next level, he needs to challenge himself with difficult things that he cannot, at the moment, do. This once again fits with the idea that expertise is not a simple matter of accumulating raw information or experience, but requires engaging with the tasks at hand in order to develop and internalize heuristics that help one solve problems. This is also why the heuristics we develop in one field usually do not apply to another one.

Scientific studies of expertise such as those discussed so far are valuable and often intriguing, and they allow us to form an empirically based mental picture of what constitutes an expert and of how expertise is actually acquired. However, this research cannot by itself settle the question of most concern to us here: how is the person on the street supposed to tell whether an alleged expert is right? Moreover, how should we make up our minds when experts seem to disagree? Imagine, for instance, that you are witnessing a debate between a creationist and an evolutionary biologist. Or between a climate "skeptic" and an atmospheric physicist. Assume for the sake of argument that both debaters have legitimate PhDs, both teach at respected universities, and both have published in peer-reviewed journals. Should we conclude from this that such matters are so much up in the air that at this moment there is no reasonable way to take sides?

3. A Five-Point Guide to Evaluate Expertise

Broadly speaking, epistemologists look at a number of aspects of the question of trust as it relates to expertise. They tend to be particularly concerned with: (i) the epistemic goods, abilities, and virtues (Stichter 2013) that one ought to possess to become an expert in a domain, and (ii) the function experts ought to fulfill within their epistemic community. As I mentioned at the beginning, however, this is not the place for a systematic review of the available literature (see Hikins and Cherwitz 2011; Origgi 2015; Quast 2018; Kilov 2020). Instead, I will focus on the best known and, in my opinion, most practically useful account of expertise and trust: the one proposed by Alvin Goldman (2018) (for some criticisms, see Quast 2018; Croce 2019; Grundmann 2021) and for an update by Goldman see Goldman (2018).

Goldman begins by departing from the classical philosophical approach to such epistemological problems. In classical epistemology, he rightly says, one makes the assumption that all agents are perfectly rational, and all have "unlimited logical competence." So what is an imperfect "epistemic agent" (i.e., all of us) to do? Goldman's answer is that it is largely evidence about the speaker (or writer), instead of evidence about the matter at hand, that allows us to reasonably make up our mind on whether, say, one alleged expert or another is more likely to be right. That's because, as Goldman puts it (18), "the novice either has no opinions in the target domain, or does not have enough confidence in his opinions in this domain." Or, I would add, *shouldn't* have confidence in his opinions, precisely because he is no expert on the matter and that matter is intricate enough to require expertise.

Before proceeding with Goldman's recommendations on how to judge whether an expert is likely to be on track, we need to take a look at his conception of expertise, which is intuitively appealing and in itself provides some useful insight into the issue. Goldman says that expertise is a question of "veritistic terms," his phrase for truth-linked assertions. In other words, someone can be considered an expert in a particular domain if two conditions are met: (i) that person holds to more true beliefs (and fewer false beliefs) about that domain than a non-expert; (ii) that person holds a "substantial body of truths" in that domain.

This may seem a bit vague, but it implies two facts that intuitively we should want to be part of the idea of expertise: the first condition means that, in order to qualify as an expert, someone has to know things better than most people when it comes to a particular domain. If I claim to be an expert in baseball (I am not), I ought to get a lot more things right about all aspects of baseball, and fewer things wrong, when compared to someone who admits he is no baseball expert. But that is not enough, because one could know more than someone else and still not know nearly enough to be an expert (one could be a good amateur or just have an occasional interest in the matter). Hence the second condition, which insures that there is a substantial body of domain-specific knowledge in the toolkit of the alleged expert.

Let us now turn to the core of Goldman's proposal, his five-point guide to figuring out whether our trust in a prospective expert is well founded or not:

- Examine the arguments presented by the expert and his rival(s);
- Look for evidence of agreement by other experts;
- Look for independent evidence that the expert is, indeed, an expert;
- Inquire into what biases the expert may have concerning the question at hand; and
- Examine the track record of the expert.

Every so often all one needs is the first criterion: we, as non-experts, are sometimes able to tell whether an alleged expert's arguments hold up to even superficial scrutiny (what Goldman calls "direct argumentative justification"), or we may be able to tell he is a phony even if we are not capable of directly refuting his arguments (Goldman's "indirect argumentative justification").

For instance, a politician is—under ideal circumstances—an expert in public policy, someone who wishes to convince us that he has a better idea than his opponents about how to run the *res publica* (the "public thing," as the ancient Romans used to call the state). The problem—as recent elections in different parts of the world have painfully reminded us—is that it isn't as easy as one might wish to spot a phony politician (or simply a politician whose ideas are confused or wrong), because most members of the public do not know enough specifics about economics, foreign affairs, law, and so on to make a judgment by direct justification (not to mention that many of us are strongly ideologically biased when making such judgments). To complicate things, judgments by indirect justification, based, for instance, on how articulate the politician in question is, are suspect because the art of politics is closely allied to the art of rhetoric, which in turn can easily slide into sophistry, so effectively criticized by philosophers ever since Socrates.

Nonetheless, an informed public may still be able to make inroads in evaluating the reliability of a particular politician by using both components of Goldman's first criterion. To begin with, many politicians are actually not more knowledgeable about, say, economics or foreign affairs than the rest of us. This means that at least some of their arguments can be evaluated directly, if only one bothers to pay attention and inform oneself a bit. Moreover, politicians can sometimes be caught in flagrant contradictions, saying different things to different audiences, such that even an indirect analysis may reveal that there is something fishy going on.

The real problem with Goldman's first approach comes when we turn to more esoteric realms, such as debates on evolution or global warming. There, most people really do not have the technical background necessary for a direct evaluation of the arguments, and indirect techniques often fail because a creationist evangelical preacher may well be more rhetorically skilled than an academic professor who rarely leaves the ivory tower. While there is much to be said for encouraging academics to get out into the real world and talk to people, they are usually not trained in rhetoric, and often (though not always) they are unable to effectively rebut so-called commonsense arguments advanced by the other side.

We move on, then, to Goldman's second criterion: look for evidence of agreement among experts. It works like this: you may be convinced that your car doesn't start because of a problem with the spark plugs, but if several car mechanics tell you that your spark plugs are fine and instead you need to replace, say, the injection pump, you'd be a fool if you went ahead and bought new spark plugs.

The said, it is unfortunately all too easy to find at least some "experts" who will defend almost any sort of nonsense. The annals of science are littered with (sometimes even prominent) scientists being taken in by one crackpot idea or another. Just think of Newton's fascination with alchemy or Alfred Russel Wallace's (the co-discoverer with Darwin of natural selection) defense of spiritualism.

This is why creationists and supporters of intelligent design often play the game of gathering signatures from "experts" willing to criticize evolutionary theory. Typically, the statements they circulate are phrased in such generic terms that even a hardcore evolutionary biologist would be hard-pressed not to sign. For instance, scientists sometimes are asked to endorse declarations to the effect that critical examination of scientific theories ought to be taught to high school students. What scientist could in good conscience object to the teaching of critical thinking in schools? But the real issue is why the authors of such petitions single out the theory of evolution for scrutiny, and not, say, the theory of gravity or the germ theory of disease. Why the focus on evolution, if there really is no underlying ideological agenda?

Goldman's third criterion may be helpful here: look at any independent evidence that the expert is, in fact, an expert. That immediately rules out most creationists and intelligent design supporters: the overwhelming majority of them do not have degrees in the sciences, and some of those who do often received those "degrees" from unaccredited universities. The few with legitimate degrees, such as Michael Behe of Lehigh University or Bill Dembski of the Discovery Institute, have often acquired them in unrelated fields (biochemistry and philosophy, respectively). A PhD in biochemistry does not make someone an expert in evolutionary biology any more than my PhD in evolutionary biology makes me an expert in biochemistry (trust me, I'm not).

Among creationists, this still leaves a small number of experts with good credentials (graduate degrees in an appropriate discipline from accredited universities). But here we can go back to Goldman's second criterion and follow his suggestion to survey the field: if the overwhelming majority of accredited experts sides with one approach, that approach is likely the best bet. This is by no means to say that the majority of experts are always right. Far from it. The history of science provides several examples where, at one time or another, most authorities in a given field endorsed the wrong theory. Almost every astronomer in Galileo's time thought the Italian scientist got it wrong, but time vindicated Galileo. However, to use this occasional turn of events to justify a systematic rejection of expert opinion is to engage in what author Nigel Warburton informally calls the "Van Gogh fallacy." It goes something like this: "Van Gogh was an artistic genius, and yet he died penniless. I am a penniless artist; therefore I am a genius." Not the best example of valid deduction, I'm sure you will agree.

Goldman's fourth criterion says that we ought to look for evidence of bias—ideological, financial, or of any other kind—that might affect an expert's opinion. His example is of a study published in 1999 in the *Journal of the American Medical Association* (*JAMA*), in which the authors surveyed research on anticancer drugs, keeping track of where the funding for the research was coming from. The results were perhaps not surprising to anyone who appreciates human nature, but embarrassing to the medical research community: 38 percent of the studies sponsored by independent organizations (such as the federal government) yielded negative assessments of the efficacy of the tested drugs, while only 5 percent of the studies sponsored by the pharmaceutical industry turned out no effect.

Of course, science is a human activity (Longino 1990), and one expects human biases to enter into it. But notice that the 1999 study in *JAMA* was done by scientists. It is science's capacity for self-evaluation, self-criticism, and self-correction that once again shows us both the power and the limits of the scientific enterprise. The fact of the matter is that it is simply not possible to find any human being entirely without biases. "Bias" is a pejorative term, but more broadly it reflects the fact that we all come with (more or less well founded) opinions and we can't help but see the world through a series of partial, often distorting, filters. The question is not whether there is bias (there always is), but how much, where it comes from, and how one can become aware of and correct it.

Following this line of reasoning, however, both Goldman and I are dangerously close to committing the "genetic fallacy," that is, rejecting someone's arguments because of who makes the argument rather than on the basis of why he makes it. I'd like to point out, for instance, that John West, a fellow of the pro-intelligent design think tank Discovery Institute, has written an entire book, Darwin Day in America, complaining about the alleged moral corrosion of modern society brought about by "Darwinism." Have you noticed that I biased you against West by mentioning first thing that he is associated with a pro-creationist think tank? Well, was it bias or fair warning? In making up your mind about West's arguments you should certainly read his book, but knowing that he has a strong ideological background that may predispose him to be anti-science, your baloney detector will be on high alert—as it should be.

Nonetheless, neither Goldman nor I are committing any fallacy, I think (Boudry, Paglieri, and Pigliucci 2015). The point is this: if one were to say that we ought to reject an argument based *solely* on the biases of who proposed it, then one would in fact be committing the genetic fallacy (one would also be left with few people to argue with, considering that pretty much everyone has one bias or another). But what is being proposed here is that it is rational to be aware of the potential effects of a variety of sources of bias, as the *JAMA* article mentioned previously clearly demonstrated. Being aware of bias is not an argument against someone's position as much as a warning signal that we need to pay particularly close attention to their reasoning and evidence because they may not be as sound or convincing as they might at first glance appear. Just a commonsense form of critical thinking.

Lastly, let's look at the fifth source of evidence available to an intelligent novice attempting to figure out how much trust to put in a self-professed expert: the track record. We do this all the time in everyday situations: we keep going to the same mechanic because he has been pretty successful at fixing problems with our car; had he not been, we would have looked for another mechanic. Of course, like every tool described so far, this one needs to be used with caution. A common misuse is related to what we might term the "Nobel effect," a tendency of some winners of Nobel Prizes to opine on anything and everything, making pronouncements that some people will take at face value, even if such pronouncements concern fields remote from the one for which the person in question won the prize.

The reasoning must go something like this: this person won the Nobel (say, for research in cancer biology); so this person must be smart; therefore, he is probably right in what he just said about, say, economics. This, obviously, does not follow. Someone most certainly must be smart (under whatever definition of intelligence we wish to use) to win the Nobel Prize. But intelligence is not the same thing as expertise, and while winning the Nobel in medicine requires expertise in that discipline, it says nothing at all about the winner's expertise in a different field, such as economics. Indeed, it suggests that the biologist in question probably is not an expert in economics because, as we have seen, it takes a lifetime of effort to be an expert in either biology or economics, and most of us don't seem to find the equivalent of two lifetimes to devote to disparate professions. On top of this, of course, past success does not guarantee future success: your mechanic may have been infallible so far, but he may yet encounter a problem with your car that he can't solve. He is still an expert, though, and it would be unwise of you to discard his advice, unless he begins *consistently* to get things wrong.

Goldman is well aware that his five recommendations do not provide us with an airtight method to tell expertise from flimflammery. Indeed, a large part of my point is that there is no such foolproof method. But Goldman (23) invites us to rely on the general approach known as inference to the best explanation, which seeks to weigh all the available evidence and then make an educated guess about which way the truth lies. More often than not, educated guesses are all we have. In which case, the more educated, the better.

4. One More Problem: Bullshit

While Goldman's analysis of trust and expertise is—I think—still the most practical one available, a number of issues have been raised concerning it as well as the broader topic of trust and expertise. For instance, Johnny Brennan (2020) argues that novices face a hard job in evaluating expertise, particularly when the novice happens to initially hold an incorrect opinion about a given subject matter. Helen De Cruz (2020) articulates a number of proposals to improve the chances of a novice to overcome ideologically based or identity-based science denialism. And Kristina Rolin (2020) has emphasized the difficult dual role of scientific experts, not just as trustworthy epistemic agents, but also as moral agents responsible for the political implications of their expert advice.

Here, however, I wish to focus our attention on a different kind of challenging problem that has arisen recently in this context and that makes it even more difficult to tell the difference between an expert and a pretender: the problem of bullshit, analyzed in detail in a paper entitled "Bullshit, Pseudoscience and Pseudophilosophy," authored by Victor Moberger.

Moberger focuses on what he thinks are some peculiar similarities between pseudoscience and pseudophilosophy, concluding that—in a technical philosophical sense—it's all bullshit. He takes his inspiration from the famous essay by Harry Frankfurt, "On Bullshit". As Frankfurt puts it: "One of the most salient features of our culture is that there is so much bullshit" (1). Crucially, Frankfurt goes on to differentiate the bullshitter from the liar:

"It is impossible for someone to lie unless he thinks he knows the truth.... A person who lies is thereby responding to the truth, and he is to that extent respectful of it. When an honest man speaks, he says only what he believes to be true; and for the liar, it is correspondingly indispensable that he consider his statements to be false. For the bullshitter, however, all these bets are off: he is neither on the side of the true nor on the side of the false. His eye is not on the facts at all, as the eyes of the honest man and of the liar are.... He does not care whether the things he says describe reality correctly" (55–6).

So, while both the honest person and the liar are concerned with the truth—though in opposite manners—the bullshitter is defined by his lack of concern for it. This lack of concern is of the culpable variety, so that it can be distinguished from other activities that involve not telling the truth, like acting. This means two important things: (i) bullshit *sensu* Frankfurt is a normative concept, i.e., it's about how one ought to behave or not to behave; and (ii) the specific type of culpability that can be attributed to the bullshitter is epistemic culpability. As Moberger puts it, "the bullshitter is assumed to be capable of responding to reasons and argument, but fails to do so" (598) because he does not care enough.

Moberger does not make the connection in his paper, but since he focuses on bullshitting as an activity carried out by particular agents, and not as a body of statements that may be true or false, his treatment falls squarely into the realm of virtue epistemology (Bhakthavatsalam and Sun 2021; Zagzebski 1996). We can all arrive at the wrong conclusion on a specific subject matter, or unwittingly defend incorrect notions. And indeed, to some extent we may all, more or less, be culpable of some degree of epistemic misconduct, because few if any people are the epistemological equivalent of sages, ideally virtuous individuals. But the bullshitter is pathologically epistemically culpable. He incurs epistemic vices, and he does not care about it, so long as he gets whatever he wants out of the deal, be that to be "right" in a discussion, or to further his favorite a priori ideological position no matter what.

Accordingly, the charge of bullshitting—in the technical sense—has to be substantiated by serious philosophical analysis. The term should not simply be thrown out there as an insult or an easy dismissal, as it so often is. For instance, when Kant famously disagreed with Hume on the role of reason (primary for Kant, subordinate to emotions for Hume) he could not just have labelled Hume's position as bullshit and move on, because Hume had articulated cogent arguments in defense of his take on the subject.

On the basis of Frankfurt's notion of bullshitting, Moberger carries out a general analysis of pseudoscience and even pseudophilosophy. He uses the term pseudoscience to refer to well-known examples of epistemic malpractice, like astrology, creationism, homeopathy, ufology, and so on. According to Moberger, the term pseudophilosophy, by contrast, picks out two distinct classes of behaviors. The first is what he refers to as "a seemingly profound type of academic discourse that is pursued primarily within the humanities and social sciences" (600), which he calls obscurantist pseudophilosophy. The second "is usually found in popular scientific contexts, where writers, typically with a background in the natural sciences, tend to wander into philosophical territory without realizing it, and again without awareness of relevant distinctions and arguments" (601). He calls this scientistic (Boudry and Pigliucci 2017) pseudophilosophy.

The bottom line is that pseudoscience is bullshit with scientific pretensions, while pseudophilosophy is bullshit with philosophical pretensions. What pseudoscience and pseudophilosophy have in common, then, is bullshit. While both pseudoscience and pseudophilosophy are rooted in a lack of epistemic conscientiousness, this lack manifests itself differently, according to Moberger. In the case of pseudoscience, we tend to see a number of classical logical fallacies and other reasoning errors at play, not to mention, of course, lack of empirical evidence in support of whatever notion is being advanced. In the case of pseudophilosophy, instead, we see "equivocation due to conceptual impressionism, whereby plausible but trivial propositions lend apparent credibility to interesting but implausible ones" (600).

Moberger's analysis provides a unified explanatory framework for otherwise seemingly disparate phenomena of interest to the general public, such as pseudoscience and pseudophilosophy. And it does so in terms of a single, more fundamental, epistemic problem: bullshitting. He then proceeds by further fleshing out the concept—for instance, differentiating pseudoscience from scientific fraud—and by responding to a range of possible objections to his thesis—for example that the demarcation of concepts like pseudoscience, pseudophilosophy, and even bullshit is vague and imprecise. It is so by nature, Moberger responds, adopting a Wittgensteinian view that complex concepts are inherently fuzzy (see Wittgenstein 2001 [1953], sections 66–8).

Importantly, Moberger reiterates a point made by other authors before, and yet very much worth reiterating: any demarcation in terms of content between science and pseudoscience (or philosophy and pseudophilosophy), cannot be timeless. Alchemy was once a science, but it is now a pseudoscience. What is timeless is the activity underlying both pseudoscience and pseudophilosophy: i.e., bullshitting.

There are several consequences of Moberger's analysis. First, that it is a mistake to focus exclusively, sometimes obsessively, on the specific claims made by proponents of pseudoscience as so many critics do. That is because occasionally even pseudoscientific practitioners get things right, and because there simply are too many such claims to be successfully challenged (a concept known informally as Brandolini's Law (Brandolini 2015). The focus should instead be on pseudoscientific practitioners 'epistemic malpractice: content vs. activity.

Second, what is bad about pseudoscience and pseudophilosophy is not that they are unscientific, because plenty of human activities are not scientific and yet are not objectionable (literature, for instance). Science is not the ultimate arbiter of what does or does not have value, after all. While this point is hardly controversial, it is worth reiterating, considering that a number of prominent science popularizers have engaged in precisely this sort of mistake (Pigliucci 2014).

Third, pseudoscience does not lack empirical content. Astrology, for one, has plenty of it. But that content does not stand up to critical scrutiny. Astrology is a pseudoscience because its practitioners do not seem to be bothered by the fact that their statements about the world do not appear to be true.

Arguably, the amount of bullshit in public discourse has significantly increased in recent years, at least in part because of the nefarious influence of social media (Haidt 2022). The latter is yet another new player in the landscape we are considering, one that was not explicitly tackled by Goldman. Though arguably Socrates understood and explicitly labelled its underlying foundation: sophistry (in the pejorative sense of the word).

5. So, What Is the Person in the Street to Do?

As we have seen, the problem of how ordinary people ought to consider claims originating from alleged experts has been with us at least for two and a half millennia. And no easy bullet proof solution is at hand, nor will it likely ever be.

Nevertheless, Goldman's analysis provides us with actionable considerations that have very wide applicability and can help each of us reduce the number of epistemic mistakes we are likely to make when confronted with potentially pseudoscientific notions.

Unfortunately, as Moberger has highlighted, we now live in a social media-fueled era of widespread bullshit, *sensu* Frankfurt, which presents us with additional complexities not contemplated by Goldman. This probably means that we need to tackle the problem not just at the level of the individual epistemic agent—as Goldman's analysis does—but at the structural level as well.

This should not be surprising. Most of our problems straddle the personal-structural divide. It is true, for instance, that lots of people are racists and/or sexists. But it is equally true that we live in societies that are characterized by structural racism and sexism at the level of laws and institutions. Likewise, it is certainly the case that the behavior of individual consumers has made issues ranging from animal suffering to human labor to climate change more difficult to tackle. But it is just as much the case that structural-level aspects of national and international politics and finance contribute to making those problems increasingly intractable.

A number of approaches may alleviate the specific structural issues underlying pseudoscience and public discourse bullshit. These will include, of course, standard suggestions like increased and better teaching of critical thinking and the nature of science, particularly at the pre-college level. We could also more aggressively regulate social media platforms while at the same time boost the efforts and visibility of fact-checking sites.

Fritts and Cabrera (2022) have even proposed to treat the related category of fake news as "noxious markets" (after the definition advanced by Satz 2010). A market of this kind may be restricted when it inhibits citizens from being able to stand in an equal relationship with one another and the problem cannot be solved without restrictions. Examples of noxious markets include human body parts, child labor, toxic waste, sex, and life-saving medicines, though of course people may reasonably disagree about whether all or only some such markets ought to be considered noxious and restricted or prohibited.

Be that as it may, there will be no major progress in the fight against pseudoscience and assorted nonsense unless we act both at the individual and the structural levels. It is therefore unfortunate that these levels are often seen in political debate as mutually exclusive, with—broadly speaking—conservatives emphasizing personal responsibility and progressives focusing on the features of the system. Human beings live in complex societies, but societies are still, in the end, groups of individuals. If we want to make things better, therefore, we ought to work at both levels. The wellbeing of future generations, and indeed of humanity at large, depends on it.

References

Bhakthavatsalam, S. and Sun, W. (2021) "A virtue epistemological approach to the demarcation problem: Implications for teaching about feng shui in science education." *Science & Education* 30, 1421–52.

Boshuizen, H. P. A., Gruber, H., and Strasser, J. (2020) "Knowledge restructuring through case processing: The key to generalise expertise development theory across domains?" *Educational Research Review* 29 (February), 100310.

Boudry, M. and Pigliucci, M. (2017) *Science Unlimited? The Challenges of Scientism.* Chicago: University of Chicago Press.

Boudry, M., Paglieri, F., and Pigliucci, M. (2015) "The fake, the flimsy, and the fallacious: Demarcating arguments in real life." *Argumentation* 29, 431–56.

Brandolini, A. (2015) "Bullshit asymmetry principle," online at https://twitter.com/ziobrando/status/289635060758507521, retrieved 29 January 2023.

Brennan, J. (2020) "Can novices trust themselves to choose trustworthy experts? Reasons for (reserved) optimism." *Social Epistemology* 34(3), 227–40.

Croce, C. (2019) "On what it takes to be an expert." *Philosophical Quarterly* 69(274), 1–21.

Daley, B. J. (1999) "Novice to expert: An exploration of how professionals learn." *Adult Education Quarterly* 49(4), 133–47.

De Cruz, H. (2020) "Believing to belong: Addressing the novice-expert problem in polarized scientific communication." *Social Epistemology* 34(5), 440–52.

Ericsson, K. A. and Smith, J. (1991) "Prospects and limits of the empirical study of expertise: An introduction," in *Toward a General Theory of Expertise: Prospects and Limits*, (eds.), K. A. Ericsson and J. Smith. Cambridge, UK: Cambridge University Press.

Frankfurt, H. (2005) *On Bullshit*. Princeton, NJ: Princeton University Press.

Fritts, M. and Cabrera, F. (2022) "Fake news and epistemic vice: Combating a uniquely noxious market." *Journal of the American Philosophical Association* (3), 1–22.

Goldman, A. I. (2006) "Experts: Which ones should you trust?," in *The Philosophy of Expertise*, (eds.), E. Selinger and R. P. Crease. New York: Columbia University Press.

Goldman, A. I. (2018) "Expertise." *Topoi* 37, 3–10.

Grundmann, T. (2021) "Preemptive authority: The challenge from outrageous expert judgments." *Episteme* 18 (3), 407–27.

Haidt, J. (2022) "Why the Past 10 Years of American Life have been Unique Stupid." *The Atlantic*, May.

Hikins, J. and Cherwitz, R. (2011) "On the ontological and epistemological dimensions of expertise: Why 'reality' and 'truth' matter and how we might find them." *Social Epistemology* 25(3), 291–308.

Kilov, D. (2020) "The brittleness of expertise and why it matters." *Synthese* 199(1–2), 3431–55.

Leinhardt, G. (1989) "Math lessons: A contrast of novice and expert competence." *Journal for Research in Mathematics Education* 20(1), 52–75.

Liebowitz, J. (ed.) (2019) *The Handbook of Applied Expert Systems*. Boca Raton, FL: CRC Press.

Longino, H. E. (1990) *Science as Social Knowledge: Values and Objectivity in Scientific Inquiry*. Princeton, NJ: Princeton University Press.

Moberger, V. (2020) "Bullshit, Pseudoscience and Pseudophilosophy." *Theoria* 86(5), 595–611.

Origgi, G. (2015) "What is an expert that a person may trust her? Towards a political epistemology of expertise." *HUMANA.MENTE Journal of Philosophical Studies* 8(28), 159–68.

Pigliucci, M. (2014) "Neil deGrasse Tyson and the value of philosophy." *Huffington Post* 16 July, online at https://www.huffpost.com/entry/neil-degrasse-tyson-and-the-value-of-philosophy_b_5330216.

Pigliucci, M. and Boudry, M. (2013) *Philosophy of Pseudoscience: Reconsidering the Demarcation Problem*. Chicago, IL: University of Chicago Press.

Plato (1985) "Charmides," in *The Collected Dialogues*, (eds.), E. Hamilton and H. Cairns, 170e–171c, 117. Princeton University Press.

Quast, C. (2018) "Expertise: A practical explication." *Topoi* 37(1), 11–27.

Quast, C. (2018) "Towards a balanced account of expertise." *Social Epistemology* 32(6), 397–418.

Rolin, K. H. (2020) "Objectivity, trust and social responsibility." *Synthese* 199(1–2), 513–33.

Ross, P. E. (2006) "The expert mind." *Scientific American*, 24 July.

Satz, D. (2010) *Why Some Things Should Not Be for Sale: The Moral Limits of Markets*. Oxford, UK: Oxford University Press.

Segalowitz, S. J. (2007) "Knowing before we know: Conscious versus preconscious top-down processing and a neuroscience of intuition." *Brain and Cognition* 65(2), 143–4.

Shanteau, J. (1992) "How much information does an expert use? Is it relevant?." *Acta Psychologica* 81, 75–86.

Sperber, M. (2000) *Beer and Circus: How Big-Time College Sports Has Crippled Undergraduate Education*. New York: Henry Holt and Co.

Stichter, M. (2013) "Virtues as skills in virtue epistemology." *Journal of Philosophical Research* 38, 333–48.

Tynjälä, P., Kallio, E. K., and Heikkinen, H. L. T. (2020) "Professional expertise, integrative thinking, wisdom, and phronēsis," in (ed.) Eeva K. Kallio, *Development of Adult Thinking*, 156–74. London: Routledge.

Wittgenstein, L. (2001) [1953] *Philosophical Investigations*. Hoboken, NJ: Blackwell Publishing.

Zagzebski, L. T. (1996) *Virtues of the Mind: An Inquiry into the Nature of Virtue and the Ethical Foundations of Knowledge*. Cambridge, UK: Cambridge University Press.

3
Authority, Legitimacy, and the Expert-Layman Problem

Gloria Origgi
Institut Nicod (ENS-PSL-EHESS), Paris

Expertise is ubiquitous in contemporary societies. The myth of autonomous citizens who can decide on their own how to act has been debunked by the massive reliance we all have on experts and policy makers to know what is good for us in many everyday matters such as which products are safe to consume, which side effects medical drugs may have, when the air is safe to breathe and so on. As Sheila Jasanoff says: "Not only do we, as adult citizens in democratic societies, not know the answers to such questions, but for the most part, we do not know how answers are produced or why we should rely on them. Citizens delegate such issues to unelected policy makers and trust the policy system on the whole to make the right choices."[1] This poses several problems for democracy. Where does the legitimacy of the experts come from if they are not elected? Why does their voice weights more than that of an average citizen? And on which grounds their opinion must be preferred to that of other citizens? I will call these problems: (1) the problem of legitimacy, (2) the problem of equality, and (3) the problem of neutrality. I will now present some reflections on these three fundamental problems with respect to the question of expertise in liberal democracies.

1. Legitimacy and Epistemic Authority

Liberal democracies draw their legitimacy on the voluntary consent to authority. This consent is based on an agreement on the procedures that transfer individual authority to the political authority of the state. In the case of transfer of *epistemic authority*, that is, the authority of knowledge, it is unclear what are the procedures we consent to when we accept a "superior knowledge" should guide our choices. Do we consent to the scientific method as part of the public reasons that make a democracy work? This is a debatable issue. Many authors disagree on this point.

---

[1] Cf. Jasanoff (2011), 307.

Gloria Origgi, *Authority, Legitimacy, and the Expert-Layman Problem* In: *Expertise: Philosophical Perspectives*. Edited by: Mirko Farina, Andrea Lavazza, and Duncan Pritchard, Oxford University Press. © Gloria Origgi 2024.
DOI: 10.1093/oso/9780198877301.003.0003

Notably, critical theorists have rejected modern science as a tool of domination. Herbert Marcuse (1964) and Jurgen Habermas (1967) have extensively criticized the neutrality of science by arguing that science is an ideology of domination of nature (Marcuse) and of instrumental rationalization (Habermas). According to Marcuse, the scientific method embodies assumptions about the exploitation of nature that shouldn't be accepted as shareable public reasons. For Habermas, although the kind of domination of nature that is intrinsic in the scientific/ technological strategic rationality is an unavoidable feature of *homo faber*, and thus, contra Marcuse, cannot be eliminated or replaced by a different science, still the purposive-rational action of the technocratic intention serves as an ideology for the new politics "which is adapted to technical problems and brackets out practical questions."[2] The epistemic authority of science is thus based for these authors on a techno-scientific instrumental ideology of the world that doesn't have a clear legitimacy for the people. Other, even more radical thinkers, such as Paul Feyerabend (1975) have argued that the scientific method is a myth and science is nothing more than an opinion among others. There is no intrinsic legitimacy of a universal method in science and believing in mainstream science is a choice of favoring an opinion that has earned authority thanks to the institutional support of governments that endorse a certain ideology. Although the intersubjectivity, the reproducibility and the verifiability of the methods and the results of science are considered today the ground of legitimacy of science, authors like Feyerabend deny the objectivity of scientific method arguing that the way in which science is funded by governments makes it an ideology deprived of objectivity. These thinkers radically reject the scientific method as an impartial tool for knowledge. But without following the radicality of these positions, other philosophers have argued that there exists a tension between science and democracy: in his 2001 book *Science, Truth and Democracy* Philip Kitcher argues that although science promotes some of the liberal values such as transparency, skepticism, and collective problem-solving, it also challenges these values through exclusivism and elitism; it contributes to the wealth and security upon which modern democracy depends and yet it creates technological risks. A new pact between science and democracy must thus be established to understand which scientific procedures can be acceptable for the democratic society. Science's aim is not only truth: it is *significant truth* that must be established in a democratic manner.

Thus, if it is not the scientific method alone, but also the values of the society that can legitimate our consent to the epistemic authority of science and expertise, where does the legitimacy of epistemic authority come from? This is an open question that is still discussed in philosophy and the sociology of science.[3] The legitimacy of the experts does not depend on the universal legitimacy of the

---

[2] Cf. Habermas (1967), 106–7.   [3] Cf. Turner (2003).

scientific method, rather, it depends on the existence of the widespread belief in a society that the aims, interests, and values of citizens are shared by the scientific experts and that these aims, interests, and values converge with the overall objectives of the society. Sometimes, interests, aims, and values do not converge, and the legitimacy of an expert advice can be challenged. Take the measures taken by various governments in the world during the COVID-19 pandemic to restrain the liberties of citizens to keep the spread of the virus under control. These policies have generated widespread debates on their legitimacy, given that they were apparently in contrast with some basic civil liberties such as autonomy and freedom of movement. There was no knockdown scientific consensus about which policies should have been adopted and different states endorsed different policies sometimes in conflict with part of the population. The legitimacy of these policies did not depend on the epistemic authority of the experts, rather, on the trust in the different governments that endorsed the policies. This shows that the legitimacy of the epistemic authority of the experts is dependent on many factors that are external to the "pure" certainty of their knowledge.

2. Equality, Democracy, and Expertise

Democracy is grounded in a principle of equality: one person, one vote, independently of the social, cultural, and cognitive differences among citizens (except for the limits of age for the vote and other exceptional restrictions). The political threat to democracy that expert knowledge poses is that it seems to grant a special class of citizens, the experts with a form of power that is not under control by the average citizen. As Stephen Turner (2003) puts it:

> Expertise is a kind of violation of the conditions of tough equality presupposed by democratic accountability. Some activities, such as genetic engineering, are apparently out of reach of democratic control, even when these activities, because of their dangerous character, ought perhaps to be subject to public scrutiny and regulation, precisely because of imbalance of knowledge. As such, we are faced with the dilemma of capitulation to the "rule of experts" or democratic rule that is "populist," that valorizes the wisdom of the people even when "the people" are ignorant and operate based on fear and rumor.[4]

Although the example of genetic engineering may be debatable, given that it is carried out either with the informed consent of patients or in laboratories, often private, using animal models, expertise seems to violate a principle of equality in

---

[4] Cf. Turner (2003), 19.

accountability that is fundamental for democracy. On the other hand, no democracy could survive without the appeal to expert knowledge in crucial issues such as health, security, new technologies, or ecology. A "division of cognitive labor" (Kitcher 1990) is thus essential for a society to thrive. As Turner points out, there is no solution to the problem of equality. A fundamental inequality of knowledge is a condition for epistemic authority: although a democracy may try to fill the knowledge gap through public education, a society of "omni-competent" citizens, in which facts are available to everyone is an impossible ideal. The condition of *epistemic dependence* is part of the human condition and the acceptability of this condition in a democratic society depends on the fairness of procedures of distribution of knowledge and the transparency on the deference cognitive relations between non-experts and experts. While we can be obliged to obey a legitimate political authority, we cannot be obliged to trust an epistemic authority. Our trust should come from an internal conviction of the superiority of a set of beliefs (the experts' beliefs) upon our set of beliefs. Epistemic trust implies an acceptance of the surrender of our beliefs to the beliefs of others (Origgi 2004). This is a choice that must depend on our trust in those to which we surrender our beliefs and cannot be a matter of coercion. Nonetheless, given the strict ties between political authority and epistemic authority in contemporary techno-scientific democracies, our trust should be based on public reasons we are able to share with others, and in the case of scientific knowledge these reasons are not "public" in the strict sense because their mastery depends on the cognitive competence of a group of citizens (the experts) that are trained to master these reasons. Turner calls the inequality of expert knowledge as the "last inequality"[5] in our societies that we are not able to eliminate.

3. Neutrality

A distinctive principle of liberalism is the "neutrality principle," that is, the idea that the State "should not reward or penalize particular conceptions of the "good life," rather, should provide a neutral framework within which different and potentially conflicting conceptions of the good life can be pursued, rather, should provide a neutral framework within which different and potentially conflicting conceptions of the good can be pursued."[6] Grounded in the works of Immanuel Kant and John Stuart Mill on individualism and autonomy, the principle of neutrality is a milestone of contemporary political liberalism. John Rawls's theory of justice is commonly thought to be committed to the principle of neutrality through the veil of ignorance, which plays a fundamental role in his argument

---

[5] Cf. Turner (2003) cit.   [6] Cf. Kymlicka (1989), 883.

from the original position.[7] Given that the parties in the original position are ignorant with respect to their conception of a good life, they will choose the principles of justice in a neutral way, that is, without presupposing any preferred conception of what a good life should be.

The Neutrality Principle limits the kinds of arguments which are acceptable in liberal dialog. "No reason is a good reason if it requires the power holder to assert: (a) that his conception of the good is better than that asserted by any of his fellow citizens, or (b) that, regardless of his conception of the good, he is intrinsically superior to one or more of his fellow citizens."[8] Do scientific reasons respect the Neutrality Principle? Does the public debate about science-based decision-making presuppose that some points of view are "intrinsically superior" to others? Of course, some *facts* need to be taken for granted for there to be a genuine political discussion. But the establishment of these facts is delegated to those experts whose opinion is "preferred" by the states. In many crucial issues that involve scientific knowledge, facts are established by a faction of experts whose point of view is considered more "objective" by the state compared to the point of view of other experts. This is not due to a malfunctioning of the scientific community, which has concepts and procedures such as "expert consensus," "peer review" that aim at increasing knowledge in an objective way. But states can choose "biased" science to pursue their interests. Morone and Woodhouse (1989), in their work on nuclear power expertise, showed how a scientific community that was bound up with the policy imposed the idea of the safety of nuclear plants based on a consensus that was created in part by dismissing critics as "not real scientists" and imposing heavy educational policies to improve the legitimacy of this point of view in the eye of the public. Scientific facts are never 100% sure and their acceptance by policy makers involves a complex of values, mutual dependence on the opinions of industry or the army, and political preferences (Wynne, 1992, 1996).

The legitimacy and authority of scientific expertise, that is, of that part of scientific research that touches citizens' and states' interests, are thus controversial, and this creates an atmosphere of distrust and skepticism toward the institutions of knowledge. Although science has its own procedures to guarantee objectivity and universality, when science is recruited by the states to make decisions that impact the citizens' life it needs a further legitimacy to be accepted by the public.

But are we passive believers toward the experts or do we actively try to find the right experts? Here I want to argue that, given that expertise is not easily perceived as legitimate, people use their cognitive capacities to evaluate expertise and decide whom to trust (Branch-Smith and Origgi 2022).

---

[7] Cf. Rawls (1971).   [8] Cf. Ackerman (1980), 11.

4. Social Indicators of Trust in Experts

Information is everywhere. Given the new technologies, the answer to every question is at few clicks of distance. False and true information are both easy to reach. The main problem of the growth of misinformation is that people do not have access to the reliability of the sources they can find on the Internet. The gap between experts and non-experts lies in the incapacity of the non-expert to recognize the trustworthiness of a source of information. A website for climate change deniers can resemble the official website of the IPCC. Yet, people use heuristics and strategies to check the validity of the expertise by browsing the social information that surrounds a piece of content found in the media or online. I will call *social indicators of trust* the social information people extract about an expert to assess their trustworthiness in absence of any detailed knowledge of the content of the expertise.

Authority is a powerful indicator of trust. The authority of an expert depends on their past records that a layman can find online by searching with *Google Scholar* the publications of the expert and their H-index, that is, the index that measures the impact of the publications. This social information is available in the online environment and easy to process.

Status is the position of the expert in a hierarchy. The layman can check the status of the expert by looking at their role in an institution and the recognition they have within the scientific community.

Influence is the exposure of the expert on the media and social networks. A layman can check the media presence of the expert and look at whom follows the expert on social media like *Twitter* to assess the influence of the expert.

Values are also important for the public to come to trust an expert. If an expert shares the values of the public, they have more credibility. Values can be checked by looking at public statements of the expert. An expert whose statements don't align with the values of the community has less chance to be believed. There are *epistemic values* shared by the scientific community, like replicability of experiments and norms against plagiarism, and *non-epistemic* values, that is, general values that a community holds, like freedom or solidarity.

How reliable are these social indicators? To maximize the rationality of our use of social information, an exercise in metacognition could be useful. We may combine these different social indicators and see whether a coherent picture of the expert emerges. A certain balance among the indicators should be a criterion for a good reputation of an expert. If, for example, an expert ranks high in influence, but low in authority, we may infer that they are able to use the social media to influence other people's behavior without being a certified expert by their peers.

People are not blind trustors: they check the reputation of the experts by using the social information available to them. The problem of trust in experts depends on the quality of the social information that surrounds the experts. "Who says what about whom" is the key ingredient to come to trust an expert, and the gap between experts and non-experts is where misinformation thrives. The gap is the "space" where the authority is claimed: charlatans or ill-intentioned informants may claim pseudo-authority on some evidence and redirect the public toward alternative authorities. The gap between experts and non-experts is not limited to the gap of knowledge about the content of expertise, that is irreconcilable given the specialization of knowledge; it is about the reputation of the expert that people actively search and that can be distorted by the information pollution that surrounds what it is said about the expert.

The information deluge that characterizes our societies comes with a price: everyone can claim authority over a piece of information and present themselves as legitimate source of information. That is why using social indicators of trust is indispensable to filter the reliable sources of information. People need to assess the reputation of an expert available to avoid false information. Of course, reputations may be misleading, but combining the social indicators of trust listed above allows to reduce the possibility of being misinformed. Using the social indicators of trust is an exercise in social cognition and a way of distributing the cognitive labor necessary to come up with trustworthy information. The search for truth is a collective endeavor that needs the cognitive work of the experts and the laymen. The expert produces the content and the layman verify its credibility by navigating the social information the surrounds the expert.

Yet, people don't easily see the collaborative aspect of the search for truth and tend to see the experts as authoritative characters who "impose" their truth to the public in a despotic way. The rejection of expertise is motivated by a demand of epistemic egalitarianism against the fundamental inequality we have seen above. This demand is entrenched in the idea that no knowledge is superior and that everyone has the right to believe what she prefers without submitting to any epistemic authority. This creates a sort of *epistemic populism* that I am going to describe in the next section.

5. Epistemic Populism

People don't like to be told what they should believe. The autonomy of our thinking is challenged by the experts, who claim to possess superior knowledge. There is a resistance toward the surrender of our beliefs in front of the expert, a sort of pride of one's own intelligence that makes us autonomous beings. This is well known by communicators and information producers of any kind who take advantage of this natural resistance to knowledge to make people believe any sort of "alternative" truths, that are not produced by the scientific establishment. The rejection of the epistemic authority of the expert in the name of a more democratic access to knowledge and of a skeptical attitude toward any sort of "mainstream" knowledge leads to what I call *epistemic populism*, that is, an ideology that promotes an egalitarian distribution of knowledge. I call it *epistemic populism* because it reminds the attitude of the political populists of renegade any statement made by the *élites* of power in the name of the commonsense of people and their capacity to choose what is better for them. Of course, this is a caricatural picture of the epistemic populist, that tries to select its most typical features. There exist other accounts of the epistemological aspects of populism that insist on the same features[9] and other ones.

The epistemic populist is convinced that there is no better knowledge than that she can appraise and understand. The epistemic populist is a naive realist: she believes that the world is described in terms that are too complex and that reality can be deciphered by simpler means than those proposed by various "pundits."

The epistemic populist is, or believes herself to be, a methodological individualist: she believes that her senses and reason are sufficient to see things as they are and that epistemic dependence on other human beings is an unbearable limitation on her cognitive abilities.

The epistemic populist is a monist: there is only one exact description of the world, the one she is able to understand, and he hates the pluralism of scientific theories that complicate a reality that is instead "in plain sight."

The epistemic populist is nostalgic: there was a simple worldview in which there were only two sexes, only four seasons, apples tasted only one way, the climate was not changing, Blacks were different from Whites, an authentic world, easy to understand with the five senses where we were not asked to "correct" the various biases that make us see things a certain way such as essentialism, tribalism, or other natural tendencies of the human spirit. She thinks that the unvarnished picture of the world would be still available if experts did not intervene on it by imposing a mainstream view about what happens around us.

The epistemic populist is not ignorant: she is an information consumerist. Her ability, mainly due to the Web, to find information that confirms her point of view gives her the arrogance to think that she does not need others to get her idea of things.

The epistemic populist is not undemocratic: he is a product of democracy. Indeed, she argues that her opinions should be as valid as everyone else's, and that it is fundamentally undemocratic for those who consider themselves in a position of epistemic superiority and lecture everyone else.

The epistemic populist (like the political populist) has a symbolic, non-delegative conception of democratic representation: those who represent her in

---

[9] Cf. Bronk and Jacoby (2020); Ylä-Anttila (2018).

ideas must be like her, must resemble her, be a mirror of her, must share the same "common sense."[10]

The epistemic populist is arrogant because she is insecure: she does not find herself in the theoretical complexities of mainstream knowledge and wants to limit her exploration of the world to what she can or wants to understand with her "common sense."

This brief ideal typic description leads us to explore an interesting epistemological problem. What is the common sense to which the epistemic populist appeals? Is there a manifest image of the world to be contrasted to a scientific image of it to which common sense refers? What kind of beliefs does common sense contain? In this next section I will try to describe what common sense is from an epistemological point of view.

6. The Appeal to Commonsense

Common sense is a concept that oscillates between two meanings: on the one hand it refers to reason: to have common sense means to be reasonable, that is, able to use reason to arrive at a conclusion. On the other it refers to the idea of "common sense" a difficult concept to define on which philosophy has spilled rivers of ink. In a famous article entitled *A Defense of Common Sense*, philosopher G.E. Moore argued that there are utterances about which no one with "common sense" can disagree. Moore's examples are such utterances as "There exists in the present moment a physical body that is my body;" "This body was born at a certain date in the past;" "The Earth existed before the birth of my body," and so on. Common sense is defined in the history of philosophy sometimes as a "sixth sense" that allows perceptions from the other senses to be integrated into a manifest view of the world (John Locke, Aristotle, Thomas Reid) and at other times as a set of practical/experiential knowledge common to all, which depends on the *koiné* of the group to which it belongs.[11] *Koiné* in Greek referred to the average language spoken by the Greeks, a mixture of dialects that established the Attica variant as the common language. Common sense is thus the wisdom of the language shared by all. The epistemic populist oscillates between these two interpretations of common sense: she appeals to her reason, an individual faculty that everyone possesses, and argues that using reason everyone should arrive at her conclusions, which are part of "common sense." To depart from this *koiné* is for the epistemic populist a dangerous abandonment of a vision shared by a people who express themselves in the same language. This call to the *koiné* soon

---

[10] For the distinction between these two conceptions of "representation" see H. Pitkin (1967) *The Concept of Representation*. University of California Press.
[11] Cf. S. Colvin (2009).

becomes a call to a tradition, to a way of thinking inherited from a glorious past in which things were said as they were without the excessive complications of contemporary technocratic language. And it is precisely against technocracy that the epistemic populist rebels. Let us now see what the reasons and wrongs of the populist in her conflicting relationship with technocracy are.

7. Democracy, Science, and Technocracy

Science is not what it used to be. A global enterprise with international standards of exactitude, science has transformed over the past fifty years into a very different business from the one evoked by the classic image of the scientist who drops an apple on his head. Scientific research has adopted a ruthless business model (publish or perish), and techno-scientific alliances in industrial production have transformed it into a capitalist activity far removed from the norms ascribed to it by sociologist Robert Merton: communalism, universalism, disinterestedness, and organized skepticism. Moreover, from the exact sciences to the social sciences, scientific knowledge is ubiquitous in political decision-making today. Democracies tend to turn into technocracies ruled by elites of experts, which creates a tension between expertise and democracy in at least two respects, as we have seen at the beginning of the chapter: expertise seems to violate on the one hand the condition of equality of subjects, because it considers some subjects cognitively superior to others, and on the other hand the condition of neutrality of liberal democracy, because it favors one opinion, the scientific one, over other opinions.

The Weberian rationalization of society under technocratic drive makes science and technology a new ideology, as Jürgen Habermas argued as early as 1967. The Weberian model still distinguished between scientific rationality and political decision-making, a distinction that according to Habermas is no longer possible because of the rationalization of decision-making processes (Rational choice theory, algorithms) that formalize power in such a way as to make political decision-making an obstacle to the complete rationalization of society. If today's technocracies find their legitimacy in science, however, science is difficult to legitimize in democracy. Experts are not elected by the people, but more than that: the procedures through which they are selected by power are often opaque, based more on mechanisms to reproduce a class of state technicians than on meritocratic selection mechanisms. This creates a detachment and distrust of the public toward a technocratic power that is distant and does not interact with citizens, nor does it bother to understand based on what values and needs expressed by the public a decision should be made.

The populist thus has apparently good reason to dislike technocracy. The epistemic populist extends this dislike to all science, arguing that scientific thinking is merely a new form of domination that does not consider her reasons. Science is also seen as an activity run by multinational elites that is no longer able to connect with the *koiné* of citizens. Although there is much common sense in this populist sentiment and it is a view shared by various thinkers and commentators, the consequences of disaffection with science can sometimes be deleterious. The epistemic populist, however, cannot do without expertise although she trusts her common sense: she will therefore seek alternative expertise to mainstream expertise to form her own opinion on a scientific issue.[12]

Ill-intentioned information producers know very well how the epistemic populist reasons: they have understood her resentment toward the mainstream institutions of knowledge. The design of an alternative source of knowledge, one that goes against the official view, is easy and cheap to create and, in the eyes of the populist, has the same legitimacy as the official one. Alternative sources proliferate on the Web by designing the ground for a new perspective on facts that satisfies the desire of autonomy and independence of the epistemic populist.

This "alternative" design was soon seen by various political and intellectual communicators as an opportunity to gain consensus online. The communicator of social networks presents herself to the audience as "one of them," writes short and clear messages, contrasts her style with the verbose style of traditional media, and creates an effect of identification with the audience. The system of likes and shares on the Web allows the effect to be amplified by turning identification into acclaim. The audience, acting on the message, appropriates it as if it were its own word, thus creating the feeling that what it believes comes from itself, whereas it comes from authoritarian and arrogant personalities that exploit the need to believe alternative authorities. The case of the French doctor Didier Raoult[13] is an example of this process. Doctor Raoult became a prominent character in the media in 2020 by recommending hydroxychloroquine as a remedy for COVID. His "anti-establishment" style was a key ingredient of his success. Raoult presents himself dressed not as a doctor, but as an average *Marseillais*, insults opponents by crying conspiracy, is arrogant and overbearing, talks about a medicine that everyone knows can cure COVID instead of citing incomprehensible statistics, uses *YouTube* instead of insiders' magazines, lashes out at the "system" of science that is cold and distant from the public, and invokes the Hippocratic Oath by saying that he is not interested in making statistics, but in saving lives. The speech is simple, the oppositions clear, the listener identifies with the character, makes his thoughts his own, and believes he himself is speaking. The typical narcissism of the network user is satisfied by a video in which he projects himself, identifies himself by making the leader's words his own without realizing that he is subservient to an authority far more intrusive than the

---

[12] Cf. Ylä-Anttila, cit.
[13] On the case of Dr Raoult, see Branch-Smith, Origgi, and Morisseau (2022).

authorities he rejects, but thinking that he is the authority. Doctor Raoult's narrative perfectly fits the mind of the epistemic populist. The complex science of random control trials is pitted against the concrete reality of a known medicine, already on the market and easily accessible to all. His arguments against the potential risks of chloroquine are personal and nostalgic: they come from his childhood in Senegal, where he had taken chloroquine against malaria, and he insists that it is a completely harmless medicine that in the good old days was taken without fuss. Consider that the debate around chloroquine is exploding at a time when very little is still known about COVID, and vaccines are not being talked about. Countering the uncertainty of science with a simple and available remedy reinforces that idea of common sense to which the epistemic populist is so attached. Raoult's arrogance toward his colleagues and the scientific establishment in general gives voice to the arrogance of the epistemic populist who sees in their ability to doubt mainstream authority a reason for cognitive empowerment. The use of non-traditionally scientific media such as social networks makes communication seemingly more democratic, although Raoult is not democratic at all and presents himself as a guru who possesses the truth and is not to be contradicted, insulting his opponents without listening to them. It is interesting to note the following paradox: the epistemic populist who rebels against authority is easily preyed upon by authoritarian personalities who reject confrontation with opponents by branding them as incompetent or malevolent.

8. Conclusion

The authority and legitimacy of expertise are problematic because they challenge the common sense of citizens and the intuition that all opinions deserve the same consideration in a democracy. People are asked to blindly trust the experts and they resist this capitulation of reason. They use social information that is available online to evaluate the reputation of the expert. The problem is that social information is uncertain and can lead to "bad beliefs"[14] about who are the reliable experts. The gap between experts and non-experts resides in the "authorities" that claim to provide exact social information about the experts. The abundance of information online empowers the laymen and may lead to an attitude of epistemic populism, that is, an overconfidence in one own's capacities to assess who are the reliable experts and an appeal to one's own commonsense against the complexities of the scientific discourse. To avoid such distortions, science communication should pay attention to the social information that surrounds an expert, that is, "who says what to whom," and reinforce the credibility and the visibility of the

---

[14] On the notion of bad beliefs see N. Levy (2022).

good informants. People do not blindly trust, nor do they trust completely rationally. They use social information as a heuristic to evaluate the prestige of the expert. That is why it is important to keep social information as clean as possible to avoid the noise produced by ill-intentioned informants.

References

Ackerman, B. (1980) *Social Justice in the Liberal State.* Yale University Press.

Branch-Smith, T., Origgi, G., and Morisseau, T. (2022) "Why trust Raoult?" *Social Epistemology*, 36(3), 299–316.

Bronk, R. and Jacoby, W. (2020) "The epistemics of populism and the politics of uncertainty." *LEQS*, 152, online at: https://eprints.lse.ac.uk/103492/1/LEQSpaper152.pdf.

Colvin, S. (2009) "The Greek *koiné* and the logic of standard language," in *Standard Languages and Language Standards: Greek, Past and Present*, (eds.), Alexandra Georgakopoulou and Michael Silk. Aldershot: Ashgate Publishing Ltd.

Habermas, J. (1967) *Towards a Rational Society*, Polity Press, Cambridge.

Jasanoff, S. (2011) "The practices of objectivity in regulatory science," in *Social Knowledge in the Making*, (eds.), C. Camic, N. Gross, and M. Lamont, 307–37. University of Chicago Press.

Kitcher, P. (1990) "*The division of cognitive labor*," Journal of Philosophy, 87(1), 5–22.

Kymlicka, W. (1989) "Liberal individualism and liberal neutrality," *Ethics*, 99(4), 883–905.

Levy, N. (2022) *Bad Beliefs.* Oxford University Press.

Marcuse, H. (1964) *One-Dimensional Man*, Beacon Press.

Morone, J. G. and Woodhouse, E. J. (1989) *The Demise of Nuclear Energy: Lessons for Democratic Control of Technology*, Yale University Press.

Origgi, G. (2004) "Is Trust an Epistemological Notion?" Episteme 1(1), 61–72. doi:10.3366/epi.2004.1.1.61.

Rawls, J. (1971) *A Theory of Justice*, Harvard University Press.

Turner, S. (2003) *Liberal Democracy 3.0. Civil Society in an Age of Experts.* New Delhi: Sage Publishing.

Ylä-Anttila, T. (2018) "Populist knowledge: Post-truth repertoires of contesting epistemic authorities." *European Journal of Cultural and Political Sociology* 5(4), 356–88.

Wynne, B. (1992) "Misunderstood misunderstanding: Social identities and the public uptake of science." *Public Understanding of Science* 1(3), 281–304.

Wynne, B. (1996) "May the sheep safely graze? A reflexive view of the expert-lay knowledge divide," in *Risk, Environment and Modernity*, (eds.), S. Lash, B. Szersynski, and B. Wynne, 27–43. Sage Publications.

4

Institutionalized Expertise

Trust, Rejection, and Ignorance

Daniel R. DeNicola

Professor *Emeritus* of Philosophy, Gettysburg College, USA

My intent is to examine two contemporary trends (or clusters of trends) that have great significance for both private and public life, and especially for professional practice. My interest in these trends arises from their collision: they appear contradictory; they pull in opposite epistemic directions; and, beyond the impact each has alone, their *collision* carries significant social and political implications. The first trend involves the expansion of the professions and increasing specialization and institutionalization of expertise; the other concerns the growing public resistance to, or at least distrust of, experts, including professionals, and their judgments and authority. Understanding the relation of these two phenomena is the problematic of this article.[1]

The tension between expert knowledge and public opinion is hardly new: Plato discourses on the opposition of *eidikoí* (those with specialized knowledge) and *hoi polloi* (the masses) and understands the special problems it poses for populist democracy (Plato 1997, esp. *Republic* IX and *Statesman*). In the Enlightenment, this tension was subsumed under the antagonism between authority and autonomy: it was seen as conflict between the epistemic authority of experts and the right-to-believe autonomy of non-expert individuals. (That interpretation still resonates despite its sloppy understanding of "authority.") What *is* new is the *intensity* of this opposition: the distrust, not only of individual experts, but of institutions that have sought, discovered, and certified expert knowledge since the Enlightenment; the elevation of "the wisdom of crowds" and public opinion; the parlous attacks on the epistemic health of the polity—all this despite increasingly profound dependence of contemporary life on hyper-specialized, technocratic expertise. We live in a time when fervent belief trumps knowledge, the very

---

[1] Each of the two trends is a cluster of developments, but each cluster forms an identifiable, coherent if complex, social development. Where confusion is at risk, I will try to make the relevant distinctions. Although I limit examples to the USA and UK, these trends appear across most liberal democracies. I first considered the topic in relation to academia and professional education (DeNicola 2015). Though that is not my focus here, I do borrow from and build on the analysis and language of that piece, especially in discussing the concept of a profession.

Daniel R. DeNicola, *Institutionalized Expertise: Trust, Rejection, and Ignorance* In: *Expertise: Philosophical Perspectives*. Edited by: Mirko Farina, Andrea Lavazza, and Duncan Pritchard, Oxford University Press. © Daniel R. DeNicola 2024. DOI: 10.1093/oso/9780198877301.003.0004

existence of truth is doubted, and ignorance is promoted and celebrated. Our polarized politics are symbiotic with our polarized epistemology.

Epistemic differences smolder beneath the societal surface all the time, but the current situation is, I would argue, an epistemic crisis on a level not seen since the seventeenth century conflicts between Roman Catholicism and Protestantism and between religion and emerging science—conflicts that propelled the modern philosophical discipline of epistemology. It is a complex crisis, and I will frame my discussion in terms of epistemic communities and their interactions. Included are various epistemic communities of experts, but I will focus on those institutionalized in what I will call "the professions," along with related cohorts of clients and the larger lay public. The first task is to set out how I use the terms *expertise* and *profession*.

1. Types of Expertise

Following on the analytical work of others (especially Watson 2020), I take *expertise* to be an epistemic good, sustained by advanced learning and epistemic virtues. It is *a demonstrably high degree of competence in a particular domain*. Such competence (1) entails specialized knowledge, skills, experience, and understanding relevant to the domain; (2) is calibrated in relation to the competence of the general population in that domain, and in relation to the state of human knowledge, skill, and understanding at the time of attribution; and (3) implies facility in the shared protocols of the epistemic community for that domain and adherence to its values. The fact that gaining expertise is a gradual process, nearly always involving intentional and rigorous study and practice, suggests that it is a scalar concept, though the term is applied only to high degrees of competence. An *expert* is one who possesses such competence in a particular *domain*, which is a topic, discipline, language, form of art, activity—in short, anything that one may pursue to gain epistemic goods. So, *Stoicism, macroeconomics, Mandarin, cello-playing,* and *chess*—all exemplify domains in which one may acquire expertise.

Domains differ, however, in the sort of epistemic good required for expertise. Being an expert in the American Civil War or fungi (mycology) is primarily about possessing certain knowledge (*knowing-that*), though there are protocols, techniques, skills, and judgment that are required for research and judging claims to knowledge. But some domains, such as the arts, crafts, sports, and games, are predominantly *performative*. For them, the dominant epistemic goods of competence are skillfulness, technique (*knowing-how*), and perhaps grace, creativity, and efficacy. So, Vladimir Horowitz and Arthur Rubenstein may be called expert pianists; Serena Williams and Roger Federer may be called expert tennis players; Garry Kasparov and Magnus Carlsen may be called chess experts. (Although, I prefer the terms *virtuoso* or *master* when one's performance record is so fully determinative of expertise.) This type of expertise is also distinctive in that mastery can often be recognized and appreciated by a wider, non-expert public through masterful public performance. That is not true for expertise in other domains. Consider the philatelic expert whose "performances" include distinguishing forged from genuine stamps; or the papyrologist who displays expertise in translating and interpreting ancient papyri. The content of their competence is arcane, inscrutable to the non-expert; the quality of their work can be accurately judged only by peer experts. Therefore, public acceptance of such specialized and esoteric expertise requires mediated credibility, formal certification, or markers of competence by appropriate institutions or community of like experts.

Setting aside the virtuosity-of-performance sort of expertise, the activities most basic to expertise in many domains are the articulation, identification, and retention of truths (and falsehoods) about the domain, and the forging of a comprehensive understanding of the domain—in short, expertise is often *descriptive*. An expert really knows the territory of the domain. But some sorts of expertise are also *normative and regulative*: these experts use their knowledge to reach value judgments, offer advice, recommend policy, and regulate practice. It is therefore useful to distinguish two types of epistemic communities of expertise: *closed-loop* communities pursue their work and advance factual claims largely in and for a context of expert peers, but do not typically advise or claim authority over external agencies or laypersons as clients. Papyrology, for example, is typically closed-loop. *Open-loop* communities use their expertise not only to describe or diagnose a client's situation, but also to advise the client on normative choices, presuming authority for doing so. Medicine, for example, is open-loop.[2] And the loop is open to client feedback that may ultimately alter practice.

These distinctions are important because the trend of suspicion and rejection of expertise is largely targeted to open-loop communities, present only in special cases of the closed-link variety,[3] while performative expertise is generally unaffected. For that reason, I want to focus in what follows on what we call "the professions"—the most fully developed and highly institutionalized set of open-loop epistemic communities.

2. Institutionalized Expertise: The Professions

There appears to be an unusual consensus about the defining features of a profession (see e.g., R. P. Wolf 1969, 9–27; D. A. Schon 1984, 22ff; L. S. Shulman 2004,

---

[2] Parallel terminology is used by Robert Pierson (1994). I have benefited from Pierson's analysis in the discussion of epistemic responsibility that follows. There are borderline and oscillating cases, which I will mention later.

[3] For example, biology in regard to evolution; geology and paleontology in regard to the age of the Earth.

520–44). The list of proper exemplars, however, is controversial. Medicine, law, and engineering, for example, are universally considered to be professions; but event planning, retail sales, travel agency, real estate, and others are contested. The issue is not whether one can acquire expertise in these domains, nor is it whether one can be paid for one's competence in these domains, for both are obviously possible.[4] The question is whether they are professions. So, let us unpack the standard view, relying on the uncontested examples as ideal types.

A profession is a domain of work structured as a practice and organized for the purpose of advancing a social good. Thus, the law aims at justice; medicine at health; the clergy at salvation; the military at victory and a secure peace. Because a profession draws on a body of theory and specialized knowledge, it requires advanced study; because it also requires proficiency in relevant technical skills, it requires mentorship and training. As Plato (1997) pointed out in the *Statesman*, this sort of expertise cannot be reduced to meticulous rule following. Because the conditions of practice involve unique cases and uncertainty, a professional needs first-hand experience (*knowing-by-acquaintance*) and the tacit knowledge it yields, along with discerning judgment. The usual means for pursuing these epistemic goods are master-supervised apprenticeships, practicums, internships, and other forms of monitored experience in the art of the practitioner. Moreover, such experiences allow for a colleagueship in which professional knowledge, technique, experience, and judgments about cases can be shared with peers.

Acquisition of all these epistemic goods is clearly a worthy achievement. They are the content of the expertise the possession of which distinguishes a professional from the lay person. Professional expertise grants delimited epistemic authority within the domain of the practice, and it imparts an identity and social regard to the practitioner. (The concern about some occupations, like event planner or estate agent is, in part, whether the content required for expertise is sufficiently complex and specialized to elevate them to professional status.)

A profession constitutes an open-loop, regulatory epistemic community; practitioners are permitted a certain range of self-regulation. Mature professions use an array of structures to regulate practice. Members of the profession, through representative bodies, set standards for certification and license others to practice. They articulate and publish codes of professional ethics, institutional standards, guidelines for best practices; and they censure or suspend or expel violators. Protocols of peer review are employed for publication of research, case reports, and innovative techniques, using shared standards of evidence. Through these devices, professionals influence the agenda of the profession. Professions evolve a

---

[4] English speakers often use the adjective "professional" to mean "receiving pay for," as in "professional shopper." Thus, the antonyms of "volunteer" or "amateur:" the amateur golfer "turns pro." But this is a weaker, more specific sense of the term than is the point here. In this case, a "professional" means *one who is engaged in a profession*.

culture with hierarchies of roles and rituals, vocabularies and implements, recognitions and honors—and, in some historic cases, distinctive dress.

Professions rest on a reliable body of knowledge deemed relevant to producing the intended social good. They cannot be grounded in mere speculation, hit-or-miss guesses, folklore, anecdotal information, unregulated experimentation, esoterica, or recipes that lack a theoretical base; genuine knowledge and understanding are essential. But this does not mean the epistemic base is fixed or frozen: it evolves with advances in knowledge (sometimes from discoveries in other domains), technological improvements, changes in social institutions, and shifts in research needs. Moreover, the bridge between theory and practice, between research and application, works in both directions: while research and current knowledge regularly inform practice, the observations of practitioners in the field can inform and sometimes alter theory and accepted truths. The reflective practitioner both applies accepted knowledge and (at least occasionally) alters it. So, although medical researchers and practicing physicians are engaged in quite different activities (and occasionally disparage each other's roles), their symbiosis ideally forms a single community of interdependent expertise, united both by its shared epistemic base and the social good it pursues.

For professional practice to be effective, it must have the respect and trust of the public, which is quite a different thing from having epistemic authority.[5] The social good which the profession serves must be widely perceived as a good by laypersons, and practitioners must be seen as generally effective in the service of that good. Should those conditions fail, respect is lost. We only get to Shakespeare's "let's kill all the lawyers"[6] if we believe that lawyers are not serving justice. And when the social good intended by a profession is prized, it is clear even to the general public that permitting anyone to practice, even the unqualified or the unscrupulous, would constitute a social danger.[7] Many of the self-regulatory

---

[5] Some claim that expertise is simply a social construct, created by respect and public reputation. This view misses the veritistic pull that is presumed by normal public respect and by the activities of experts.

[6] "The first thing we do, let's kill all the lawyers." Shakespeare, *Henry VI*, Part 2, Act IV, Scene 2.

[7] These are precisely the points made in the thought-experiment of *Statesman* (Plato 1997, 298a–e). Thinking of those whose expertise we value, like doctors and ship navigators, imagine we establish a rule that

> both laymen and craftsmen other than steersmen and doctors would be permitted to contribute an opinion, whether about sailing or about diseases, as to be basis on which drugs and tools of the doctor's art should be used on patients, and even how to employ ships themselves, and the tools of the sailor's art for operating them, for facing not only the dangers affecting the voyage itself from winds and sea, but encounters with pirates....

Then, recording both the recommendations of the non-experts and the experts, suppose we stipulate that *the majority opinion* would determine the required policy. Or, what if such crucial roles were simply assigned by *drawing lots*, "and that those who take office should execute it by steering the ships and healing patients according to the written rules." The point of this bizarre scenario is to make four aspects of human societies vivid: (1) our deep dependence on trustworthy experts; (2) our vulnerability to malevolent experts; (3) the dangers of substituting public opinion for genuine knowledge and skill; and (4) the fallacious view that expertise can be reduced to following a detailed instruction manual.

and quality-assurance devices put in place by professions are also aimed at reassuring the public and their lay clients. Trust and confidence of non-competent citizens in a profession and its practitioners is bolstered by the knowledge that there are rigorous tests of competence for licensure, that there is an enforced code of ethics and standards of practice, that malpractice is punished, and that continuing education is required to maintain currency. Visual symbols may have a similar effect: hanging out one's shingle, posting diplomas or certificates or honors. Universities, especially their graduate professional schools, host the training of professionals, sustain the research, certify the epistemic content of professional expertise, and provide a basic level of professional credentialing. These vital roles feed public confidence (only) to the extent that the public respects institutions of higher education.

In all the ways presented in this account of the ideal type, professions are *institutionalized* forms of expertise. This ideal with its generous idealism has long been challenged with a darker vision: professions are self-serving clubs, thriving on exclusivity, disposed to close ranks, jealously guarding self-regulation by obscuring their activities and speaking in needless jargon. Post-modernist analyses have deepened this critique: deconstruction has revealed the power relations that define the professions, the conceit of expertise to claim authority over the public, and the fabrication of an ideology that manufactures the need for the professional. Like nearly every other institution, the critic asserts, professions are political at their core, not merely epistemic. They lobby for legislation and public policy that serve their interests; moreover, they promote specific constructions of the social good and a doctrinaire view of appropriate techniques, marginalizing alternative methods. Think how the medical profession defines "health," medicalizes normal conditions[8] like shyness (C. Ireland 2009), and limits mainstream treatment—and how, traditionally, it has disparaged forms of alternative medicine. In short, professionals define what counts as knowledge in the field, and, therefore, what counts as truth.

These are trenchant observations, but I cannot accept them as "the whole picture" or as a rejection of the ideal. We cannot claim corruption of an ideal without affirming the ideal itself; we cannot claim epistemic injustice without an implicit ideal of epistemic justice. Yet clearly, these critiques may be pertinent to understanding the two contemporary trends regarding institutional expertise. I will turn next to the first trend, and the ways in which professionalization is changing.

3. Evolving Professionalization

The phenomenon of expanding professionalization began slowly in the nineteenth century but has increased rapidly since the post-war boom of the 1950s. One

---

[8] My thanks to D. Pritchard for this example, and more for several penetrating questions and helpful comments on a draft of this article.

measure is the increase in the number of recognized professions along with the number of occupations seeking professional recognition. "Recognition" has several dimensions: (1) practitioners organize as an epistemic community around a substantial body of specialized knowledge and techniques, and adopt the culture of a professional; (2) the lay public regards the practitioners as professionals, with the rights, obligations, and social status that carries; (3) governmental agencies require licensure for practice, bowing to practitioners for determining criteria and standards; and (4) universities establish graduate schools with degree programs devoted to preparatory education for professional practice in the field.

Medicine, law, and theology have long traditions of professional recognition. During the nineteenth century, dentistry, music, engineering, education, agriculture, forestry, architecture, business, dramatic arts, mortuary science, and other fields became professional in the ways described. By the close of the twentieth century in the United States, public policy, accountancy, library science, social work, (graphic) design, dance, environmental sciences, hotel administration, computer science, recreation management, tourism, folklore, and many others had become professions with graduate professional schools. As professions mature, it is common for the educational requirements for licensure to grow. In 1836, Abraham Lincoln became a lawyer simply by intensive self-study and standing examination by the Illinois Supreme Court; in most US states today, he would need to earn a Doctorate in Jurisprudence from an accredited law school and pass a bar exam. Another example: a field that currently seems to be moving toward increased professionalism is physiotherapy, where passing a national exam is required for licensure, and training expectations are shifting to graduate programs and continuing education.

Though occupational prestige, limitation of professional competition, and economic advantages are contributing motivations to a quest for professional recognition, it is true that many traditional jobs have simply become more complex. Expertise in the field may now rest on a larger, more specialized epistemic foundation, and the practitioner may face a more complicated policy environment. In addition, professions are increasingly global, responding to transnational standards, practicing in multicultural contexts, and researching with diverse and far-flung peers. Today, more of us seek expertise in more areas of daily living: we consult experts to organize our closets, pursue an exercise program, to plan a wedding, appraise our heirlooms. Professionals in one field are part of the lay public for other fields, of course, and they are likely to carry an understanding of the value of credentialed expertise.

Another factor is the increasing specialization within professions, resulting in narrower specialties and subspecialties. Given its rich base of scientific knowledge, continual research, and advanced techniques, the field of medicine has long required a specialization of roles. Now, even the medical generalist, the primary care family physician, is a specialty. Any complex operation, such as surgery for a brain tumor, now requires a team that includes a neurosurgeon, an oncologist or pediatric oncologist, a radiologist, an anesthesiologist, surgical nurses, and several others. Recently, in mature professions, "para-professionals" such as physician's assistant and paralegal have become a job specialty requiring advanced training. Perhaps the final step in medical specialties is the establishment of the medical ombudsman who helps patients navigate the opaque medical and insurance bureaucracies. Increasing specialization renders an epistemic community of peers both smaller and siloed; professional conflict and rivalry can become dysfunctional, as it is harder to maintain the unity of the profession as an epistemic community. In addition, as preparatory professional education claims more time within a student's educational program, it loses connection with its roots in liberal education and a broader, richer understanding of the social good.

Moreover, professional programs within universities have now constructed an increasingly complex network of relationships with other elements and agencies of the community. For example, medical schools now own or are affiliated with hospitals, free clinics, pharmacological and biotech research corporations, rehabilitation centers, proprietary software for professional use, patented products of research, etc. One consequence of this is that the allegiance of faculty and staff is not solely to the university or the profession; there are many implied commitments in this network that engage the time, attention, and loyalty of the faculty. Those active primarily as practicing professionals often seek formal university affiliations that carry titles, benefits, and privileges.

Professional associations now wield enormous influence over licensure requirements and program accreditation, often stipulating curricular content, adding requirements, and raising minimal levels of study. Their reaccreditation review expectations are more specific and usually more demanding than those of the institution's own regional, state, or national accrediting agency. The culture of assessment and ranking has now overtaken the professions—and not just from the accrediting agencies. All professional education programs are thoroughly engaged in formal assessment, but they are also now subject to less formal public rankings (as is almost every institutionalized enterprise). Individual practitioners themselves are also assessed and rated, *not just by peers but by their clients* through websites, social media, and publications. Self-regulation and peer review processes are increasingly supplemented by government oversight and independent assessment. Naturally, such external assessments influence professional performance and presentation. Hospitals now promote their low average waiting time in the emergency room. Private practices publicize their "voted best" ratings. There is a tendency to be drawn away from the valuable assessment of factors that might improve practice toward boasts that are more useful in image marketing.

Finally, it is noteworthy that professional practice in all domains is devolving to a business. What I mean is that the business elements of a practice are looming larger within the daily routine of the practice. Operating a practice has indeed become more complicated, but especially in aspects that are peripheral to the core art of serving the client. Compliance with government policy and reporting, personnel issues and payroll, patient/client billing, service documentation, insurance and other third-party interaction, liability concerns, complex scheduling, and so on—these require added specialized staff. The novice professional who "just wants to practice and serve clients" soon finds that private practicing is owning a business requiring efficient management. It has become common in many professions for offices to have fewer full professionals than administrative support staff. The realization that a professional is an entrepreneur, an owner of or partner in a business, a managerial role for which many feel unprepared, has spurred demand in numerous fields for professional degrees to include training in managing a practice.[9] But such a fundamental shift in one's vision of professional life carries risks. Medicine can seem less about patient health, more about the finances of the practice, insurance coding, or the liability of care. Law can seem less about justice, more about billable hours and profits for the partners. When a practice becomes distant or detached from its social good, the means may overwhelm the end.

In summary, this narrative portrays the professions as forms of institutionalized expertise that connect knowledge to practice, as growing in extent and diversity, and gaining in complexity, power, reach, and influence, yet also as struggling to prevent epistemic silos and to remain faithful to the human good at which they aim. All professions (and indeed all of higher education) are subject to these developments: curricular explosion, specialization, globalization, assessment and ranking, and the "mission creep" of the practice-as-business model. These developments also suggest some of the factors that may contribute to the second trend: the public resistance to the epistemic authority of experts.

4. The Rejection of Expertise

The proper place of expertise has always been problematic for democracy. It seems common sense that one who does not understand a matter should heed the claims of one who does. Still, the notion that recommendations of *cognoscenti* should be treated as privileged testimony that ought to govern conduct of others violates the egalitarianism of democratic epistemology. But the faith that every human being has the capacity to recognize the truth, the right to think for oneself and believe what one chooses, crashes into hard facts of social life, especially in a complex,

---

[9] For example, a recent issue of *Dentistry iQ*, a leading information source for English-language dentists, a printed piece headlined "The Desperate Need for Business Education in Dentistry." Among the advice to practicing dentists: "You must work *on* the business, not just *in* the business." (A. Greer 2018).

advanced technological society. Constructing knowledge is a communal endeavor: we cannot verify every fact for ourselves; most of what we know, we learn through some form of testimony. We don't have the time or desire or often the ability to learn many things, even many things that impinge on our own vital interests. In these matters, we must trust those who know.

Today, however, there is a crisis of confidence in expertise and the traditional institutions that create and sustain it. In the United States, surveys by the Pew Research Center (B. Kennedy, A. Tyson, and C. Funk 2022), and Gallup (M. Brenan 2021), and other highly regarded independent surveyors indicate a significant decline in public confidence in public schools, higher education, the medical community, mainstream media, economists, and scientists. A major survey in the UK (British Social Attitudes 37, 2020 and 39, 2022) finds that trust in British government has dropped to lowest level in decades.[10] These surveys also reveal an increasing partisan gap in confidence: in the US, Republicans express far less confidence in major institutions than Democrats, except for religious organizations, financial institutions, and the police. In Britain, there are parallel gaps in similar directions between Brexit Leavers and Remainers (respectively).

This distrust, resistance, and even rejection is especially strong for open-loop professions, which engage both descriptive and normative roles, and address the public. As professional practices adopt a business model, attention shifts to the "market response." The clients are the market, and their satisfaction is crucial. The clients become the experts, and homage is paid to the wisdom of crowds.[11] Who needs a film critic with a degree in cinema studies when one can easily see how an audience of thousands judged the film? Who needs a restaurant critic's pronouncements when there are dozens of websites that tally the evaluations of those who have dined there? Assessment from the clients' perspective becomes the standard, and quality seems displayed by public ratings, the number of "likes" or "thumbs up" (or down), the content of the customer's comments. Since patients are the market, not the health professionals, pharmaceutical companies now pitch

---

[10] Some examples: Gallup reports US confidence levels were only 23% for public schools, 44% for the medical system, and a woeful average of 18.5% for print and TV news (M. Brenan 2021). Pew reports: "Overall, 29% of U.S. adults say they have a great deal of confidence in medical scientists to act in the best interests of the public, down from 40% who said this in November 2020. Similarly, the share with a great deal of confidence in scientists to act in the public's best interests is down by 10 percentage points (from 39% to 29%)" (B. Kennedy, A. Tyson, and C. Funk 2022). The British Standards Agency (BSA) reports that only 15% of those surveyed said they trust the UK government either "most of the time" or "just about always"—the lowest level in forty years. Reported satisfaction with the NHS, long a point of pride, stood at only 36% in 2022, a drop of twenty-four points since before the pandemic. Comparative data for forty-two countries may be found in reports of the International Social Survey Programme (ISSP).

[11] While I believe large groups can sometimes reach better solutions than individuals, I do not accept that this applies to all or even most problems; and when it does apply, the composition and size of the group become relevant. The well-known experiments are intriguing, but application to many other situations seems an excited over-generalization. A popular synthesis of this research is the work of J. Surowiecki (2005).

their pills directly to the public, by-passing physicians and creating a self-diagnosed demand for prescriptions. Patients are pressed to rate their "experience" and their physician. The value of peer review can soon be submerged by the larger phenomenon of public response.

For open-loop professions, clients and patients traditionally interact individually with professionals; many of these relationships are protected by privacy and client confidentiality. I noted earlier that social networking has facilitated *client-to-client* relationships and the creation of groups whose commonality is their relationship with a particular professional, firm, or establishment. This allows for sharing experiences *as* clients or patients. One could argue this is clearly a positive development: to some extent, this trend corrects for the self-absorption of professional practice; it reminds professionals that the good at which they aim is a social good, owned and interpreted by all; it revolts against the hegemony of experts and empowers the client as the best judge of professional services the client has received. The emerging epistemology, however, is more radical: it suggests that the truth is a matter of vote, not justified belief. And as such, it is a head-on challenge to professionalism and the epistemic authority of expertise.

Education is perhaps the most open of open-loop professions and, especially in the United States, education at every level is now bedeviled by this trend. In many states, teachers, librarians, and school administrators have been harangued by activist parents seeking to control the curricular content their students will encounter in schools. Books, especially textbooks, have become weaponized. Politicians have attempted to use legislation to regulate classroom interaction and discussion topics, especially in subjects like history, sexuality and sexual preference, racism, and religion. A third-time US Presidential candidate supports the election of school principals by parents. An emerging and contagious political rallying cry is *parent rights*. Taken as a whole, these events effectively deny or discard the value of educational expertise, the professionalism of educators, and the principle of academic freedom. The very concept of education depends on the hope of coming to *know* and on the value of knowledge over raw opinion. The attitude might be summarized in this way: expertise is merely ideology; since everything is ideology, we should adopt mine.

Because we cannot function in the world without reliance on testimony, the politically disaffected and the ignorant often disown established institutions and professionals and place their trust in other sources of information. Hyperpartisan media, conspiracy theorists, celebrities, and internet "influencers" displace genuine experts and institutionalized expertise. Their unjustified, sometimes dangerous, claims are received as reliable testimony by their followers. It is an epistemology built on a strange combination of gullibility and skepticism, and willful ignorance, claiming to offer "the real truth" behind the established institutional façade.

What has precipitated this erosion of trust and epistemic authority? Social networks have enabled and structured it, but they are not the only cause. Some blame is to be placed on the experts themselves. Widely circulated reports of professional malpractice and fraud, especially when unpunished, contribute; so do the juicy scandals in which established experts are hoodwinked by cunning pranksters or pretenders. Violations of codes of ethics and illegal activities erode the integrity and authority of the profession—not just of the individual violator. Another detriment is the showcasing of experts who are mercenary, endorsing whatever claim is to be made for a fee. The rush to publicize breakthroughs and discoveries strains research protocols and increases the probability of reporting false information. The public display of persistent disagreement among experts and frequent changes in professional consensus, especially on advice important to the public—on economic policy, diet regimens, or pandemic conduct, for example—sow confusion, doubt, frustration, and ultimately a sense that experts really know less than they profess. Those who are notable experts in regard to a specific domain contribute to the epistemic problem when they fail to "stay in their lane" and make pronouncements about a quite different domain in which they have no expertise—a behavior aptly called *epistemic trespassing* (N. Ballantyne 2019a, 2019b). No doubt, as evidenced by the partisan disparity in trust, the recent attacks on established institutions by the political right have played a major role in fanning distrust. When expertise is seen as political, a polarized politics correlates with polarized attitudes toward expertise. Intellectual groundwork for the critiques on established institutions and the undermining of the concept of truth was, however, offered up by the post-modernist left, then exploited by the political right. Though I will not recite the dismaying litany of ways in which our public discourse has been corrupted (it is so familiar by now, and so distressingly unhelpful), but all those epistemic distortions contribute to the skepticism and suspicion of institutionalized expertise.

Social science disciplines have, from their nineteenth-century inceptions, embraced both descriptive and normative aims, often drawing research-based conclusions that were promulgated as recommendations for social policy. The desire to understand human social life is naturally paired with the desire to improve society. Social science practitioners—economists, political theorists, sociologists, some historians, social psychologists, ethnic studies specialists, anthropologists, and others—sometimes oscillate between closed- and open-loop models of their profession. But how could we expect those who are expert to keep silent when their knowledge is relevant to vital social issues? We cannot and should not. Wise social policy is *informed* by deep knowledge of human affairs and systems of human interaction. Yet, when member of parliament (MP) Michael Gove famously inveighed against experts in the Brexit debates of 2016, it was this social science sort of expert he had in mind: Gove declared: "I think the people in this country have had enough of experts from organisations with acronyms saying that they know what is best and getting it consistently wrong" (Steerpike 2021). Indeed, one might think of political pollsters, economic forecasters, monetary policy authorities, writers of social science curricula, and recent turmoils that have surrounded their judgments. The root of their difficulty is that human social phenomena are causally overdetermined and complex; analyses are inevitably selective or reductive; and their predictive application is vulnerable to unintended consequences and dramatic disruptions. As a result, these disciplines are open not just to methodological differences, but also to ideological and activist approaches. This also applies to those fields grounded in the natural sciences but having strong social implications (such as public health and nutrition, environmental studies, urban planning); and a few fields are inherently activist (such as peace and justice studies). For the conscientious practitioner, intellectual caution and humility can seem politically ineffective or cowardly when the social consequences of an issue within their domain of expertise are significant and deep values are at stake. Unfortunately, public recommendations from these sorts of experts can seem less like informing and more like convincing or proselytizing, especially to non-experts who resist the conclusion or who prefer willful ignorance.

5. The Collision of Two Trends

The two trends I have described—the expansion and deepening of institutionalized expertise and the rise of public disfavor and abandonment of expertise—pull in opposite directions and reflect a bifurcated epistemology. But I argue they are also correlated: my claim is that *what is happening with institutionalized expertise, especially in the open-loop professions, feeds the rebellion against expertise.* The causes of large-scale social trends are likely over-determined. I don't mean to discount other factors I noted earlier—malpractice, mercenary practitioners, professional disagreements, effects of client-client social networking, postmodernist deconstruction, the polluted epistemic environment, and so on. But the argument that one trend feeds the other, if sound, has profound implications for the effective role of expertise in liberal democracies, for professional practice, and for professional education.

Professions aim at a recognized social good. But as professions become more specialized and subdivided, more complex and bureaucratic, more entrepreneurial and business-like, more siloed units within large-scale corporations, the social good for which the profession was organized becomes more distant and difficult to serve. Yet that good is the reason individuals become clients. If we erect a bafflingly complex, wildly expensive, profit-oriented or budget-constricted health care system, with the unjust outcome that public health is quite uneven; if we establish a many-layered legal system that is costly, mercenary, and capricious, unable to assure justice or even the appearance of justice; if we have a banking system so byzantine that it can plan to reap profits from mortgages known to be imprudent; if we deploy a system of education that is so over-burdened with non-academic duties and policy requirements, and so over-staffed with administrators and part-time adjuncts that educational quality is not a priority—then we can expect a loss of public faith in the integrity of these institutions and the professionals who staff them, and hence a contempt for their expertise and the authority it has carried.

Today, there is a tendency for many open-loop professionals to behave in close-loop ways: intra-professional concerns take the fore. The practitioner's focus is drawn toward the firm or institution, reputation, liability, compliance, the bottom line—not the client or the social good. The expert's need to stay current in mastery of the relevant domain can easily be swamped by the administration of the practice. From the client's perspective, negotiating a large-scale, specialized, bureaucratic system requires more time and effort than arranging a one-on-one professional consultation. Even determining just what sort of expert one needs can be difficult, and the more challenging it is to select and obtain professional services, the more frustrated and disillusioned with the practice the potential clients become. Of course, these sorts of institutional barriers always affect most severely the poorest, the most vulnerable, the least technologically proficient, those most in need of services.

Bureaucracy is often in the service of capitalism and the *homo economicus* model, and the mission creep of business concepts is notorious. When applied broadly in open-loop professions, clients and patients become customers or consumers—and they come to see themselves that way. The relationships become transactional, fee-for-service. Sadly, that model fails to grasp the implicit moral contract between physician and patient, attorney and client, professor and student. It reduces the epistemic community to an exchange market, and promotes the commodification of the product or service rendered: medical services are coded as procedures and products that fit the insurance company's approved list; education becomes the consumer-student's purchase of discrete courses and the accumulation of credit units; legal services are charted as billable hours. The exchange market model (drawn in classical economics) involves self-interested transactions with both parties understanding the product or service to be bought and sold. The disanalogy with expertise consultations is clear: when a lay person consults a professional, the non-expert—by definition—does not (and perhaps cannot) fully understand the content of the expertise required, or perhaps even specify precisely the service or product or requirements prescribed. Their relationship is epistemologically asymmetrical. An effective relationship requires the non-expert to have some level of confidence or trust (faith, if you will) in the expertise of the professional—grounded not in cordiality or popularity, but in competence.

To bestow trust is to enact one's willingness to remain ignorant, to abandon the need for firsthand verification, in regard to a specific agent and a given domain. When enlisting the services of an expert, the non-expert *is unable to validate the expert's services and recommendations*, and therefore *must* tolerate a considerable level of ignorance and extend trust. But trust in professional authority does not imply that the layperson simply be in thrall to someone who knows more.[12] There are, of course, public *indicators* of professional competence: the diploma or certificate on the wall, the awards or plaques on display, and yes, the client ratings or rankings. But the client still has an epistemic responsibility to judge for herself the worth of professional advice being given for her own case. Seeking expert advice does not entail loss of autonomy, absolute control, or abdication of responsibility. Clients may not be in a position to evaluate a lawyer's claims about court procedure or a doctor's knowledge of pharmaceutical interactions, but each remains the proper judge of whether that advice should be followed in their own case. Each client is free to reject expert advice or even to withdraw trust.

6. Coping with the Epistemic Asymmetry of Expertise

Certain basic hard facts confront us: (1) The scope of knowledge, technique, and understanding to sustain a complex human society requires the division of epistemic labor, and thus we need and value expertise. (2) The epistemic asymmetry that exists between experts and the lay public, between professionals and clients, is ineluctable. (3) That epistemic gap is rapidly increasing due to specialization, expansion of knowledge, and professionalization, trends which are likely to continue. (4) Trust—earned *and* given—is essential to bridge this asymmetry. (5) Contemporary liberal democracies are experiencing increased distrust of experts and established epistemic institutions, and a turn to alternative sources of information. I have argued that the third fact is one factor contributing significantly to the fifth.

There are plausible ways to ameliorate the inevitable epistemic gap, to earn trust, and perhaps to reclaim the valid role of experts. Unfortunately, most seem obvious or commonplace, platitudes that are significant if taken seriously, but easily shelved. Nonetheless, they are worth highlighting in conclusion. Most are implicit in my earlier comments. Some are steps that require commitment by the profession as a whole; others fall to individual professionals. Those have

---

[12] A sizeable literature now exists on the issue of trust and autonomy in consulting experts in which this point is affirmed (e.g., E. Fricker 2006; J. Hardwig 2006; L. T. Zagzebski 2012; C. T. Nguyen 2018).

significant implications for professional education. Inevitably, other steps are the responsibility of clients and the lay public.

The most obvious step experts can take (also the quickest for others to propose) is to maintain or tighten *strict quality management* of the professions and their practices. This covers standards for professional admission, ethical conduct, meaningful peer review, promulgation of "best practices" and ethical standards, and so on—measures that are intended to protect the integrity of the profession and, for open-loop professions, the client relations of practitioners. This need not require additional layers of bureaucracy—indeed, it might involve dismantling or reducing some existing agencies. (Many professionals would retort that such measures are already in place, and occasional publication of violations only proves they are already sufficiently "tight.")

To be reflective and trustworthy practitioners, professionals need to reach beyond competency to acquire practical wisdom, beyond legal compliance to moral understanding; beyond a ritualistic acknowledgment of the social good that defines their profession to a broadly humanistic understanding of and commitment to that good. This means placing limits on the practice-as-business model and its effects on service. So far as possible, every important decision made by professionals in a practice should be decided by reference to the central social good that is the *raison d'être* of that profession. Just imagine if every academic institutional decision were subjected to the test: *is it good education?* Unfortunately, some professions suffer from a confusion of social goods, such as journalism (especially broadcast, cable, and web) in which reporting news, and providing entertainment, and bolstering an ideology have become conflicting aims. Other professions operate on the principle that working for the interests of individual clients structures a system that works for the social good, such as lawyers who, serving their innocent or guilty clients with the strongest defense, claim the system ultimately serves justice. Clarifying for the public the values on which a particular practice makes its decisions can be quite helpful; for example, brokerage firms that now pledge their expert financial advisers and brokers will always act *in the interests of the individual client,* not for their own financial benefit, are addressing the issue of client trust.

Experts in open-loop professions should attempt to ameliorate the asymmetry of epistemic authority and eliminate an attitude of *power over* clients; that is, they should offer clients an understandable rationale for the course of action they recommend, and should regard such explanatory communication as essential to their service. This need for the professional to inform, even educate, clients should be addressed in professional education programs—not to better market services, but to help the client make informed decisions. Plato, we remember, doubted that those who are wise could (or should attempt to) fully explain the rationale for their decisions to those who were ignorant, so he resorted to lies, parables, and propaganda. But liberal democracies try, not always successfully, to rely on transparency, public education, and open access to information. These elements may help reduce the epistemic gap, but they cannot close it. Busy, finite creatures, however able, cannot know everything.

There are epistemic virtues and vices that are particularly salient for experts. Among the virtues are the many forms of *knowing the limits of one's competence*: showing intellectual humility, not professional arrogance; avoiding epistemic trespassing; attending to one's own ignorance, even "mapping one's ignorance" (DeNicola 2017), identifying factors that, even as expert, one does not know about the domain or specific cases; listening to the client, not silencing or stereotyping; avoiding epistemic injustice.

An epistemic community includes many roles: knowers, believers and disbelievers, skeptics; experts and non-experts, the ignorant; witnesses, reporters, those who give testimony, those who listen and receive testimony; critics, speculators, fabulists, and many others. We move easily in and out of these roles. What helps dampen the effects of the asymmetry is that our division of epistemic labor is not fixed into classes as Plato's ideal *polis* imagined, but constantly shifts across individuals—as when a lawyer needs a surgeon, or the doctor seeks the services of her lawyer patient. This phenomenon is spread more broadly, as when an engineer consults an arborist, a professor consults an auto mechanic, or when a patient reports side effects to his doctor. In this way, individuals alternate between asking and giving information, between engaging experts and offering informed advice regarding different domains. So, one commonly experiences both sides of many different asymmetric epistemic relationships.

Ultimately, there are epistemic responsibilities for the lay public, and especially for those seeking expertise or advice. One commonsense principle is that the greater the negative consequences of bad judgment or failure for the non-expert in a case, the greater the responsibility the non-expert has for careful epistemic evaluation. A diabetic, for example, has a special call to consult with qualified medical professionals, and to understand the disease, its implications, and the appropriate regimen. The catastrophic implications of climate crisis burden all citizens with the responsibility of some level of understanding and informed consideration of proposed responses. But even such selectivity of focus must be coupled with the display of epistemic virtues like open-mindedness, giving due weight to evidence, alertness to cognitive biases, giving judicious deference to expertise, along with the avoidance of epistemic vices like intellectual apathy, willful ignorance, and commitment to totalizing beliefs that preempt evidence and block inquiry. The challenge for the lay public is not to protect autonomy, but to develop effective epistemic agency.

References

Ballantyne, N. (2019a) "Epistemic Injustice." in *Mind* 128, 367–95.

Ballantyne, N. (2019b) *Knowing Our Limits.* Oxford: Oxford University Press.

Brenan, M. (2021) "Americans' Confidence in Major U.S. Institutions Dips," on *Gallup.com*, January 14, 2021 (Washington, DC: Gallup). Available at: https://news.gallup.com/poll/352316/americans-confidence-major-institutions-dips.aspx

British Social Attitudes (2020, 2022) *British Social Attitudes* 37, 39 London: National Centre for Social Research, available at https://www.bsa.natcen.ac.uk.

DeNicola, D. (2015) "Higher education, the professions, and the place of expertise." in *Higher Education and Society*, (eds.), J. DeVitis and P. Sasso, 25–40. Lausanne, Switzerland, and New York: Peter Lang Press.

DeNicola, D. (2017) *Understanding Ignorance: The Surprising Impact of What We Don't Know.* Cambridge, MA: MIT Press.

Fricker, E. (2006) "Testimony and epistemic autonomy," in *The Epistemology of Testimony*, (eds.), J. Lackey and E. Sosa, 225–50. Oxford: Clarendon Press.

Greer, A. (2018) "The desperate need for business education in dentistry." (*Dentistry iQ*, June 20, 2018; reprint from *Principles of Practice Management* e-newsletter), available at https://www.dentistryiq.com/practice-management/practice-management-tips/.

Hardwig, J. (2006) "Epistemic dependence," in *The Philosophy of Expertise*, (eds.), E. Selinger and R. Crease, 328–41. New York: Columbia University Press.

Ireland, C. (2009) "Scholars discuss 'medicalization' of formerly normal characteristics." Cambridge, MA: *The Harvard Gazette*, April 28, 2008, available at https://news.harvard.edu/gazette/story/2009/04/.

Kennedy, B., A. Tyson, and C. Funk (2022) *Americans' Trust in Scientists, Other Groups Decline.* Washington, DC: Pew Research Center.

Nguyen, T. C. (2018) "Expertise and the fragmentation of intellectual autonomy." *Philosophical Inquiries* 6(2), 107–24.

Pierson, R. (1994) "The epistemic authority of expertise," in *Proceedings of the Biennial Meeting of the Philosophy of Science Association*, vol. 1: Contributed Papers (PSA), 398–405.

Plato (1997) *Statesman*, in *Plato: Complete Works*, (eds.), J. M. Cooper and D. S. Hutchinson, 294–358. Indianapolis: Hackett Publishing.

Plato (1997) *Republic*, in *Plato: Complete Works*, (eds.), J. M. Cooper and D. S. Hutchinson, 426–35. Indianapolis: Hackett Publishing.

Schon, D. A. (1984) *The Reflective Practitioner: How Professionals Think in Action.* New York: Basic Books.

Shulman, L. S. (2004) "Theory, practice, and the education of professionals," in *The Wisdom of Practice: Essays in Teaching, Learning, and Learning to Teach*, (ed.), Shulman, 520–44. Hoboken, NJ: Jossey-Bass.

Steerpike (2021) "Fact check: What did Michael Gove actually say about 'experts'?" *Spectator*, September 2, available at https://www.spectator.co.uk/?s=Fact+check%3A+What+did+Michael+Gove+actually+say+about+%E2%80%98experts%E2%80%99%3F.

Surowiecki, J. (2005) *The Wisdom of Crowds*. New York: Anchor Press.

Watson, J. C. (2020) *Expertise: A Philosophical Introduction*. London and New York: Bloomsbury Academic.

Wolff, R. P. (1969) *The Ideal of the University*. Boston, MA: Beacon Press.

Zagzebski, L. T. (2012) *Epistemic Authority: A Theory of Trust, Authority, and Autonomy in Belief*. Oxford: Oxford University Press.

PART 3
SITUATED AND GROUP EXPERTISE

5
Expertise as Perspectives in Dialogue

Michael Larkin
Aston University, UK

Lisa Bortolotti
University of Birmingham, UK

Michele Lim
University College London, UK

1. What is Expertise?

Philosophers and social scientists have recently turned their attention to the notion of *expertise by experience* (e.g. Dings and Tekin 2022; Castro et al. 2019), a notion[1] that has been formulated to account for a form of expertise that does not depend on studying a particular issue but by having a particular experience.

> If those who are not experts can have expertise, what special reference does expertise have? It might seem that anyone can be an expert. We say that those referred to by some other analysts as 'lay experts' are just plain 'experts'—albeit their expertise has not been recognized by certification; crucially, they are not spread throughout the population, but are found in small specialist groups. [W]e will refer to members of the public who have special technical expertise in virtue of experience that is not recognized by degrees or other certificates as 'experience-based experts'. (Collins and Evans 2002, 238)

In the background of many discussions about expertise the following question seems to emerge: are experts by experience *real* experts? But in this formulation the question raises a number of concerns: why do we need to demarcate expertise

---

[1] We are aware that the phrase 'experts by experience' is problematic and controversial and that in some contexts other phrases are preferred, such as 'lived experience advisors'. In this chapter, we use the phrase 'experts by experience' because we are interested in addressing recent literature that discusses the strengths and the limitations of that role, and we want to focus on the notion of expertise. However, the arguments presented in the chapter can be interpreted as an invitation to reflect on whether the phrase 'experts by experience' is adequate to the purposes for which it is used.

Michael Larkin, Lisa Bortolotti, and Michele Lim, *Expertise as Perspectives in Dialogue* In: *Expertise: Philosophical Perspectives*. Edited by: Mirko Farina, Andrea Lavazza, and Duncan Pritchard, Oxford University Press. © Michael Larkin, Lisa Bortolotti, and Michele Lim 2024. DOI: 10.1093/oso/9780198877301.003.0005

in that way? How does power affect the process by which lived experience is recognised as a legitimate source of expertise? Is there an underlying assumption that there can only be a limited number of experts and thus some gatekeeping is required? Or that experts by experience can only have a well-defined, limited role in their contribution to shared knowledge and decision making? Such concerns resonate with our experience in meetings with patient and public involvement groups, where people with lived experience are often assigned a mere consultancy role, and asked to comment on a small number of issues, but are not actively engaged in the more general discussions that occur between professionals within the same meetings. This is an illustration of the idea that experts by experience are not considered as *real* experts and that their involvement is often tokenistic.

In the literature, objections raised to the notion of expertise by experience have started to be identified and addressed. These are presented either as general worries about expertise that are made more salient by the case of expertise by experience, or as concerns that are specific to the acceptance of lived experience as an epistemic resource. The general issues debated in this literature are the need to ensure the *objectivity* of knowledge, and the importance of finding effective ways to deal with *disagreement* among experts, combined with the need to integrate various perspectives in decision making. Such challenges are not unique to expertise by experience but can arise in encounters between experts trained in different disciplines, belonging to different schools of thought, or using different methodologies. The more specific issues with the legitimacy of expertise by experience can be seen as at least partially prompted by a threat to the authority of experts by training: such experts might worry that their more 'traditional' expertise is at risk of devaluation if other routes to expertise are validated. However, such issues are also conceptually grounded. One of these issues is whether contributions by experts by experience are a source of *evidence* for the claims and decisions they support. In other words, lived experience is not always recognised as a legitimate epistemic resource.

In this chapter, we acknowledge the urgency of clarifying the notion of expertise, but we consider questions about the legitimacy of expertise by experience as a lens through which key aspects of our epistemic practices come into clearer focus rather than as a problem in itself. In the course of the discussion, we start developing what we take to be a more productive way to look at the challenges of group decision-making: expertise is something that emerges from a collective and dialogical pursuit that involves a group reflection on the available evidence. That is, it arises out of collaborative discussion and critical consideration. One way to bypass the constant requirement to justify a seat at the table for experiential insights is to think of contributions as *perspectives in dialogue* rather than independent claims to expertise.

Our aim is threefold. In Section 2, we will revisit the notion of expertise and ask how individuals and groups come to be seen and treated as having expertise, independent of the source of their expertise. We argue that expertise requires scaffolding and is perspectival. It is not a property of an individual or a group but a function of the relationship between an individual or a group and the world around them. As Matt Stichter says:

> Deciding who to credit as an expert is not like describing a natural kind. We credit people with expert status because it serves a useful function, and we decide who to confer this status upon depending on what we want from them, and this also determines what power we choose to grant to them. So we want to avoid what some have termed an 'immaculate' conception of the expert—someone who counts as an expert whether we like it or not, and to whom we must defer judgment.   (Stichter 2015, 126)

In Section 3, we will consider and address some of the most common objections to the idea that lived experience is a source of expertise. How can experience give rise to knowledge if it is subjective? Aren't experts supposed to provide evidence to reach the best decisions for a group of people, whereas lived experience is always an individual experience of something in particular? What type of evidence can experts by experience provide? And what is the value of this type of evidence? We respond to these challenges, arguing that in some contexts subjective experience is exactly the type of perspective we need, and highlighting that narratives reporting personal experiences can be evidence of something worth knowing, but are not the only contribution that experts by experience can offer. Moreover, we reiterate how being able to access and use information is necessary but not sufficient for any form of expertise. If it is reflection on experience that generates expertise as opposed to experience alone (e.g. Castro et al. 2019), the complaint that all that experts by experience can offer is a first-person account of their own experience misses the point. The point is that what the experience offers is a unique perspective on the problem that deserves to be reflected upon.

In Section 4, we offer some an example of how distinct forms of attaining knowledge, including expertise by experience, can be integrated in the mental health context and reflect on the process of co-design and co-production before making a concrete proposal about how the promotion of different perspectives in dialogue may be facilitated.

2. Expertise as Scaffolded and Perspectival

'Expertise' is usually defined in terms of a high level of skill or knowledge in a given field where the skill or knowledge can be attained by training, study, or practice. In this chapter, we address the notion of expertise *as knowledge* and leave aside the notion of expertise *as skill*. In addition, to make the discussion more manageable, we focus on situations where experts meet to make an informed decision or solve a problem.

Most definitions of expertise (such as the one to be found in the dictionary of the American Psychological Association, where expertise is defined as 'a high level of domain-specific knowledge and skills accumulated with age or experience') present expertise as a property of an individual or a group rather than a function of the relationship between the individual or group and the world around them. Most commonly, the property referred to is *knowledge*. However, it is unhelpful to define expertise as a high level of knowledge where knowledge stands for an individual's or group's capacity to gather information pertinent to a given topic and draw upon this information when appropriate. That is because, if the availability and use of information were sufficient, then it would be very difficult not to attribute expertise to anybody who has access to a source of information, such as someone sitting in a library or searching the internet.

The literature on *distributed cognition* (e.g. Hollan et al. 2000) and *communities of practice* (Wenger 2000) are powerful reminders that an intrapsychic view of expertise is inadequate. Both of these notions, and the research developed from them, emphasise the means through which intrapersonal knowledge is boosted, scaffolded, and augmented through dialogue with others, through structured interactions within knowledge systems (as informal as team meetings, or as formal as peer review), and opportunities to 'offload' tasks to technology (e.g. see Hollan and Hutchins 2009). These processes apply to domains as varied as flying a plane (Hutchins and Klausen 1996) and working in public health (Barbour et al. 2018).

An alternative, more enactive, formulation suggests that expertise is *afforded*—by people, questions, culture, expectations, power relations, and systemic structures. For example, an academic in their university office, with their colleagues across the corridor, their student asking the 'right' kinds of questions, the spines of their books on the shelves to prompt recall, and the internet at their fingertips, may feel very comfortable in their professional expertise. But the same academic may feel considerably less able to draw upon a sense of their own expertise when they are caught off guard and asked the 'wrong' sort of question in a media interview, while at a bar with friends, or in a taxi. These examples underscore that expertise is usually defined and performed in a given environment, and in that environment, it is accessed and benefits someone, either another individual or a community. Expertise is a form of *social contribution*, which is—by nature of its recognition *as* expertise—valued and, correspondingly, afforded by others.

If expertise is afforded, then this has important consequences, because affordances can be changed at the interpersonal level through skilful facilitation, preparation, and expectation-setting. A well-planned and well-chaired research meeting allows for different kinds of expertise to be heard, for example. Expertise might also be afforded via thoughtful configuration of the tools, environments, systems, and structures through which different kinds of expertise are recognised and warranted. Think for example about the conventional structures of the 'ward round' in which the patient's expertise is barely heard, while a range of professional forms of expertise are granted more time, space, and authority. Now contrast this with the kinds of patient-and-family-first discussions which characterise the Open Dialogue approach (Seikkula 2011). In the traditional ward round, professionals talk *about* the patient. In the Open Dialogue model, the patient's concerns are the *central* focus—and no one talks about the patient unless the patient is present.

Moreover, in the complex environments in which we live, most decision-making tasks require inputs from distinct sources of information and information is attained via a multiplicity of methods, so each contribution needs to be integrated in a wider context. Consultations which incorporate perspectives from sales, user experience, marketing, and product design are needed to make decisions about new products in the business boardroom. Problem-solving in technological industries requires the input of designers, developers, software engineers, programmers, hardware managers, and moderators. Emergency responses in public health draw on the insights of epidemiologists, infectious disease specialists, psychologists, and health communication experts. Each party in these situations offers valuable information despite their distinct perspectives and areas of specialty—all of which need to be carefully balanced, navigated, and integrated through dialogue, to reach a comprehensive and suitable plan. The key claim here is that people's capacity for providing expert insight can either be scaffolded or obstructed. The recognition that expertise is situated in this way opens up the possibility of creating environments and relationships in which different kinds of expertise can be heard and understood. Keeping these contextual and ethical matters in mind enables us to formulate a more useful *applied* concept of expertise.

Individuals or groups we consider as experts do not merely gather and use information but understand the information at their disposal and know how to apply it in order to engage with problems and processes, or to come to a decision. They know something about *how* to use what they know. Such 'meta-knowledge' arises from prior experience and is situated. As with many situated phenomena, it will reflect certain social inequalities: you may have learned how to successfully chair a committee in such a way that it reaches decisions that you approve of; or you may have learned what information to share with (and withhold from) a health professional, in order that they suggest a treatment that you will be able to accept. In both cases, you may find that what you have learned in one domain has some utility in the other.

However, if we make the leap too quickly from saying that expertise is just 'high level knowledge of information pertinent to a topic' to saying that it is 'the capacity to understand and apply high level knowledge', then we may overlook an important feature that is especially salient in some contexts, such as health and mental health, where professionals work in multi-disciplinary teams. This critical feature is that the information *pertinent* to any topic can come from many possible sources (e.g. distinct professional and disciplinary knowledge; distinct methodologies and epistemologies), and can take many possible forms (for example, statistical patterns, experiential narratives, clinical images, organisational logics, professional standards, treatment guidelines, case studies, and so on). There are many perspectives.

Expertise is *perspectival* in a very important sense: it implies a view which is situated *somewhere*, and which provides—under the right conditions—*depth* of insight. Resolution of differences between expert views is something which occurs *dialogically* (e.g. a team talks it through and decides on balance what the best course of action is); through *consultation* (e.g. options are shared with a wider community and the members of that community provide feedback); or through *consensus* (e.g. stakeholders work together to co-design solutions that meet everybody's needs). The capacity and opportunity to engage in such processes is indicative of some degree of expertise, of *knowing how to use what one knows*. Thus, expertise is not just specialist knowledge. It is often built upon such knowledge, but in practice it is a form of reflection which arises from the extensive experience of seeking to understand something—it is a hard-earned familiarity with the contours of a particular set of problems. The ability and opportunity of an expert to draw upon familiarity with a problem *in a way which is valued by others* comes from the expert being in a situation where they can *reflect* upon the knowledge they have, for the benefit of others. This reflective thinking is a product of an interaction between the person's potential and capacity, and the context's affordances and scaffolding, and it is shaped to some extent by the values of the other parties who are present. In the context of health and mental health, these dynamics are captured in the popular meme:

DOCTOR: don't confuse your Google search with my six years at medical school.
PATIENT: don't confuse the one-hour lecture you had on my condition with my twenty years of living with it. (Greenhalgh 2018)

If we conceive of expertise as situated in this way, as something that is always scaffolded and perspectival, we appreciate a number of things: that one can become an expert by many routes; that expertise is generally performed for some purpose to benefit someone; that it can be afforded or obstructed by features of the context in which that performance is sought; and that it involves offering reflective and perspectival insight into a particular problem or topic. This understanding of expertise can apply to expertise derived from (personal or clinical) experience, just as well as to expertise stemming from specialist or technical knowledge (via scholarship or research).

3. Subjectivity and Disagreement

Expertise primarily derived from specialist or technical knowledge is commonly assumed to be genuine expertise or a superior form of expertise, whereas expertise by experience is often thought to be a sort of pretend expertise or a second-class form of expertise, if recognised as expertise at all. In many disciplines dealing with the real-world effects of complexity and uncertainty (such as social science and medicine), the admission of multiple and perspectival sources of expertise is considered to be both pragmatically and conceptually sensible. However, expertise by experience is rarely acknowledged as an example of this acceptable heterogeneity, and that is due to the assumption that expertise by experience is not based on objective knowledge.

How generalisable is the significance of expertise by experience for decision making and problem solving? Expertise by experience grants us insight into things which are 'human-scaled' and thus is more likely to make an important contribution to mental health research than to climate science. But acknowledging the value of the 'human-scaled' perspective can be a useful route to reflecting on the bigger picture. Not only can a climate scientist persuade us of the urgency of the problem of climate change by sharing the intensity of their personal concerns, but experiential insights can help us better understand the complexities of the problem. For instance, they can get us to appreciate how climate change does not affect all human lives equally. There are some populations and communities who will feel its effects at an earlier stage and more dramatically than others. Experiential insights tell us something about the *human systems* aspect of climate change, even if they do not add much to our understanding of the *physical systems* aspect.

In domains like social science or medicine, practitioners often encounter situations where objectivity is unattainable or undesirable, or face problems that have irreconcilable features—under these conditions, collaborating to develop a nuanced and negotiated formulation of the problem is the way in which consensus is reached and an 'expert' view emerges. In what comes next, we focus on a specific context to make our discussion more manageable. In the recent literature in philosophy of science and medical ethics, there has been a genuine attempt to legitimise expertise by experience in the context of mental health, integrating the experience of people whose health is at stake with the views of experts by training. It is overwhelmingly plausible that a person who lives with a specific health condition acquires knowledge about the ways in which the health condition affects them (via lived experience), and also acquires knowledge about the ways in which various medical interventions impact on their health (also via lived experience). This knowledge can contribute to practical decisions about treatment but also to a better theoretical understanding of the health condition as such.

3.1 Objectivity

There are several notions of objectivity that have been discussed in the philosophy of science (Reiss and Sprenger 2020). Objectivity has been interpreted as: (1) a way of looking at the world that does not depend on any perspective, but is neutral—as in Nagel's (1986) *View from Nowhere*; (2) a way of looking at the world that is not affected by any value or normative commitment—what Kuhn (1977) denies when he talks about the *incommensurability of rival paradigms*; (3) a way of looking at the world that is not influenced by personal bias—what Feyerabend (1978) deems a misleading ideal in science. In feminist approaches to the philosophy of science and in epistemology, such accounts of objectivity are rejected on the basis of the value-ladenness of science, (Longino 1990) and the perspectival and situated character of all forms of human knowledge (Haraway 1988; Harding 1993).

In a recent paper, Şerife Tekin describes how patient perspectives have been excluded from the wide range of consultations leading up to the revised version of the *Diagnostic and Statistical Manual of Mental Disorders* (DSM-5) for fear that they would compromise the objectivity of the process.

> It is also important to note that in DSM-III and DSM-IV, the APA highlights that it sought the 'advice of experts in each specific area under consideration' (APA 1994, xv). 'Experts' in the DSM language refer to scientifically trained researchers or clinicians with recognized degrees who treat individuals with mental disorders. During the DSM-5 revision process, there were calls to the APA to include patients, by involving them in the decision-making process about the diagnostic criteria. However, this was not accepted by the DSM-5 Task Force, on the grounds that patient inclusion would reduce objectivity in the scientific process. (Tekin, forthcoming)

Tekin argues that the decision to exclude patients from the consultation was misguided because it was based on the wrong notion of objectivity for psychiatry, a positivistic notion where objectivity (as in the accounts listed previously) is characterised either as *the opposite of subjectivity* or as *concordance*. For Tekin, there is a more useful notion of objectivity, what she calls *participatory interactive objectivity*. Inspired by feminist philosophy of science and social epistemology, Tekin sees objectivity as a product of a 'participatory and interactive negotiation process' bringing together people with different forms of expertise, 'the professionals trained in psychopathology and those with experiential knowledge'. For Tekin, this notion of objectivity is not opposed to subjectivity and is not to be assimilated to concordance.

Tekin's contribution to the debate helps us see how key philosophical notions such as expertise and objectivity should be contextualised to the domains where they are applied. However, as a defence of expertise by experience, Tekin's proposal does not go far enough. Rather than redefining what objectivity means to legitimise the contribution of people with lived experience to the production of shared knowledge, couldn't we simply acknowledge that, in some contexts, the type of knowledge we need to access is *subjective*? If the reason why we find the contribution of experts by experience intrinsically valuable is that we need an understanding of illness as it is lived by those who are ill, what we need is knowledge of their subjective experience, a type of knowledge which training alone cannot offer.

Psychiatry and psychology are examples of the kinds of applied domains which deal with very few matters that can be known objectively. In these domains, subjectivity is a feature of *many* or *all* perspectives, and the thing which distinguishes one perspective from the other is not the presence or absence of objectivity, but the kind of evidence which one draws upon. Reflection on personal experience is a source of knowledge which has an especially important contribution to make when the other available resources (e.g. clinical experience, qualitative and quantitative research evidence, professional training and ethics, pragmatic and local knowledge) are in need of a contextual interpretation.

Tekin observes that objectivity is usually posited as the opposite of subjectivity:

> Such detachment or distance is thought to keep the individual observer from being overly invested in a particular outcome or fearing another, thus biasing/spoiling her understanding of the phenomenon in question. In the positivistic view, science should be based on impartial or third-person observations, not 'subjective' or first-person perspectives.   (Tekin, forthcoming)

If we are investigating the causes of a mental health condition or the centrality of a specific symptom for that condition, a person's report about their experience of illness will not provide all of the evidence required to come to a reasoned conclusion. But in the context of other research questions, such as 'how does it feel to have depression?' or 'how do people with unusual beliefs experience clinical encounters?', then we do not need evidence that is detached from the observer's standpoint, and we should not be concerned about biases spoiling the person's 'understanding of the phenomenon'. On the contrary, it is the person's standpoint that we need to have an understanding of, and potential biases will not spoil the object of investigation but will be integral to it. The value of lived experience comes from its subjectivity, because it provides a deeply personal, visceral understanding of the illness that goes beyond (or is at least qualitatively distinct from) any form of academic or scientific knowledge.

What is the relationship between objectivity and concordance? Again, if we are investigating the causes of a mental health condition or the centrality of a specific symptom for that condition, a person's report about their experience of illness can only be part of the evidence that we need. It will have to be integrated with other types of evidence, and gathered across a range of patient reports to identify patterns, that is, similarities and differences between individual perspectives. Discordance is not a weakness and does not give us a reason to doubt either the importance of the reports as evidence or the competence of the people who share those reports. Rather, those discordances are precisely the data we need in order to come to a better understanding of how the health condition manifests: the differences do not need to be eliminated but appreciated in a dialogic process where multiple perspectives coexist. As Tekin also suggests, when the phenomenon is complex, the 'inclusion of differently trained experts', and we would add, *people with different experiences*, make the process of inquiry 'interactional, collaborative, and critical'.

Thus, for at least some of the relevant research questions, we need neither unbiased information from a neutral standpoint nor perfect agreement among the people who play the role of experts. We need, as Tekin puts it, a 'participatory and interactive negotiation process' and there should be no pressure to brand it as a way to achieve objectivity. The view of expertise emerging from this discussion is a form of a dynamic and collaborative dialogue. Interactive participation and the ensuing negotiation processes won't guarantee objectivity but can ensure that subjective information is reflected upon and complements the existing body of knowledge on a particular issue in a way that is coherent and considered. For some research questions, objectivity is not the gold standard. In the process by which we attempt to understand and enhance patient experience of healthcare (e.g. see Green et al. 2020), for instance, the value of lived experience lies precisely in harnessing *subjectivity*, that is, harnessing knowledge of the experience of approaching services with a health condition that is perspectival, value-laden, and idiosyncratic.

3.2 Experience as Evidence

Objectivity and evidence are intimately related in the philosophical literature: for the logical positivists, and more generally in the philosophy of science, an objective inquiry is an inquiry that is driven by evidence and that makes it possible to reach agreement across multiple perspectives (Feigl 1953). So, it should not be a surprise that a standard objection to thinking of lived experience as a source of expertise, and to drawing upon the first-person accounts of people with lived experience, is that lived experience can only contribute personal narratives of illness. Given that these narratives, by being personal, are both partial and subjective, they are thought to be of limited relevance to understanding the illness more generally. The traditional approach is to emphasise the sharp distinction between the *narrative paradigm* and the *argumentative paradigm*, where only the latter provides evidence and contributes to rational argumentation (see Fisher

1984). But this dichotomy neglects the possibility that stories can serve as evidence for rational arguments and as a window into another person's or another group's valuable perspective.

In the domains of qualitative research, phenomenology, narrative psychology, and intersectional epistemology, there are a number of counter arguments rejecting claims that experiences can only produce narratives (e.g. see Høffding et al. 2023); that the narratives (and other structures of experiential knowledge) of one person cannot have relevance to other people's experiences (e.g. see Finfgeld-Connett 2010); that partiality can be avoided in any form of knowledge; and that subjectivity or perspectivity is always a form of epistemic deficit (e.g. see Grant and Kara 2021). Some of these defences of lived experience as a form of expertise arise from the literature which explores the different forms through which experience can be accessed and represented by research participants and researchers (e.g. see Thomsen and Brinkmann 2009; Boden and Eatough 2014); from methodological work on 'transferability' and 'theoretical generalisability' (Kuper et al. 2008) as tools for conceptualising the ways in which knowledge always requires contextualisation and interpretation in order to be applied in a new domain (e.g. see Konradson et al. 2013); and from the examination of the inescapability of a 'standpoint' for human knowledge (e.g. Bacevic 2023), or of the way in which specific standpoints can grant insights and capabilities that others may not.

These responses suggest that the objections to lived experience as a source of evidence are based on misleading assumptions. We already saw in the previous section that a description of the events can be valuable, and sometimes also essential to answering some research questions, even when it is not obtained from a neutral point of view or is not agreed upon. Arguably, the value of lived experience is precisely that it enables a particular perspective to be shared with people who would not otherwise have access to it. People are often biased when they self-report and confabulate reasons for their actions in contexts as diverse as consumer choice, politics, and morality (see for instance, Bortolotti 2018; Stammers, 2020; Murphy-Hollies 2022), and this is especially true when people report events that are emotionally salient and personally significant to them. This applies to experts and novices, and experts by training, and experts by experience in equal measure.

However, people's 'subjective' ways of telling their stories, sharing their knowledge, or explaining their actions are not a confounding variable, but part of the evidence we need when we appeal to lived experience. Stories are *curated* versions of the events and the curation itself can be evidence of the impact of the events on that person. The effects of the curation may be more obvious when information is presented in a narrative form, but other expert views are also the product of a process of selection and reconstruction of the data that can add or detract from their value depending on the aims and context of the exercise. When experts by training offer their advice, for instance, some information may be foregrounded and other information passed on quickly, and these curatorial choices can determine how the expert view is received and assessed and what impact it has on decision making.

Another related assumption is that lived experience can *only* produce personal narratives, where the narrative is the beginning and the end of the process. Instead, personal narratives are only one way to communicate a person's experience, and often they are just the starting point of a process of reflection and elaboration. Experiential insights may well be communicated via narrative, but qualitative and phenomenological researchers know that they can also be expressed and represented in the form of metaphors, concepts, images, emotion, or network maps, atmospheres, sounds, invariant structures, themes, and touchpoints. Many of these forms do not rely on core features of a narrative (such as temporality and causality), and they allow us to understand experiential knowledge along a number of different dimensions (such as intensity, quality, tone, connectedness, pattern, and essence). Substantive bodies of literature in qualitative research draw upon these sources of experiential insight as *data*. Thus, the claim that experience is only communicated via narratives cannot be supported.

Narratives can be a helpful form of knowledge. The blanket claim that personal narratives cannot be evidence is deeply unsatisfactory. In deciding whether a piece of information can be evidence, we need to agree on what we are gathering evidence *of*. A narrative of a person's experience of depression may not be a reason to believe that the next person with depression will have a similar experience, but it is evidence that depression can be experienced in that way. Although personal narratives may not always help us understand what caused an event or predict how likely that event is to happen again in the future, the elements of curation that make the story unique, how emphasis is used, what roles different actors and events play in the story, and what the narrative arc includes and excludes, provide powerful evidence of how the narrators see and interpret being in certain situations, of how those situations make them feel (Murphy-Hollies and Bortolotti 2021; Bortolotti and Jefferson 2019).

4. Expertise by Experience in Mental Health

There is an extensive literature which describes the harnessing of lived experience expertise for conceptual and evidential insight in mental health. As a source of data, participants in qualitative research studies, or in projects which use 'expert consensus' methods (such as Delphi surveys, or nominal group techniques), provide accounts of—and from—their lived experience. These accounts inform the theoretical and conceptual understanding of topics right across the disciplinary spectrum.

As a source of direct expertise, people with lived experience may lead and publish their own academic scholarship (e.g. Milton 2012), may provide insightful personal reflections (e.g. Barnsley 2023), or may work as part of a collaborative research team (e.g. see Bond et al., 2022). They may be the source of insights driving an experience-based co-design process (EBCD) (e.g. Springham and Robert 2015), or inform the general direction and ethos of a large project, through their contributions to an advisory group (e.g. Weich et al. 2020). Across all these roles, as we have discussed previously, people's experiential expertise, and their capacity for reflection, can be supported by the surrounding context, or obstructed. Considerable guidance exists to help researchers and other decision-makers to create supportive contexts (e.g. see NSUN 2015). Much of the guidance focuses on practical matters such as establishing clear aims; allowing adequate time and preparation for people to contribute; agreeing on processes and terminology which are inclusive; providing payment or other forms of appropriate recognition and recompense, and so on.

4.1 An Example from Our Work: Developing the Agential Stance

In some of our own recent work, we have been reflecting on how best to support experts by experience in engaging with the substance and detail of research data, in order to develop interpretations arising from, and useful for, a range of disciplinary perspectives.

Our project aimed to investigate whether the sense of agency of young people was safeguarded in mental health clinical encounters. To address this issue, two groups of researchers worked together to analyse the research data. One was a group of six people with backgrounds in philosophy, psychology, psychiatry, clinical communication, clinical practice, and public involvement in research. The other was a group of five young people with experience of mental healthcare services (the 'experts by experience'). As it happens, there were experts by experience in the first group and experts by training in the second group, so the divide was not as neat as it may sound. All the researchers brought their training and experience to the table.

What was the data to be analysed? The data consisted of filmed interactions between young people in a mental health crisis and practitioners in mental health emergency services. We watched the videos and with the help of the members of the team who are trained in conversation analysis, we examined interactions that went well, where the young person felt heard and shared information with the practitioner in a fruitful exchange; and interactions that did not go well, where the practitioner either dismissed the young person's perspective or excluded the young person from the decision-making process leading to the identification of further support.

To begin with, both groups met separately, facilitated by the member of the team specialising in public involvement in research, who also had lived experience of receiving mental health care. The analysis conducted by the group with lived experience was fed back to the academic group. The analysis informed the selection of subsequent video clips in an interactive research process. In turn, observations from the group with academic expertise informed selection of video clips and the agenda of the meetings of the group with lived experience. Preliminary themes were identified by the group of researchers with academic expertise based on the data analysis sessions and these themes were then brought to the group of researchers with lived experience for open discussion and further interpretation of meaning and significance.

This dialogue *first* within groups, and *second* between them, and *third* as one more integrated group, led us to identify a particular attitude, *the agential stance* that practitioners can adopt toward young people in a mental health crisis to make sure that the young person feels heard and has an opportunity to participate in the decisions to be made about further support (Bergen et al. 2022). The agential stance is made of five key steps: validation, legitimisation, avoidance of objectification, affirmation of the capacity for positive change, and involvement in decision making. Agreement on the key steps of the agential stance—including what the steps are, how best to describe them, what benefits and risks they have, how best to implement them in communication—was achieved via a process of reflection, discussion, and negotiation occurring within each group of researchers and across the two groups.

For instance, a choice driven by considerations brought to the table by experts by experience was to focus on how the practitioner can affirm the young person's capacity to contribute to positive change rather than talking about the young person's taking responsibility for their own health journey. In the psychological and philosophical literature, agency is usually associated with the notion of responsibility, and responsibility is in turn associated with blame and praise. However, in the context of a young person who is experiencing a mental health crisis, it was found that it is better to acknowledge that there is something meaningful and helpful that the young person can do to contribute to their progress rather than describe them as responsible for what happened to them or for how their health journey is developing. It is good to stress the young person's agency, but it is also important to identify the pressures and constraints they are experiencing due to the surrounding environment. Future health prospects are never entirely in the agent's own hands, and this becomes even more salient in the case of young people seeking emergency support for their mental health.

This is just one illustration of the many negotiations that occurred in the course of the analysis. Once the analysis was completed, members of both groups participated actively in public engagement and dissemination activities, and co-authored the project's outputs.

4.2 Integrating Perspectives in Dialogue

With our collaborators, we have learned a lot about how to elicit, integrate, and apply different kinds of expertise within a project. Here we mention four useful places to start.

a. Providing scaffolding
People need an accessible route into the problem. This is as true for an expert by training who is working outside of their disciplinary comfort zone as it is for an expert by experience working with academics and clinicians. Effective collaboration is often scaffolded when all parties are willing to use terminology and concepts in ways that are respectful and support mutual understanding. In well-managed projects where experts by experience advise on the management of programmatic research, for example, this arises from taking care to introduce key concepts, and then agreeing on shared terminology.

In collaborations focused on the interpretation of data—as in our own previous brief example—this may arise from finding an accessible means of *presenting* those data (for example, video recordings rather than transcripts). In each case, it is important to think ahead, and discuss *how* the next steps will work. 'Talking about how we will talk about it' is thus a simple and important component of supporting the kinds of collaboration which can afford the expression of expertise from a range of perspectives.

b. Accepting partiality
Empirical approaches which accept reflexive practices as responses to the 'problem of reflexivity' are helpful here (Jamieson et al. 2023). That is, if all parties begin from the position that *everyone* is likely to have some bias, and that they will all have some learning to do, then it becomes possible to begin to reflect together on what each party can bring to the project (both in terms of bias and insight), and what they have yet to understand. In experience-based co-design work, for example, these reflections often arise at the 'feedback stage' of a project, where different stakeholder groups learn about the differences and overlaps between their own insights, and those that have arisen from *other* stakeholder groups. These kinds of conversations encourage perspective taking and can make people aware of their own assumptions and preconceptions (Larkin et al. 2015).

c. Cultivating epistemic humility and curiosity
One practical feature which links epistemic humility with scaffolding is the *timing* and *scope* of the involvement of experts by experience. If all parties accept that they have something to learn from each other (and we argued that they should), then it is important to ask experts by experience the right questions at the right time. This means moving away from a model of occasional consultation, and

instead sustaining a dialogue with experts by experience *throughout all the stages of the project*.

This can ensure that research plans are developed for projects that are meaningful to all parties, and decisions are reached as a result of the collaboration. Cultivating genuine curiosity in the other parties' perspectives can contribute to avoiding the dangers of merely bringing expertise by experience in to validate or rationalise plans and decisions that have already been made. Openness and flexibility can make space for a more bottom-up methodology in the identification of both key issues and solutions.

d. Supporting constructive dialogue

In our experience, the process of 'building up a picture' of a problem and its possible solutions, arises from relationships of trust.[2] It takes time, goodwill, resources, and effort to develop trusting relationships between parties who may experience inequities of power and social-political capital. This is one important dimension on which collaboration with experts by experience is often *not* like collaboration with experts from other disciplines.

Here, collaborations benefit from researchers first taking care to think about—for example—how less powerful people will be heard, and how differences will be accommodated. This will include consideration of practical issues too, for example, time, money, childcare, and safety. In user-led or user-centered research, trusting relationships can be nurtured when other parties provide skilful facilitation, additional support, learning opportunities, supervision, problem-solving, and access to resources. In more traditionally structured research, we may need to think harder—and use more resources—to create safe and sustainable spaces in which experts by experience can both make their contributions and develop a dialogue with other experts.

To sum up, in this chapter we took the lively discussion about the merits and limitations of expertise by experience as an invitation to reflect on expertise more generally. We argued that expertise is scaffolded and perspectival, and we offered some examples in the area of mental health research where different forms of expertise have been brought together as perspectives in dialogue. In order for people to be experts, knowledge of the subject matter, attained via training or experience, and the capacity and opportunity to reflect on their knowledge are both required. But they are not sufficient: people also need to be in an environment where their knowledge can be accessed, used, and valued by others. To make

---

[2] In political philosophy, there is a literature on trust and expertise, focusing on the conditions in which the public trusts the experts, especially epistemic authorities with scientific expertise (see for instance, Whyte and Crease 2010). We recognise the relevance of this literature for the notion of expertise more generally, but also think that there are special considerations concerning trust that apply to the way in which experts whose source of expertise is different interact with each other in group discussion and decision making and that is our focus here.

the exercise of integrating expertise from multiple sources genuinely productive, we need our social environment to be structured in a such a way as to support constructive dialogue, and the people involved to be open to alternative perspectives, exercising their humility and curiosity.

Acknowledgements

We wish to thank our collaborators on the Agency-in-Practice project (peer researchers, Young Person's Advisory Group members, lived experience research managers, and academic researchers) for their feedback and discussion on an earlier draft of this chapter.

This work was funded by UK Research and Innovation (UKRI) under a Methodology call for the Medical Research Council (MRC)/the Arts and Humanities Research Council (AHRC)/the Economic and Social Research Council (ESRC) 'Adolescence, Mental Health and the Developing Mind' programme. The short-form title 'Agency-in-Practice' was chosen by the Young Person's Advisory Group. The funder's project reference is MR/X003108/1.

References

Bacevic, J. (2022) "'Philosophy herself': An essay." *The Philosopher* 110, 1, available at https://www.thephilosopher1923.org/post/philosophy-herself, accessed March 2023.

Barbour, L., Armstrong, R., Condron, P., and Palermo, C. (2018) "Communities of practice to improve public health outcomes: A systematic review." *Journal of Knowledge Management* 22(2), 326–343, available at https://doi.org/10.1108/JKM-03-2017-0111.

Barnsley, S. (2023) "OCD, metaphor, and me: The horse that can't eat apples." *The Lancet Psychiatry* 10(3), 170–71.

Bergen, C., Bortolotti, L., Tallent, K., Broome, M., et al. (2022) "Communication in youth mental health clinical encounters: Introducing the agential stance." *Theory & Psychology* 32(5), 667–90.

Boden, Z. and Eatough, V. (2014) "Understanding more fully: A multimodal hermeneutic-phenomenological approach." *Qualitative Research in Psychology* 11(2), 160–77, available at doi: 10.1080/14780887.2013.853854.

Bond J., Kenny A., Mesaric A., Wilson N., et al. (2022) "A life more ordinary: A peer research method qualitative study of the Feeling Safe Programme for persecutory delusions." *Psychology and Psychotherapy* 95(4), 1108–1125, available at doi: 10.1111/papt.12421.

Bortolotti, L. (2018) "Stranger than fiction: Costs and benefits of everyday confabulation." *Review of Philosophy and Psychology* 9(2), 227–49.

Bortolotti, L. and Jefferson, A. (2019) "The power of stories: Responsibility for the use of autobiographical stories in mental health debates." *Diametros* 60, 18–33.

Castro, E. M., Van Regenmortel, T., Sermeus, W., and Vanhaecht, K. (2019) "Patients' experiential knowledge and expertise in health care: A hybrid concept analysis." *Social Theory & Health* 17(3), 307–30, available at https://doi.org/10.1057/s41285-018-0081-6.

Collins, H. M. and Evans, R. (2002) "The third wave of science studies: Studies of expertise and experience." *Social Studies of Science* 32(2), 235–96.

Dings, R. and Tekin, S. (2022) "A philosophical exploration of experience-based expertise in mental health care." *Philosophical Psychology*, DOI: 10.1080/09515089.2022.2132926.

DSM-IV. (1994) *Diagnostic and Statistical Manual of Mental Disorders* (4th ed.). Washington, DC: American Psychiatric Association.

DSM-V. (2013) *Diagnostic and Statistical Manual of Mental Sisorders* (5th ed.). Washington, DC: American Psychiatric Association.

Feigl, H. (1953) "The scientific outlook: Naturalism and humanism," in *Readings in the Philosophy of Science*, (eds.), H. Feigl and M. Brodbeck. New York: Appleton-Century-Croft.

Feyerabend, P. K. (1978) *Science in a Free Society*. London: New Left Books.

Finfgeld-Connett, D. (2010) "Generalizability and transferability of meta-synthesis research findings." *Journal of Advanced Nursing* 66, 246–54, available at https://doi.org/10.1111/j.1365-2648.2009.05250.x.

Fisher, W. R. (1984) "Narration as a human communication paradigm: The case of public moral argument." *Communication Monographs* 51(1), 1–22.

Grant, A. and Kara, H. (2021) "Considering the Autistic advantage in qualitative research: the strengths of Autistic researchers." *Contemporary Social Science* 16(5), 589–603, available at doi: 10.1080/21582041.2021.1998589.

Green, T., Bonner, A., Teleni, L., Bradford, N., et al. (2020) "Use and reporting of experience-based codesign studies in the healthcare setting: A systematic review." *BMJ Quality & Safety* 29(1), 64–76, available at doi: 10.1136/bmjqs-2019-009570. Epub 2019 Sep 23. PMID: 31548278.

Greenhalgh, T. (2018) Twitter status on 26th May, 10:30am, available at https://twitter.com/trishgreenhalgh/status/1000308119115915264?lang=en.

Haraway, D. (1988) "Situated knowledges: The science question in feminism and the privilege of partial perspective." *Feminist Studies* 14(3), 575–99, available at doi:10.2307/3178066.

Harding, S. (1993) "Rethinking standpoint epistemology: What is strong objectivity?," in *Feminist Epistemologies*, (eds.), L. Alcoff and E. Potter, 49–82. New York, NY: Routledge.

Høffding, S., Heimann, K., and Martiny, K. (2023) "Editorial: Working with others' experience." *Phenomenology & the Cognitive Sciences* 22, 1–24, available at https://doi.org/10.1007/s11097-022-09873-z.

Hollan, J., Hutchins, E., and Kirsh, D. (2000) "Distributed cognition: Toward a new foundation for human-computer interaction research. *ACM Transactions on*

Computer-Human Interaction 7(2), 174–196, available at https://doi.org/10.1145/353485.353487.

Hollan, J. D. and Hutchins, E. L. (2009, August) "Opportunities and challenges for augmented environments: A distributed cognition perspective," in *Designing User Friendly Augmented Work Environments: From Meeting Rooms to Digital Collaborative Spaces* (pp. 237–59). London: Springer.

Hutchins, E. and Klausen, T. (1996) "Distributed cognition in an airline cockpit," in *Cognition and Communication at Work*, (eds.), Y. Engeström and D. Middleton, 15–34. Cambridge: Cambridge University Press, available at doi:10.1017/CBO9781139174077.002.

Jamieson, M. K., Govaart, G. H., and Pownall, M. (2023) "Reflexivity in quantitative research: A rationale and beginner's guide." *Social and Personality Psychology Compass* e12735, available at https://doi.org/10.1111/spc3.12735.

Konradsen, H., Kirkevold, M., and Olson, K. (2013) "Recognizability: A strategy for assessing external validity and for facilitating knowledge transfer in qualitative research." *Advances in Nursing Science* 36(2), E66–E76.

Kuhn, T. (1977) "Objectivity, value judgment, and theory choice," in *The Essential Tension. Selected Studies in Scientific Tradition and Change*, (ed.), T. Kuhn, 320–339. Chicago: University of Chicago Press.

Kuper, A., Lingard, L., and Levinson, W. (2008) "Critically appraising qualitative research." *British Medical Journal* 337, a1035, available at doi:10.1136/bmj.a1035.

Larkin, M., Newton, E., and Boden, Z. (2015) "On the brink of genuinely collaborative care: Reflections on the use of an adapted form of 'experience-based design' as a mechanism for translating qualitative research into service development." *Qualitative Health Research*, 11, 1463–1476, available at doi: 1049732315576494.

Longino, H. E. (1990) *Science as Social Knowledge: Values and Objectivity in Scientific Inquiry*. Princeton, NY: Princeton University Press.

Milton, D. E. (2012) "On the ontological status of autism: The 'double empathy problem'." *Disability & Society* 27(6), 883–887.

Murphy-Hollies, K. (2022) "Self-Regulation and political confabulation." *Royal Institute of Philosophy Supplements* 92, 111–128, available at doi:10.1017/S1358246122000170.

Murphy-Hollies, K. and Bortolotti, L. (2021) "Stories as evidence." *Memory, Mind & Media* 1(E3), available at doi:10.1017/mem.2021.5.

Nagel, T. (1986) *The View From Nowhere*. New York, NY: Oxford University Press.

NSUN National Involvement Team. (2015) "Involvement for influence: 4Pi National Involvement Standards," available at https://www.nsun.org.uk/projects/4pi-involvement-standards/

Reiss, J. and Sprenger, J. (2020) "Scientific objectivity," in *The Stanford Encyclopedia of Philosophy*, (ed.), E. Zalta, available at https://plato.stanford.edu/archives/win2020/entries/scientific-objectivity/.

Seikkula, J. (2011) "Becoming dialogical: Psychotherapy or a way of life?" *The Australian and New Zealand Journal of Family Therapy* 32(3), 179–193.

Springham, N. and Robert, G. (2015) "Experience based co-design reduces formal complaints on an acute mental health ward." *BMJ Open Quality* 4, u209153.w3970, available at doi: 10.1136/bmjquality.u209153.w3970.

Stammers, S. (2020) "Confabulation, explanation, and the pursuit of resonant meaning." *Topoi* 39, 177–187, available at https://doi.org/10.1007/s11245-018-9616-7.

Stichter, M. (2015) "Philosophical and psychological accounts of expertise and experts." *HUMANA.MENTE Journal of Philosophical Studies* 8 (28), 105–128, available at https://www.humanamente.eu/index.php/HM/article/view/83.

Thomsen, D. K. and Brinkmann, S. (2009) "An interviewer's guide to autobiographical memory: Ways to elicit concrete experiences and to avoid pitfalls in interpreting them." *Qualitative Research in Psychology* 6(4), 294–312, available at https://doi.org/10.1080/14780880802396806

Weich S., Fenton S-J., Staniszewska S., Canaway A., et al. (2020) "Using patient experience data to support improvements in inpatient mental health care: The EURIPIDES multimethod study." *Health Service and Delivery Research* 8 (21).

Wenger, E. (2000) "Communities of practice and social learning systems." *Organization* 7(2), 225–46.

Whyte, K. P. and Crease, R. P. (2010) "Trust, expertise, and the philosophy of science." *Synthese* 177, 411–425, available at https://doi.org/10.1007/s11229-010-9786-3.

6
Affective, Cognitive, and Ecological Components of Joint Expertise in Collaborative Embodied Skills

John Sutton
Macquarie University

1. Introduction: Collaborative Embodied Skills

Many striking cases of expert performance are undertaken in and by small groups. Ethnographer Natasha Iskander describes one of the small teams of migrant construction workers in Qatar who built vast modernist edifices for the football World Cup under challenging conditions: they 'hang from the sky' on unprecedented, vertiginous scaffolding several stories high, working smoothly together, with no common language (Iskander 2021, 115).

> Each scaffolder handled one ton of material every day.... In the heat and din of the construction site, the pipes, planks, joints, and spanners were all manipulated wordlessly. The men relied on hand gestures and manual signals, like a twist to the pipe to convey the soundness of their grip to those passing the material or a tug to signal the direction in which they were moving the material.

This initial case reminds us that joint expertise need not be developed or exercised in the service of self-generated goals or as a shared expression of autonomy. Many phenomena and processes of social ontology are non-ideal, and even the smoothest coping can facilitate politically or morally unpalatable or troubling outcomes.

The primary focus of this essay, though, is on less risky environments in sport and the arts where we also engage in or observe joint intelligence in action. In the middle of a song they have played together many times, one of four professional musicians in a touring band suddenly extends its structure, improvising two transformed repetitions of the chorus line: despite the lack of warning, the other three musicians stay synchronized in their music and coordinated in their movements, continuing the performance more or less seamlessly (Geeves, McIlwain, and Sutton 2014). Experienced tango dancers move as one, constantly extending and playing with the basic 'leader-follower' dynamics of the genre: two dancers'

John Sutton, *Affective, Cognitive, and Ecological Components of Joint Expertise in Collaborative Embodied Skills*
In: *Expertise: Philosophical Perspectives*. Edited by: Mirko Farina, Andrea Lavazza, and Duncan Pritchard,
Oxford University Press. © John Sutton 2024. DOI: 10.1093/oso/9780198877301.003.0006

individual skills micro-modulate and mesh with each other's movements across fluctuations in the embodied dynamics of the joint action (Kimmel 2016). Japan's unfancied rugby union team beat mighty South Africa at the 2015 World Cup with extraordinary tries in which almost all fifteen team members created space from nothing over a few gripping seconds of action: such tightly knit teams coordinate movements and decisions perfectly, acting for a time as if of one mind (Sutton and Tribble 2014).

In such collaborative performance, team members' shared experience, intense practice regimes, and complementary skills animate joint action on the fly. Success is precarious: things can go wrong at any point. But outsiders marvel at the dynamic, improvisatory intelligence of such small groups. It is the envy of corporate culture and management science, where explicit training practices often fail to elicit innovative outcomes. Here I focus on such forms of skilful embodied interaction in small groups. While we can learn much from recent attempts to *analyse* joint know-how or collective practical knowledge, so as to be able to distinguish genuine cases from other distinct phenomena, my project here is different. I hope instead to delineate some of the component processes involved in clear cases like those just described, to encourage diverse further research on collaborative performance in context at multiple timescales, from fast embodied coordination and alignment, through explicit collaboration when intentions in action can be shared and negotiated, to slower kinds of cooperation formed over careers in shared histories of culturally embedded practice (Williamson and Sutton 2014; Bietti and Sutton 2015).

Philosophical and scientific theories of mind and action should throw light on the nature and mechanisms of shared skills in real-world practices. But work on 'embodied cognition' often proceeds at an abstract level, neglecting both the experience of specialists who commit themselves together over decades of practice, and rich bodies of research in fields like sport psychology and music cognition. It is a good starting point to mention striking examples and anecdotes like those previously given, but it is not enough: in order to grasp the phenomena of joint expertise in collaborative embodied skills in their depth and complexity, we must tap more systematically into the abundance of applied work on collaborative skills that uses experimental, ethnographic, qualitative, practice-based, and phenomenological methods (McIlwain and Sutton 2015; Sutton and Bicknell 2020). I can't live up to this demand in this short essay, but I can point to a recent collection in which contributors deploy and integrate tools and methods from just this range of fields in examining case studies of thinking in action (Bicknell and Sutton 2022). In this essay, building on the momentum of that interdisciplinary work, I aim to pull out and identify a range of components of joint expertise which vary in telling ways across contexts, and to which it will be productive to attend as we seek to extend research on skill to address collaboration, and to apply research on collaboration to domains of embodied skill. In the embodied joint action of expert performers, the mind-body problem comes to life in social practice.

Some clarifications will sharpen the focus. First, I am thinking here primarily of small groups doing things together that cannot be done by individuals, as in the previous examples. I do not feel the force of the charge that an account of joint expertise in dyads and small groups acting together is incomplete if not clearly applicable to larger-scale and diffuse groups (Habgood-Coote 2022, 187). Of course understanding much larger corporations or social movements is also important, but the dynamics and mechanisms may be so different as to require another kind of social ontology. Second, similar processes may be involved when activities or tasks that *could* also be accomplished by an individual are in fact jointly performed, and I will mention such cases: but for the most part it is clearer to focus on 'Hutchins-style' tasks for which more than one person is needed (Hutchins 1995). Third, I concentrate on cases involving embodied activity because it's easier to see the range and variety of the component processes that participants have to deploy and mesh. This does not, however, assume a sharp distinction between cognitive skills and motor skills, and I do draw lessons from forms of collaborative cognition in other domains. Fourth, remembering the previous example from Iskander's work in Qatar, I am aware that my focus on sport and the arts runs the risk of reinforcing what Aagaard (2021) calls 'the dogma of harmony' in research on embodied and distributed cognition, the tendency to offer 'an overly idealized picture' of coordination and collaboration, and to neglect cases of socio-technical or socio-cultural interaction in the service of oppressive or brutal enterprises (compare Slaby 2016); it will be a task for other occasions to show that the same conceptual tools and frameworks can also throw useful light on such cases. Finally, I am not interested here in problems about the attribution or recognition of expertise, but in examining domains where the existence of expert performance is manifest and uncontested, to try to improve our understanding of its nature and basis. Neither in ordinary language nor in the current state of science do we have anything like natural kind-like terms to pick out our topics in these domains—we don't need to treat 'expertise,' or 'know-how,' or 'practical knowledge' as labelling well-defined phenomena or abilities, or to be delayed by worrying at the edges of applicability of these labels, in order to get on with the job of saying something illuminating about cases where they clearly do apply.

To proceed, I first assess some recent philosophical accounts of joint know-how, to help orient my different kind of enquiry. In Section 3, to address the challenges of integrating research on skill with research on collaboration, I consider concepts and lessons from the philosophy and psychology of collaborative cognition and small group research that can inform our thinking on joint know-how. The rest of the essay selectively surveys features and dimensions of joint expertise that are embodied in the kinds of professional, artistic, and sporting performance contexts mentioned previously. In Section 4 I discuss situational components of joint know-how in practice in the form of ecological and bodily factors. Section 5 examines some of the affective and socio-cognitive capacities and processes salient when collaboration succeeds or fails. I conclude in Section 6 with some programmatic notes on what the approach suggests about the mechanisms of complementarity operative among heterogeneous group members. This is an impressionistic, high-level survey of rich, under-explored but challenging research terrain, and all of my pointers and recommendations need to be tightened and deepened in application to specific contexts and cases.

2. Joint Know-How: Analyses and Approaches

S. Orestis Palermos and Deborah Tollefsen (2018) offer a pair of non-reductive proposals of joint know-how, analyzing it by reference to joint intentionality and distributed cognition respectively. While they take inspiration for these two approaches from distinct accounts of individual know-how, they plausibly suggest that pluralism may be appropriate here, to accommodate a range of cases and of forms of mutual responsiveness. Sometimes group members do make and rely on explicit or semi-contractual commitments which then guide their actions in the form of 'collectively known propositional knowledge' (2018, 119, 127); on other occasions, more implicit or dynamic interdependence between individuals, involving reciprocal and continuous interactivity, allows them to keep monitoring performance (2018, 123–4). While noting that the hypothesis of distributed cognition on which the latter account rests 'is being widely debated', Palermos and Tollefsen hope that it opens paths towards more detailed explanations of the sense in which group know-how can emerge from the complex interactions between individual members rather than resulting from the mere summation of all the individuals' knowledge states (2018, 117, 125).

In Jonathan Birch's 'active mutual enablement' account (2019), joint know-how in a dyad is 'an interlocking package of individual know-how states ... each agent knows how to perform his role in an actively coordination-enabling way for the other agent while predicting, monitoring, and making responsive adjustments in response to the other agent's behaviour' (2019, 3339). Aiming to put flesh on Bratman's (2014) notion of 'mutual responsiveness in action', and on psychological statements of minimal conditions for joint action (Vesper et al. 2010), Birch suggests that such monitoring and adjusting, on the part of each individual in such a dyad (or group), actively enables their coordination: in particular, key kinds of adjustments arise when one agent—as a result of predicting and monitoring the other agent's actions in real time—can avert the failure of the joint action (2019, 3336–8). In one sense this is a reductive account, in that joint know-how is analyzed by reference to the mutually compatible knowledge states of the distinct individuals in the dyad or group. But the relations between individual know-how and joint know-how are not simple or summative: Birch stresses that neither individual need know how to perform the action in question as a whole or to perform *both* component roles, and that thus the joint know-how may be genuinely distributed across the dyad (2019, 3333, 3346). For our purposes, it is notable that Birch acknowledges that the exact forms that such 'actively coordination-enabling performance' will take 'will depend a great deal on the nature of the action in question' (2019, 3338).

Finally, Joshua Habgood-Coote (2022) develops a broader view intended to apply to larger collectives as well as dyads and small groups, while acknowledging that the approaches provided by Birch and by Palermos and Tollefsen have some appeal when we consider small-scale groups 'with a tight pattern of mutual engagement' (2022, 186–7). His view is inspired by considerations about the semantics of how-to questions. Group members may each have fragmented practical knowledge of answers to parts of larger practical how-to questions: group know-how then requires them to be able to exercise their interrogative capacities together to generate effective answers to such questions in the course of acting together (2022, 191–3). Habgood-Coote also notes that in the case of larger groups, some such answers may be embodied in the institutional design and structure, or distributed across sub-groups who do not need to collaborate directly (2022, 194).

In developing these distinctive and promising accounts of joint know-how, these authors all acknowledge its context-specificity (compare also Martens 2021; Pino 2021). Yet they aim, like most philosophers, at analyses that are general enough to cover many different cases and kinds of case. Such abstraction gives these analyses their generalizing power, in applying across contexts. But the concrete and specific can't be left behind if we want to know more about just *what* experts know, what some of the answers to difficult why-questions might be, or just *how* experts generate answers to them or coordinate by interactively monitoring each other such that their sub-plans continue to mesh. While I too have generalizing ambition, mine is directed not towards analysis but at establishing toolkits to be applied differently in different cases and contexts—sets of factors and dimensions to which we as researchers can develop sensitivity as we examine varying forms of group expertise.

In hoping to motivate attention to under-noticed features of joint know-how, I note that accounts of the sort just sketched operate at a fairly abstract level, deliberately abstracting away from aspects of collaborative performance that (I suggest) matter greatly. My alternative approach aims not to replace but to supplement and complement these more standard philosophical projects, in the three specific directions that I address in turn in the next three sections. While they are rightly concerned to align their proposals with current approaches to individual skill, whether intellectualist or not, they are less engaged with research on collaboration. While they do acknowledge in principle the importance of systemic or structural support for group performance, they do not directly focus on technological or ecological scaffolding (for example) in shaping what group members do together. And since these analyses are not intended to dig down into the specific interacting cognitive and affective processes which individuals engage in together, there is ample room to point to research on the entangled roles of emotion, perception, attention, memory, and more in dynamic interaction. Overall, the aim is gently but firmly to nudge philosophers of skill and collaboration towards a more integrative and insistent interdisciplinarity in which experimental and ethnographic methods and results both hold a significant place.

3. Collaborative Cognition and Group Process

Our target phenomena are cases of clear expertise in joint performance, in domains including sport, the arts, and professional teamwork in action domains. Working together successfully is far from easy: this is not only because the individual skills involved are challenging, but also because of the extra demands of collaboration, encapsulated in the slogan that a team of experts is not inevitably an expert team (Eccles and Tenenbaum 2004; Gaffney 2015). This is a natural point on which to look to research on collaborative cognition in domains other than movement or action. People often hope that sharing cognitive resources with others will be beneficial: we brainstorm in search of creative ideas, we pool our views in committees or with friends in making important decisions, and we often enjoy remembering past experiences together. If collaborative success required only that the group would perform better (according to any agreed metric in assessing either process or output) than any one individual working alone, we would be justified in identifying regular 'process gains' through working together. But in most contexts that is too low a threshold. Instead researchers typically evaluate group performance while collaborating against the *sum* of the same number of individuals working alone and then pooling their (non-redundant) outputs, in what is known as a *nominal* group, 'a group in name only' (Weldon 2000, 92). Against this more challenging bar, we find that the conditions for collaborative success are surprisingly fragile, and that 'collaborative inhibition' and other forms of 'process loss' are more typical (Harris, Paterson, and Kemp 2008; Meade et al. 2018).

Many research traditions across the disciplines engage with such findings and seek to identify factors that may disrupt or enhance group performance. One useful broad-brush way to arrange such factors is by asking how 'higher-level' and 'lower-level' cognitive states and processes interact. As I have argued in the context of sport (Williamson and Sutton 2014; Sutton and McIlwain 2015), higher-level states including beliefs and attitudes, intentions and plans, and 'shared mental models' of a changing task domain are:

> to a first approximation, the kinds of things that team or group members can talk about, ... that can be rendered explicit, as in the use of written or verbal information sharing, or even deliberate, iconic bodily cueing, like pointing or hand waving. These processes can be plans, strategies or instructions made and shared before or after a match, or changed and adapted during play, but they can also include more immediate verbal cues or directions used on the fly to signify an intention or to influence the attention of a team member. In some contexts, they can also include the use of formalized or formalizable game plans, visually represented for instance through diagrams, video footage or on-field/court reenactment.

In contrast,

> lower-level processes are those that are not immediately, easily or perhaps ever able to be tapped by talk. They include gestural, bodily and movement-based forms of information-sharing and cueing, often driven by skillful and honed perceptual and attentional processes. These processes are often thought of as implicit and non-deliberative. They can be fast and adaptive, but they are also developed and shaped through practice and performance history. Broadly, lower-level processes are those processes that rely on non-verbal forms of communication and information-sharing: anticipating and responding to the bodily presence of a team member, the direction, speed and shape of a team member's run, the feel and rhythm of the team's movement. (Williamson and Sutton 2014)

The roles of higher-level cognitive states and processes in collaboration are highlighted in various ways in formal and conceptual approaches in social ontology in philosophy, in behavioural economics, and in experimental research on collective intelligence, collaborative recall, and transactive memory systems in cognitive and organizational psychology. I will draw on results from these fields in pointing to some natural features of collaborative process to which research on joint expertise might look. But like Palermos and Tollefsen (2018), I want in pluralist spirit both to acknowledge that such higher-level states can play significant roles even in dynamic action contexts, and to deny that they are sufficient to grounds and explain successful performance in embodied collaborative movement. Groups often have to innovate together in responding spontaneously to unpredictable circumstances (Preston 2012): these on-the-fly adjustments occur at rapid timescales that do not allow for explicit modification of shared strategies, but seem to involve interactions driven by dynamical factors that are harder to access, articulate, and reflect on. Such factors include patterns of gaze and fast information pickup or changes in body movement, rhythm, or affective expression, as typically highlighted in 'alignment studies' and research on joint improvisation in ecological psychology, phenomenology, and sports psychology (Tollefsen, Dale, and Paxton 2013; Williamson and Cox 2014). Such lower-level processes are of course also operating when people collaborate in making decisions or remembering the past, but the salience of subtle, multi-channel modes of communication in joint know-how is even more obvious in sport and other action contexts where slower, more measured plan execution and verbal interaction can play only a minor role if any. In mining studies of collaborative cognition in other domains, we want to track anything that shapes the dense and complex interplay between higher- and lower-level processes (Williamson and Sutton 2014).

In cognitive domains like memory and decision-making, researchers systematically vary the kinds of tasks employed and the measures used to assess performance in groups of different sizes, durations, and structures (Larson 2010; Laughlin 2011). They want to see if the significance of materials or tasks to group members, or their motivation and interest, may affect collaborative processes and results (Marion and Thorley 2016). Yet cases of collaborative *facilitation*—when working together produces benefits beyond the pooled performance of individuals working alone—remain rare. In the case of memory, for example, disruption to cognitive mechanisms seems to occur even if groups are offered extra incentives for success. But when we review the experimental studies that have produced these robust results, we find that many use convenience groups of strangers rather than real groups with common interests or shared history, and that many do not encourage densely interactive processes of communication in working together (Harris et al. 2014). Looking for lessons to bring back to research on collaborative embodied skills, we can briefly examine factors relating to the nature of groups and the microprocesses of communication in turn.

Michelle Meade and colleagues (2009) first showed collaborative facilitation among domain experts. The key result is not that expert pilots perform better than non-experts when (and only when) remembering material related to their domain of expertise, but that expert pilots do better when working *together*, in a collaborative group, than when left to work individually and having their performance pooled as a nominal group. The specific members of the collaborating expert groups in this study did not have a history of working together, though they had been trained in communication as well as in the first-order skills of aviation. This result casts doubt on a recent claim by Katherine Sweet that 'participants must *already* have an existing relationship to collaborate well' (2023, 2, original emphasis). Where a group has shared history, as I note in a moment, this *may* indeed support collaborative benefits. But this is not inevitable. Such shared history is neither sufficient nor necessary for good collaboration, as Meade's study confirms: where it does make a positive difference, it likely does so by way of other factors such as shared domain knowledge and effective collaborative process. Alongside ongoing empirical research on the effects of expertise on collaborative creativity, problem-solving, and memory (Nokes-Malach et al. 2012; Malone and Woolley 2020; Rosenberg et al. 2022), and with dyads and small groups like long-term couples who have built rich systems of cognitive and affective interdependence over time (Harris et al. 2017, 2019), philosophers of expertise and skill can consider broader cognitive, emotional, and communicative processes that may be affected or transformed in the course of long-standing interactions between familiar individuals in shared social or professional contexts and ecologies.

In these domains, certain specific microprocesses of communication have been identified as among the mechanisms by which shared history animates collaborative performance. Celia Harris and colleagues, for example, find factors consistently associated with effective and ineffective collaboration in memory (Harris et al. 2011, 2019). Group-enhancing factors include the provision of cues—even if they do not successfully stimulate retrieval—and the acknowledgement or mirrored repetition of another member's contributions, whereas strategy disagreements, asymmetric assignments of expertise, and corrections are among factors that seem to diminish group performance. These results are based primarily on analyses of verbal interaction and are not intended to capture embodied or environmental aspects of communication between group members. Preliminary ethnographic work in our experiments with older couples shows that even in tasks where only verbal output is typically counted towards measured success, a range of factors like gaze, touch, posture, humour, and shared responsiveness to familiar environments mediate the operation of such communicative interaction (Bicknell, Harris, and Sutton, MS). The significance of bodily and non-verbal processes is likely even greater when we consider the fast-action domains in which many kinds of collaborative skill are exercised. The remainder of this essay aims at the selective initial identification of some of the diverse ecological, social, bodily, cognitive, and affective resources and factors that seem likely to be involved in the successful group performance of collaborative embodied skills over time.

4. Cognitive Ecologies of Skill: Ecological and Bodily Components

Expert individuals and expert teams alike, in many skill domains, seek to extend their capacities to perform successfully in different, challenging conditions. Going, as they say, beyond their comfort zones, they resist excessive automation or proceduralization of their grooved skills (Ericsson 2006, 687), and instead aim to expand their region of expected expertise (Sutton et al. 2011; Christensen et al. 2019). They can transfer their techniques and skills to novel situations, within an envelope of performance possibility, even though development is fluctuating and uneven, given human imperfection and the vanishingly rare chances in many domains of winning every time (Christensen et al. 2016; Bicknell 2021). If such ongoing development is one mark of expertise over time, if skill acquisition never stops, then the environments of performance—sometimes and in some fields more reliable or stable, sometimes more volatile and unpredictable—are not merely external settings in which the real expert action unfolds, not just stimuli to performance, but in many respects are intrinsic and active components of complex and heterogeneous systems: *ecologies* of skill (Bicknell and Sutton 2022).

Some domains require a wider range of such forms of scaffolding, and within each domain individuals and groups may vary greatly in the extent to which, and the ways in which, they interact with these resources. But attention on behalf of skill researchers to such material, environmental, technological, and institutional factors is not an optional sociological extra on top of direct investigation of embodied decision-making and performance (compare Becker 2008, 8). Just as, in sport, music, and professional action domains, equipment and terrain, crowds or observers, locations and weather, can all disrupt effective performance, so coping with troubles or unexpected changes in these components strongly contributes to (or sometimes just *is*) success in action. Each item of technology and each physical space of performance has its own history, its own dynamics, its own ways of challenging or facilitating skilled coping. Many elite performers across domains reasonably try to focus only on what they can control, so as not to waste energy or intensify anxiety by worrying over aspects of the performance ecology that are outside their spheres of influence. But over time some also seek to expand that sphere of influence and to gain at least some familiarity with or even control over a broader range of ecological components. Elite sports teams may not be able to regulate the weather or to choose whether they're playing at home or away, but in some contexts they can and do seek indirectly (and fallibly) to affect what the media focuses on, the mood of the crowd, or the specifications of or regulations on the available equipment.

In teamwork and joint expertise in particular, interactions with the resources of these ecologies of performance occur in various forms of combination within the group. One of the great potential benefits of collaboration, when things go well, is to be able to develop and tap specialization, to cultivate and implement effective divisions of labour among group members. Such specialization may emerge in relation to the component actions each member performs within broader group endeavours, and may also involve cognitive and emotional divisions of labour. Over time, the pooling of distinctive capacities can help generate a more explicit 'we-awareness' and a sense of the collective (Sutton and Tribble 2014). Even in those action domains like certain forms of rowing which appear to require more homogeneity among team members and their activities, the sense of togetherness has to be laboriously nurtured and sustained, as rhythm can remain elusive across a group of individuals with subtly distinctive physical, stylistic, and emotional profiles (King and De Rond 2011).

In the professional worlds of many elite team sports, institutions and training systems are exquisitely calibrated to respect, hone, and support the unique physiological and technical needs of each team member. There is no perceived tension between the driving overarching performance goals of shared success as an integrated team, and the highly differentiated bodily regimes—spanning diet, pain management and medical care, strength and conditioning programmes, and aspects of technical support—required by each distinctive team member. In the next section I shift attention to examine affective and cognitive components of joint expertise, domains in which individual differences are often not so easily acknowledged, despite lip-service paid to the importance of 'the mental game'. But first I note the bodily and technical factors which differentiate collaborating performers in many domains. In classical orchestras and jazz bands, for example, and in some team sports like cricket and rugby, it is an essential feature of joint expert performance that the individuals involved are—often dramatically—distinctive, both in terms of the capacities open to them by way of their physical make-up (their height, for example), and in the roles, and styles, and skills they bring to their different specialist parts in the shared activity as a result of their backgrounds, their training, and their experience.

Individual team members with these distinctive embodied capacities thus each *bring* something different *to* group performance. Interacting with each other in the course of action, in many domains they each also therefore *do* something different *in* that group performance. As we learned from considering collaboration in other cognitive domains previously, interaction does not inevitably bring benefits: some groups are notoriously *less* in action than the sum of their parts, as when groups of highly skilled individuals fail to gel with or complement each other, or to settle in effectively to a novel performance ecology. But the hope is always that beneficial forms of emergent outcome will result over time, that members mesh well with each other and together develop the right networks of resilient interaction across whole systems.

5. Cognitive Ecologies of Skill: Affective and Cognitive Components

We can start with affective components of the ecologies of collaborative embodied skills. As with all the factors I discuss in this section, different domains of expertise have many different norms and requirements on the roles of emotions, moods, motivation, and arousal in practice and in performance. And within single domains, different cultures enforce or recommend different ways of regulating or harnessing affective dynamics. Though it's clear that the possibilities for and operations of emotion expression in action for professional tennis players are quite unlike those at play among members of a symphony orchestra or in the cockpit of a long-haul aircraft, we can still be surprised at just how much variability there is in these domains in practice across styles and cultures. Not all orchestras conduct their business in the same way; you don't always have smooth 'plug-and-play' modularity, even if experienced musicians can often quickly adapt to a new setting. The point of surveying these components of joint expertise is obviously not that they all play equally important roles in all cases, but that sensitivity to their presence and dynamics, and to the shifting balances among them within and across contexts, improves our understanding of effective and ineffective collaboration in action.

Operating at a range of timescales (from fleeting occurrent affective processes through to explicit emotional work over a lifetime) and at a range of levels, affective phenomena play vital roles in many group performances. As for all the factors in this section, there are rich bodies of theory, backed by extensive empirical data, on affect, mood, and emotion in individuals, but considerably less research on how they work in small groups such as expert teams. One striking evocation of the affective dimensions of team dynamics in elite football has recently been offered by John Protevi. In '*Esprit de Corps* and thinking on (and with) your feet' (2023), Protevi works outwards from a detailed analysis of one celebrated, marvellous goal in the 2011 Women's World Cup. Alongside an inventive integration of conceptual frameworks that tap relational autonomy, collective intentionality, and enactive phenomenology into 'a bio-neuro-social-subjective approach' to team performance, Protevi pushes each such framework towards firmer and fuller acknowledgement of the centrality of affect in action. At the general level of mutual engagement and awareness, 'the constraints on player action—what constitutes them as team players and allows the emergence of a team—are primarily affective: the players must reward the trust their teammates put in them as team players, or they risk criticism' (2023, 30–31). More specifically, the looping resonance that emerges among players who know each other so well is underpinned or constituted primarily by a variety of affective embodied interactions over time: both through direct interactions in touch, drill, rhythmic movement, and shared effort, and in the joyous affective mediation of successful joint attention and joint commitment (2023, 31). Protevi brings this framework to life in a phenomenological analysis of a retrospective report by Megan Rapinoe, who provided the inch-perfect time-pressured cross from the left for Abby Wambach to score. In reading Rapinoe, Protevi catches the shifting affective intensities involved at each stage of this incredibly fast team movement, from awareness of the crisis situation in which a goal was urgently needed, through the embodied anticipation driving the selection of an action to afford opportunity to a deeply trusted teammate, to the disbelieving joyous rush as Wambach scores (2023, 34–36; compare Bicknell's analysis of a report by cyclist Chloe Hosking in Sutton and Bicknell 2020, 200–202).

Of course not all emotions in performance are positive or easy to manage. Teamwork can also elicit, and can be supported by, the full spectrum of feelings, from boredom to rage, uneasiness to misery. In professional practice the challenges of mood- and emotion-regulation, which also take on a different cast at the group level, are as readily acknowledged as positive emotions. Recent accounts of socially distributed affectivity and of emotion-regulation as extended across brains, bodies, and world can inform applied research on emotion in team performance (Colombetti and Krueger 2015; Salmela and Nagatsu 2017; Thonhauser 2022; Rimé and Páez 2023). Some discussions of emotion in sport, music, and organizational psychology focus heavily on cases of trouble and breakdown, as the mental health challenges of elite performance—and the corresponding need for improved systems of emotional support—are at last more widely acknowledged. This significant cultural shift should not, however, sanction ongoing neglect of emotion-regulation when things are going well. It can be a highly sophisticated skill to tune and regulate emotion experience and emotion expression in and around performance: this is more challenging in group contexts where experts who know each other well are constantly influencing each other. Sometimes, subcultural norms require strong feelings, especially negative ones, to be masked or muted to protect fellow performers, but over time many practitioners come to be able to adjust their own affective dynamics more or less effectively in and through the social and institutional contexts of performance.

Deep entanglements between emotion, perception, and attention are in play here. Depending on the domain, experts harness and work *with* emotions to help them shift energy and attention as required to what is salient, when needed. Managing difficult feelings effectively helps both in switching off to recharge and in then tuning back in to the cues that matter: knowing what, when, and how to feel can be a vital component of the capacities to pick up task-relevant information and think on your feet in selecting and performing actions in response. The years-long processes of enculturation into the world of a specific professional action domain involve an education and reorientation of perceptual attention so as to attune to cues across multiple modalities at a range of timescales—to be able to detect them, help set them up, and respond fast to them when necessary (Goodwin 1994; Grasseni 2004; Bicknell 2021). When it is a domain of *group* performance, the sociality of such enculturation is intensified: not just in apprenticeship learning and in ongoing openness to instructional nudges and other input from peers and trusted coaches or leaders, but also in explicit reflection on and redesign of interaction processes, and then faster, in effortlessly distributing attention to team mates or colleagues within the larger ecology of performance in the moment.

With more space we could work through more of the other cognitive processes which drive group action, and which may themselves be transformed in the context of ongoing collaboration. I mention just a few here. Distinct forms of *memory* operate to track and direct joint action over time in light of shared past embodied experience (Sutton and Williamson 2014). Action *control* can be more or less distributed: as individual experts may rapidly adjust different meshing combinations of integrated cognitive and automatic control processes (Christensen and Sutton 2019), so teams of experts may be able rapidly to reallocate the weights or influence of different members' contributions as situations change and evolve. And—again, depending on the domain and the task structure—more implicit and more explicit forms of shared or joint *metacognition* may operate to monitor performance over time: the group must *together* be able to discriminate among emerging evaluative feelings about ongoing activity, to interpret those feelings, and as appropriate to broadcast them to all and only those who need to know (cf. Shea et al. 2014).

All of these affective and cognitive processes, considered at the level of the small group, involve various forms of communicative interaction. Our pluralism requires us not to overemphasize either the slower, more explicit forms which are harder to deploy on the fly in some fast action skills, but which may be vital in managing trouble or realigning grooved patterns, nor the dynamic, implicit forms which operate seamlessly when all is going smoothly together. It is easy to overegg or romanticize the intensity and bandwidth of intrateam dynamics when we consider those rare occasions on which group behaviour seems to arise fully formed in perfect unfolding sequence just as the changing conditions demand. Not only are there many familiar occasions on which things do *not* fall into place quite so easily: even when there is emergent harmony in joint action, this need not require entirely mutual participatory co-regulation equally among all members. Some studies of the reciprocity of attention in effective football and basketball teams, for example, suggest that direct real-time cognitive-perceptual couplings between team mates are relatively rare: shared experience may mean that 'expert interactors probably do not need to pay as much attention to their co-agents during ongoing task performance', instead allowing them 'to adopt a parsimonious but effective structure of regulation of the intra-team coordination' (Bourbousson and Bourbousson 2016, 3). Indeed co-regulation may best loop out in 'indirect' or 'extrapersonal' modes as individual performers attune not to *all* of their own familiar colleagues, but to instructive features of the ecology of stimuli from the cue environment or the competing players (Millar, Oldham, and Renshaw 2013; Gesbert, Durny, and Hauw 2017). Likewise, it's not essential that the affective bonds that underpin trust in action must extend beyond the task domain. After examining two very particular collaborations—Watson and Crick in science, and Rodgers and Hammerstein in music—Sweet argues that 'if collaborators are constantly on the verge of estrangement from one another, then it is unlikely that they are collaborating well' (2023, 19). This is too strong: in sport, the arts, and in professional action domains alike, there are cases in which team members are indeed emotionally or personally estranged from one another yet retain the requisite mutual responsiveness in action.

The point of this tentative sketch of some affective and cognitive components of joint expertise was to begin identifying some of the ways that shared history can colour the operation of collaborative embodied skills. It is not by any means inevitable that small groups or teams will develop more effective modes of interaction over time. But when they do, the idea is, such benefits are likely mediated by transformations in some of these emotional and cognitive processes across the interacting members.

6. Mechanisms of Complementarity

In line with our pluralist stress on multiple levels of processing in joint expertise, I finish with a brief cautionary note on the complexity of interdependence in typical human collaborative embodied skills. It is tempting to think of group interaction, especially in action contexts, on models deriving from the range of wonderful, relatively simple processes of coordination operative in many non-human animal groups. Mechanisms relevantly similar to those at work in animal swarms, flocks, and herds may well operate in many human cases too. This focus, across a number of fields and research traditions, encourages us to think of the kinds of affective and cognitive coordination involved in interdependent groups of the kind I've been discussing as based in or requiring convergence. The idea is that—whether in action, navigation, emotion, or memory, for example—what is shared among interdependent group members is something similar: synchronous or entrained actions (Mogan, Fischer, and Bulbulia, 2017; Paxton and Dale 2017), combined estimations of orientation or direction (Fernandez Velasco 2022), or a 'shared rendering' of the past (Coman and Hirst 2015).

When philosophers apply these ideas to human interaction, they sometimes describe the groups in question as operating by way of a kind of 'social parity principle', understood on the model of Clark and Chalmers' (1998) parity principle for understanding relationships between brains and artifacts. Deborah Tollefsen (2006), for example, noted that when a relevant 'external' cognitive or affective resource is another person rather than an artifact, there are no deep differences of kind, mechanism, and process between the parts of the distributed system, such that there may be (or develop) relevant forms of functional similarity (or, in the extreme, kinds of merger) between members of an interdependent team or couple.

But the approach I have sketched here, involving shifting balances among many heterogeneous components in cognitive ecologies of skill, suggests that convergence and similarity are not the only or the most interesting forms of interdependence in cognition, affect, and action, even when we are considering human-human relationships rather than the material culture of cognitive artifacts. I have stressed the importance of specialization and of cognitive and affective division of labour among interacting groups, in which individual members bring distinctive capacities into the interaction and perform often very different tasks during interaction. Agreement and convergence are not required for effective group performance. If we think of a group as a certain kind of mechanism, in which distinctive components interact in characteristic operations in producing novel outputs (Theiner 2013), we do not need to assume homogeneity among those components. Unlike many of the relevant processes in non-human animal coordination, the effect of shared history within a human group is often to magnify differences and specialization, rather than reduce to a set of common or similar processes which are then additively combined. More explicit or reflective processes can play significant roles in human groups, in ways that are again unlike many in non-human animals, in allowing shared awareness of the group's history and performance, and in bringing that awareness into the open among group members for adjustment or refinement or critique (Sutton and Tribble 2014; Sutton 2018).

Aiming to complement philosophical analyses of joint know-how and group expertise, in this essay I have built on recent integrative interdisciplinary approaches to collaborative embodied skill in real performance contexts. I have sought to identify, and to offer some initial directions for research into, an expanded array of resources or components that can scaffold or partly constitute the remarkable capacities of some small teams and groups to think effectively on their feet in challenging situations, exhibiting various forms of joint intelligence in action.

Acknowledgements

Many thanks to the editors and the two reviewers. Thanks to all the participants in two focused events on group know-how and joint expertise (at the 2022 conference of the International Social Ontology Society (ISOS), and at a dedicated workshop in Antwerp in 2023 co-organized by Judith Martens), and to all who have commented so helpfully on this material, especially Mason Cash, Judith Martens, John Protevi, Axel Seemann, and David Spurrett. The writing of this chapter was supported by a Marie Sklodowska-Curie fellowship at the Paris Institute for Advanced Study and by a Leverhulme Trust visiting professorship at the University of Stirling, for both which I am very grateful. The longer-term work on which it is based was supported by the Australian Research Council through Discovery Project DP180100107 'Cognitive Ecologies: a philosophical study of collaborative embodied skills' awarded to John Sutton, and a Templeton Foundation grant 'Concepts in Dynamic Assemblages: cultural evolution and the human way of being' awarded to Agustín Fuentes, Greg Downey, and team. Huge thanks to all my collaborators on this longer-term skill research, especially Kath Bicknell, Wayne Christensen, Andrew Geeves, Doris McIlwain, Lyn Tribble, and Kellie Williamson.

References

Aagaard, J. (2021) "4E cognition and the dogma of harmony." *Philosophical Psychology* 34(2), 165–81.

Becker, H. (2008) *Art Worlds* (updated and expanded). Berkeley: University of California Press.

Bicknell, K. (2021) "Embodied intelligence and self-regulation in skilled performance: Or, two anxious moments on the static trapeze," *Review of Philosophy and Psychology* 12(3), 595–614.

Bicknell, K. and Sutton, J. (eds.) (2022) *Collaborative Embodied Performance: Ecologies of Skill.* London: Bloomsbury.

Bicknell, K., Sutton, J., and Harris, C. B. (MS) "The Wisconsin moment: Embodied interaction in collaborative recall experiments."

Bietti, L. and Sutton, J. (2015) "Interacting to remember at multiple timescales: Coordination, collaboration, cooperation and culture in joint remembering." *Interaction Studies* 16(3), 419–50.

Birch, J. (2019) "Joint know-how." *Philosophical Studies* 176, 3329–52.

Bourbousson, J. and Fortes-Bourbousson, M. (2016) "How do co-agents actively regulate their collective behavior states?." *Frontiers in Psychology* 7, 1732.

Bratman, M. E. (2014) *Shared Agency: A Planning Theory of Acting Together.* Oxford: Oxford University Press.

Christensen, W., Sutton, J., and McIlwain, D. J. F. (2016) "Cognition in skilled action: Meshed control and the varieties of skill experience." *Mind & Language* 31(1), 37–66.

Christensen, W. and Sutton, J. (2019) "Mesh: Cognition, body, and environment in skilled action," in *Handbook of Embodied Cognition and Sport Psychology*, (ed.), M. L. Cappuccio, 157–64. Cambridge, MA: MIT Press.

Christensen, W., Sutton, J., and Bicknell, K. (2019) "Memory systems and the control of skilled action." *Philosophical Psychology* 32(5), 693–719.

Clark, A. and Chalmers, D. (1998) "The extended mind," *Analysis* 58(1), 7–19.

Colombetti, G. and Krueger, J. (2015) "Scaffoldings of the affective mind." *Philosophical Psychology* 28(8), 1157–76.

Coman, A. and Hirst, W. (2015) "Social identity and socially shared retrieval-induced forgetting: The effects of group membership." *Journal of Experimental Psychology: General* 144(4), 717–22.

Eccles, D. W. and Tenenbaum, G. (2004) "Why an expert team is more than a team of experts: A socio-cognitive conceptualization of team coordination and communication in sport." *Journal of Sport and Exercise Psychology* 26, 542–60.

Ericsson, K. A. (2006) "The influence of experience and deliberate practice on the development of superior expert performance," in *The Cambridge Handbook of Expertise and Expert Performance*, (eds.), K. A. Ericsson, R. R. Hoffmann, and A. Kozbelt, 685–705. Cambridge University Press.

Fernández Velasco, P. (2022) "Group navigation and procedural metacognition." *Philosophical Psychology*.

Gaffney, P. (2015) "The nature and meaning of teamwork," *Journal of the Philosophy of Sport* 42(1) 1–22.

Geeves, A., McIlwain, D. J. F., and Sutton, J. (2014) "The performative pleasure of imprecision: A diachronic study of entrainment in music performance." *Frontiers in Human Neuroscience* 8: 863.

Gesbert, V., Durny, A., and Hauw, D. (2017) "How do soccer players adjust their activity in team coordination? An enactive phenomenological analysis." *Frontiers in Psychology* 8, 854.

Goodwin, C. (1994) "Professional vision." *American Anthropologist* 96(3), 606–33.

Grasseni, C. (2004) "Skilled vision: An apprenticeship in breeding aesthetics." *Social Anthropology* 12(1), 41–55.

Habgood-Coote, J. (2022) "Collective practical knowledge is a fragmented interrogative capacity." *Philosophical Issues* 32, 180–99.

Harris, C. B., Paterson, H. M., and Kemp, R. I. (2008) "Collaborative recall and collective memory: What happens when we remember together?." *Memory* 16(3), 213–30.

Harris, C. B., Keil, P. G., Sutton, J., Barnier, A. J., et al. (2011) '"We remember, we forget:' Collaborative remembering in older couples." *Discourse Processes* 48(4), 267–303.

Harris, C. B., Barnier, A. J., Sutton, J., and Keil, P. G. (2014) "Couples as socially distributed cognitive systems: Remembering in everyday social and material contexts." *Memory Studies* 7(3), 285–97.

Harris, C. B., Barnier, A. J., Sutton, J., Keil, P. G., et al. (2017) "'Going episodic:' Collaborative inhibition and facilitation when long-married couples remember together." *Memory* 25(8), 1148–59.

Harris, C. B., Barnier, A. J., Sutton, J., and Savage, G. (2019) "Features of successful and unsuccessful collaborative memory conversations in long-married couples." *Topics in Cognitive Science* 11(4), 668–86.

Hutchins, E. (1995) *Cognition in the Wild*. Cambridge, MA: MIT Press.

Iskander, N. (2021) *Does Skill Make Us Human? Migrant Workers In 21st-Century Qatar and beyond.* Princeton University Press.

Kimmel, M. (2016) "Embodied (micro-)skills in tango improvisation," in *Das Entgegenkommende Denken*, (ed.), F. Engel, 57–74. Berlin: De Gruyter.

King, A. and de Rond, M. (2011) "Boat race: Rhythm and the possibility of collective performance." *British Journal of Sociology* 62(4), 565–85.

Larson, J. R. (2010) *In Search of Synergy in Small Group Performance*. London: Taylor & Francis.

Laughlin, P. R. (2011) *Group Problem Solving*. Princeton, NJ: Princeton University Press.

Malone, T. W. and Woolley, A. W. (2020) "Collective intelligence," in *The Cambridge Handbook of Intelligence*, (ed.), R. J. Sternberg, 780–801. Cambridge: Cambridge University Press.

Marion, S. B. and Thorley, C. (2016) "A meta-analytic review of collaborative inhibition and post-collaborative memory: Testing the predictions of the retrieval strategy disruption hypothesis." *Psychological Bulletin* 142(11), 1141–64.

Martens, J. H. (2021) "Habit and skill in the domain of joint action," *Topoi* 40(3), 663–75.

McIlwain, D. J. F. and Sutton, J. (2015) "Methods for measuring breadth and depth of knowledge," in *The Routledge Handbook of Sport Expertise*," (eds.), D. Farrow and J. Baker, 221–31. London: Routledge.

Meade, M. L., Nokes, T. J., and Morrow, D. G. (2009) "Expertise promotes facilitation on a collaborative memory task." *Memory* 17(1), 39–48.

Meade, M. L, Harris, C. B., van Bergen, P., Sutton, J., et al. (2018) *Collaborative Recall: Theories, Research, and Applications*. Oxford: Oxford University Press.

Millar, S. K., Oldham, A., and Renshaw, I. (2013) "Interpersonal, intrapersonal, extrapersonal? Qualitatively investigating coordinative couplings between rowers in Olympic sculling." *Nonlinear Dynamics, Psychology, and Life Sciences* 17(3), 425–43.

Mogan, R., Fischer, R., and Bulbulia, J. A. (2017) "To be in synchrony or not? A metaanalysis of synchrony's effects on behavior, perception, cognition and affect." *Journal of Experimental Social Psychology* 72, 13–20.

Nokes-Malach, T. J., Meade, M. L., and Morrow, D. G. (2012) "The effect of expertise on collaborative problem solving." *Thinking & Reasoning* 18(1), 32–58.

Palermos, S. O. and Tollefsen, D. P. (2018) "Group know-how," in *Socially Extended Epistemology*, (eds.), J. A. Carter, A. Clark, J. Kallestrup, S. O. Palermos, et al., 112–31. Oxford: Oxford University Press.

Paxton, A. and Dale, R. (2017) "Interpersonal movement synchrony responds to high- and low-level conversational constraints." *Frontiers in Psychology* 8, 1135.

Pino, D. (2021) "Group (epistemic) competence." *Synthese* 199, 11377–96.

Preston, B. (2012) *A Philosophy of Material Culture: Action, Function, and Mind*. London: Routledge.

Protevi, J. (2023) "*Esprit de Corps* and thinking on (and with) your feet: Standard, enactive, and poststructuralist aspects of relational autonomy and collective intentionality in team sports." *Southern Journal of Philosophy* 61(S1), 24–38.

Rimé, B. and Páez, D. (2023) "Why we gather: A new look, empirically documented, at Émile Durkheim's theory of collective assemblies and collective effervescence." *Perspectives on Psychological Science* 18(6), 1306–1330.

Rosenberg, M., Gordon, G., Noy, L., and Tylen, K. (2022) "Social interaction dynamics modulates collective creativity." *Proceedings of the Annual Meeting of the Cognitive Science Society* 44.

Salmela, M. and Nagatsu, M. (2017) "How does it really feel to act together? Shared emotions and the phenomenology of we-agency." *Phenomenology and the Cognitive Sciences* 16(3), 449–70.

Shea, N., Boldt, A., Bang, D., Yeung, N., et al. (2014) "Supra-personal cognitive control and metacognition." *Trends in Cognitive Sciences* 18(4), 186–93.

Slaby, J. (2016) "Mind invasion: Situated affectivity and the corporate life hack." *Frontiers in Psychology* 7, 266.

Sutton, J. (2018) "Shared remembering and distributed affect: Varieties of psychological Interdependence," in *New Directions in the Philosophy of Memory*, (eds.), K. Michaelian, D. Debus, and D. Perrin, 181–99. London: Routledge.

Sutton, J. and Bicknell, K. (2020) "Embodied experience in the cognitive ecologies of skilled performance," in *The Routledge Handbook of the Philosophy of Skill and Expertise*, (eds.), E. Fridland and C. Pavese, 194–205. London: Routledge.

Sutton, J. and McIlwain, D. J. F. (2015) "Breadth and depth of knowledge in expert versus novice athletes," in *The Routledge Handbook of Sport Expertise*, (eds.), D. Farrow and J. Baker, 95–105. London: Routledge.

Sutton, J., McIlwain, D. J. F., Christensen, W., and Geeves, A. (2011) "Applying intelligence to the reflexes: Embodied skills and habits between Dreyfus and Descartes." *JBSP: Journal of the British Society for Phenomenology* 42(1), 78–103.

Sutton, J. and Tribble, E. (2014) "The creation of space: Narrative strategies, group agency, and skill in Lloyd Jones's *The Book of Fame*," in *Mindful Aesthetics: Literature and the Sciences of Mind*, (eds.), C. Danta and H. Groth, 141–60. London: Bloomsbury.

Sutton, J. and Williamson, K. (2014) "Embodied remembering," in *The Routledge Handbook of Embodied Cognition*, (ed.), L. Shapiro, 315–25. London: Routledge.

Sweet, K. (2023) "How to collaborate well." *Pacific Philosophical Quarterly* 104(2), 252–273.

Theiner, G. (2013) "Transactive memory systems: A mechanistic analysis of emergent group memory." *Review of Philosophy and Psychology* 4(1), 65–89.

Thonhauser, G. (2022) "Towards a taxonomy of collective emotions." *Emotion Review* 14(1), 31–42.

Tollefsen, D. P. (2006) "From extended mind to collective mind." *Cognitive Systems Research* 7(2), 140–50.

Tollefsen, D. P., Dale, R., and Paxton, A. (2013) "Alignment, transactive memory, and collective cognitive systems." *Review of Philosophy and Psychology* 4(1), 49–64.

Vesper, C., Butterfill, S., Knoblich, G., and Sebanz, N. (2010) "A minimal architecture for joint action." *Neural Networks* 23(8–9), 998–1003.

Weldon, M. S. (2000) "Remembering as a social process." *Psychology of Learning and Motivation* 40: 67–120.

Williamson, K. and Cox, R. (2014) "Distributed cognition in sports teams," *Educational Philosophy and Theory* 46(6), 640–54.

Williamson, K. and Sutton, J. (2014) "Embodied collaboration in small groups," in *Brain Theory: Essays in Critical Neurophilosophy*, (ed.), C. T. Wolfe, 107–33. London: Palgrave.

PART 4
EXPERTISE AND PUBLIC POLICY

7

Expert Judgement without Values

Credences not Inductive Risks

Rivkah Hatchwell
Department of Philosophy, King's College London

David Papineau
Department of Philosophy, King's College London

1. Initial Meta-Ethical Observations

We shall be arguing that values should play no role in deciding scientific claims. But let us start by distancing ourselves from a more general notion of scientific 'value freedom'.

It is often said that science should be 'value-free'. As it is usually meant, we do not think this dictum is defensible. That is, we do not think that scientific claims should be free of evaluative implications. In our view, such implications are both inevitable and desirable.

The dictum that science should be 'value-free' seems to us to be premised on outmoded meta-ethical views. We have in mind vulgar relativist views on which there is some unbridgeable gap between facts and evaluations. On such views, any descriptive claim leaves it quite open what evaluations should follow. Different people are free to adopt different evaluative responses to the facts. Given this, the business of science is solely to determine descriptive information, not to prescribe any evaluative add-ons. Not only will science be exceeding its authority if it adds evaluative glosses to its claims, but it will run the risk of alienating audiences with different evaluative reactions.

This view of values might still be common in certain scientific circles, but from a meta-ethical perspective it is nearly a century out of date. Contemporary meta-ethical views are generally agreed in rejecting relativism. This covers both realist and non-realist meta-ethical stances. While these options disagree about the import of ethical claims, they coincide in their resistance to relativism. Thus moral realists hold that the natural facts in and of themselves metaphysically entail evaluative claims, and perhaps also conceptually entail them. And most moral non-realists, while denying that moral attitudes answer to substantial moral facts, still incline to the non-relativist view that humans will generally be drawn to similar evaluative conclusions given the same descriptive facts. On all these views, then, we can expect given factual claims to have agreed evaluative consequences.

There is of course much more to be said here. We shall content ourselves with pointing to the oddity of relativism. Is it really optional how you should react evaluatively to news that, say, an infectious agent is causing children to go blind, or that a breakdown in medical supplies is resulting in painful deaths? These are not unusual examples. Many factual descriptions of real-life situations have obvious ethical implications. In consequence, any search for 'value-free' terminology that is somehow insulated from evaluative connotations is doomed to failure. Given that many descriptions of natural facts automatically dictate evaluative judgements, science cannot possibly be 'value-free'. What we want from science is the facts, not the facts shorn—*per impossibile*—of evaluative significance.[1]

Even if people agree generally about values, this does not mean that they will always agree about how to weigh them against each other. In rejecting relativism, we take it that humans will generally agree about which features of a situation have positive value and which negative value. Suffering and death are bad, health, prosperity, and justice are good. But this leaves plenty of room for even fully informed people to disagree about the best course of action, all things considered. For example, some might attach more weight to prosperity, others to justice. Agreement on pro tanto values does not therefore ensure a universally shared calculus for decisions. Even reasonable people in full possession of the facts can diverge in their choices of political and practical policies. This point will be central to what follows.

2. Evidence not Expedience

We say it is futile for science to aspire to 'value-freedom'. This does not mean, however, that it is acceptable to adopt scientific claims *because* of your values. Scientific claims may carry inevitable evaluative consequences. But this by no means implies that our scientific beliefs should be tailored to pre-existing values, rather than to evidence for their truth.

After all, why would anybody want to adopt scientific claims in the absence of good evidence for them? Whatever values are in play, it would seem to be in everybody's interests to believe truly if they can. If you harbour false views of the world, you will be ill-placed to select effective means to your ends and your projects will fail. Tailoring your beliefs to pre-existing values rather than good evidence does not seem a sensible strategy for anybody to adopt.

Of course, some people might have an interest in *persuading others* to adopt false beliefs. If I am rich, I might promulgate the theory that high marginal taxes have a

---

[1] In line with this, we do not take 'thick' concepts (Anderson 2002; Kirchin 2013) or 'mixed' concepts (Alexandrova 2018) to be incompatible with good science.

negative impact on the overall economy, in the hope this will stop my own taxes being raised. Again, if I grow crops, I might try to persuade the public that neonicotinoids don't poison bees, so as to lessen the danger of a ban on my using them. Even so, it would be silly for me to believe these things myself if the evidence doesn't support them. As before, it will only reduce the chance of my being able to select the appropriate means to my ends. In particular, if I care at all about the general economy, or the bees, as well as about my income and my crops, then I will be ill-advised to embrace my own propaganda, for it will only lead me to underestimate the dangers to some of the things I value, and so to advocate policies ill-suited my overall interests. (And if I don't care at all about the general economy or the bees, then why should I want to fool myself about the impact my selfish interests will have on them?)

As a matter of fact, of course, people with sectional interests do often delude themselves in just this way. In addition to promulgating factual views that favour policies that serve their own interests, they tend to believe those views themselves. It is an interesting question why this happens. After all, it is not only epistemically but also prudentially irrational, for the reasons just given. Still, people are not always rational. Perhaps an aversion to 'cognitive dissonance' is operative: rather than accepting that your different ends are in conflict, you come to persuade yourself, against the evidence, that they are in truth jointly achievable. Or perhaps a simpler form of self-deception is responsible: if you are going to espouse propaganda that serves your sectional interests, it is psychologically more comfortable, and indeed more effective, to believe in it yourself.

But we can leave these psychological conundrums for others. The nether regions of the human mind are not our present subject. Whatever irrational processes sometimes influence humans, we take it to be uncontroversial that it's bad to induce others to form unevidenced beliefs in order to further our sectional interests, and it's not made any better if we end up duping ourselves into those beliefs too.

Surprisingly, an influential strain in contemporary philosophy of science comes close to arguing that it is the duty of scientists to do just these things.

3. Rudner and 'Inductive Risk'

The line of thought goes back to the 1950s. In 'The scientist qua scientist makes value judgements' (1953), Richard Rudner argued that existing evidence can never conclusively prove or falsify any scientific claim. As he saw it, there will always be some inductive risk[2] involved in definitely accepting or rejecting any claim on the basis of finite evidence. In the face of this risk, said Rudner, scientists ought to

---

[2] As it happens, the now-standard phrase 'inductive risk' does not appear in Rudner's original paper. It was introduced by Carl Hempel in 'Science and Human Values' (1965) when discussing commitment to universal generalisations.

compare the relative dangers of falsely accepting or rejecting the claim. To the extent that false acceptance will have significant adverse consequences, they should demand stronger evidence before endorsing it. And the same applies in reverse. If mistaken rejection will lead to costly results, they should wait on strong negative evidence before rejecting the claim.[3]

Let us illustrate with neonicotinoids and the bees. Suppose that our hypothesis is:

neonicotinoid (N) spraying reduces bee populations by more than 20% each year.

And suppose that neonicotinoids will be banned if and only if this is accepted.

So if we reject this hypothesis when it is true, we will carry on spraying neonicotinoids and the bee population will be inadvertently depleted. On the other hand, if we accept the hypothesis when it is false, we will stop spraying and crop yields will decrease to no good effect. Suppose now that scientists agree that a significant fall-off in the bee population would be a worse outcome that reduced crop yields. According to Rudner's analysis, they should therefore make it easier to accept that neonicotinoids significantly affect bee populations. They should be ready to embrace N on weaker evidence than they'd require to reject it. This would have the happy result of reducing the risk of continuing to spray neonicotinoids when it will be hurting the bees.[4]

4. Whose Values?

One obvious problem with Rudner's recommendation is: 'whose evaluations of the costs and benefits should go into the calculation?'

Rudner's suggestion is that the scientists should look to their own values to tell them when the evidence is sufficient to warrant accepting or rejecting a hypothesis. In our example, the idea was that, given that scientists will value bee populations more than crop yields, they should be relatively quick to accept that neonicotinoids significantly affect bee populations. Still, there is no reason why everyone should share these values with the scientists. Rural communities, or indeed the general public, might well have different evaluative priorities. They might attach higher value than the scientists to food production, and so be inclined to demand more evidence before banning neonicotinoid spraying on the grounds that is has been shown to significantly affects bee populations.

---

[3] Rudner in fact only claims that scientists *will* compare these dangers, not that they *should*. In line with most subsequent literature, we shall skip over this. Given that a Rudnerian appeal to values is not the only possible response to inductive risk (see Levi 1967; John 2015a), the philosophically pressing issue is whether it is the best response. The actual practice of scientists is a secondary matter.

[4] See John (2015a) for further discussion of this example.

Relying on the values of scientists would thus seem to be in danger of pre-empting decisions that should properly be subject to democratic debate. At bottom, the Rudnerian line means that neonicotinoids will be banned more readily if the scientists prioritise bees over crops more than the wider public do. However, put like that, it seems clearly wrong. Why ever should the scientists' values carry more weight in these matters than those of the wider public?

What about modifying the Rudnerian line so that the scientists don't use their own values in deciding how to set evidential thresholds, but rather those of the democratically answerable politicians they are advising?[5]

That strikes us as better, but not a lot better. If the values determining evidential thresholds are those of the democratic majority, then at least the policies that come to be adopted will reflect those majority values. However, there still seems to be something wrong with allowing even majority values to influence decisions on scientific hypotheses. After all, not everybody shares those majority values, and such dissenters might reasonably take exception to scientific experts categorically asserting claims on relatively weak evidence, just because those claims align with majority preferences.

5. Which Choices?

Our main concern in this paper will be with the way that Rudnerian choices are sensitive to *whose values* are in play, in the way just explained. But before proceeding we would like to observe that Rudnerian choices are also sensitive to *which choices* are being addressed.

At bottom, the Rudnerians are making a choice aimed at certain non-epistemic ends—in our example, food production and bee preservation. Their eventual aim is to maximise expected values over those ends. True, it is not a simple expected value maximisation calculation, because their framework requires them to find a route to these ends that goes via a choice of an evidential threshold and subsequent formation of a categorical judgement. Even so, if we ignore the bells and whistles, the Rudnerians are in effect making a choice aimed at certain non-epistemic ends.

One consequence is that, in a different context where different ends are being pursued, Rudnerians might end up setting a different evidential threshold for some hypothesis, and so might reach a different categorical judgement, even given the same evidence. To illustrate, imagine that we are still interested in:

---

[5] This is advocated by Heather Douglas (2000, 2009). She argues that evidential thresholds ought to be determined, not by the values of scientists themselves, but by democratic values generated by involving the public in science.

neonicotinoids (N) reduce bee populations by more than 20% each year.

But now imagine a context where we are concerned, not with bee conservation and crop yields, but about the best way to deal with an infestation of bees that is disrupting a school. Should we use neonicotinoids or some more proven but more expensive insecticide? Assuming it would be much worse to fail to deal with the infestation than to spend a bit more money, Rudnerians will now have reason to set a higher threshold for accepting N, the better to guard against using the cheaper neonicotinoids when they are in truth ineffective. So they might end up categorically rejecting N on the basis of evidence which would have led them to accept it when they were concerned with bee conservation and crop yields.

In the last section we pointed out that the Rudnerian setting of evidential thresholds is sensitive to whose values are at issue: the thresholds favoured by ecologists will be different from those favoured by agriculturalists. We now see that, even when we are dealing with one fixed set of values, the setting of thresholds is also sensitive to which choices are being addressed.

It might have been thought that it is the responsibility of science to issue judgements that all people can trust to inform whichever choices they need to make. Rudnerian science, however, turns out not to be able to fulfil this responsibility. The categorical judgements it delivers are inevitably tailored to specific values and specific choices.

6. The Current Debate

The debate about values and inductive risk has had a had a new lease of life over the past couple of decades, no doubt due to the increased awareness among philosophers of the political implications of scientific claims. A surprising number of philosophers endorse the Rudnerian line that values should influence commitment to scientific claims, among others Philip Kitcher, Heather Douglas, Katie Steele, Eric Winsberg, and Torsten Wilholt.[6]

We are going to argue against the Rudnerian line. Our objection will not be to his proposal for setting evidential thresholds by reference to values, but to the whole idea that we need such thresholds at all. We think that Rudner goes wrong from the start, in assuming that science needs to overcome 'inductive risks'. Once this assumption is made, the intrusion of values into scientific theory choice becomes hard to avoid. Still, as we shall explain in the next section, we don't need to start where Rudner does. A better account of science is available, one in which worries about the role of values fall away.

---

[6] See Kitcher 2001; Douglas 2000, 2009; Wilholt 2009; Steele 2012; Winsberg 2012.

We should say at this point that we will not be able to address all the issues raised in the contemporary debate about values and inductive risk. These are many and varied. For a start, different authors locate the role of values at different stages of the scientific process. Thus Douglas (2000) argues that they are needed at intermediate levels of research, as for example when a tumour is categorised as 'cancerous' or not. Others have suggested we should be concerned, not with the claims scientists themselves believe, but rather with those that they communicate publicly (John 2015a). Yet others have focused on particular cases, especially involving climate change. In this last context some have advocated a system of intervals to represent uncertainty (Betz 2013) while others (John 2015b; Frank 2017) have responded that this will not eliminate the need to address inductive risks.

Rather than attempting to engage with all these detailed concerns, our focus in this paper will be on more foundational matters.

7. Credences versus Significance Tests

Our view, as we said, is that Rudner goes wrong from the start in the way he thinks about the scientific enterprise. He assumes without argument that the job of science is to reach *categorical judgements* about the truth or falsity of scientific claims. However, it is contentious that we should think of science in this way. The alternative would be for science to advise us on what *credences* (equivalently *degree of belief*, or *subjective probabilities*) in scientific claims are appropriate given the evidence. After all, Rudner himself insists that existing evidence will never compel a definite decision on any scientific claim. Given this, why assume that science must issue in definite categorical judgements,[7] rather that indicating appropriate degrees of belief?

This response to Rudner's line is of long standing. Rudner's original paper in 1953 soon prompted a response from the early Bayesian Richard Jeffrey, who observed that there would be no need to invoke values to fix evidential thresholds for categorical commitments if scientists simply eschewed such commitments (Jeffrey 1956). Instead they could just report the degree of belief warranted by the evidence. If scientists didn't try to bridge the 'inductive risk gap' by setting evidential thresholds, Jeffrey observed, they wouldn't need evaluate the costs and

---

[7] How exactly should we understand 'categorical judgement' in the present context? Given that the attitude is supposed to be a response to non-zero inductive risks, it can scarcely be required to exclude any possibility of error. What it does require, however, is unclear. However, we shall say little about this issue here (though see Section 11). Since we deny that categorical judgements play any serious intellectual role in science or policy-making, we regard the issue as a problem for our opponents rather than ourselves.

benefits of doing so. As Jeffrey saw it, instead of forming categorical judgements, 'the activity proper to the scientist is the assignment of probabilities' (1956, 237).

At the time he wrote, Jeffrey's Bayesian line was fighting against the tide of scientific orthodoxy. For better or worse, scientists were committed to arriving at categorical conclusions, and they standardly appealed to the apparatus of *significance testing* to facilitate this.

We take it that this apparatus will be familiar, but it will be convenient to rehearse the basics briefly. The central idea of significance testing is that, when presented with some hypothesis and some evidence supporting it, scientists should assess how likely it is that we would have found supporting evidence of that strength if the hypothesis were not true. If statistical analysis shows that this probability is small enough, say less than 5%, or less than 1%, then we should conclude that the hypothesis is indeed true. The probability required is called the *significance level*, or the *p-value*, and the rationale for this procedure is that it means we will only rarely end up endorsing a hypothesis when it is in fact false—5% of the time, say, or 1%, depending on what *significance level* is chosen.

The Rudnerian line on value-ladenness goes hand-in-hand with this methodology of significance testing. Indeed Rudner and many others in the ongoing debate explicitly refer to significance levels as what they mean when they say thresholds for acceptance should be sensitive to the costs of inductive mistakes. As they see it, if there is a high cost to rejecting the neonicotinoid hypothesis N when it is actually true, then we should choose a relatively undemanding significance level, with a 5% p-value or higher, rather than 1%, so as to lessen the risk of wrongly going on spraying.

Still, this affinity with the methodology of significance tests lends no real support to the Rudnerian approach. This is because the whole logic of significance testing is itself deeply flawed. When properly examined, the idea that we should categorically accept a hypothesis H whenever we observe a significant result makes little sense. Sure, if we do that we'll only end up accepting H *when it is false* one time in twenty (or one hundred, or whatever). But that's no reason at all to think H is likely to be true—which one might have thought would be a minimum requirement for categorically accepting it.

To see the point, imagine that scientists are indefatigable assessors of unlikely hypotheses, and that only one in one hundred of the hypotheses that they test are in fact true. Now focus on the ninety-nine of every one hundred hypotheses that are false. We can expect about five of them to show a significant result at the 5% level, and about one to do so even at the 1% level. After all, that is precisely what 'significance level' means—how often you will get a significant result if your hypothesis is false. So, even if we take it for granted that true hypotheses will always generate significant results themselves, a significant result at the 5% level will only warrant about a 1:6 credence in the hypothesis under test—since we will get a significant result from false hypotheses five times for every one from a true hypothesis—and even at the 1% level only a ½ credence will be warranted. This would seem a terrible basis for the categorical commitment to the hypothesis that is mandated by the logic of significance testing.[8]

Much of the motivation for the traditional 'frequentist'[9] approach to statistical inference lay in the way that the alternative 'Bayesian' credences depend on 'prior probabilities', that is, initial pre-testing degrees of belief in the hypotheses under test. For example, the toy analysis of the preceding paragraph hinged on the prior assumption that the hypothesis under test has a one in a hundred probability of being true. Since this assumption preceded the gathering of any evidence, it would seem to lack any proper empirical backing. Reliance on such prior probabilities thus seems in tension with a commitment to evidence-based science. The attraction of frequentist significance testing was that it seemed to offer a way of bypassing any dependence on arbitrary initial ideas.

We shall come back to the supposed arbitrariness of prior probabilities in Section 10. Still, whatever view we take on that issue, we trust that it is already clear that the methodology of significance testing is itself fatally flawed. The previous toy analysis was not meant to be fanciful. Science is currently beset by a 'replication crisis'. In areas that rely on statistical evidence, surveys indicate that most published results accepted on the basis of significant evidence turn out not to be replicable in repeat studies (Open Science Collaboration 2015; Baker 2016). No doubt a number of different factors have contributed to this, but a growing consensus now takes the view that the main cause has simply been the misplaced faith in significance testing. As we have seen, failures of replication are exactly what we would expect if the preponderance of hypotheses subject to test are in fact false. From this perspective, the replication crisis is simply the chickens of significance testing coming home to roost.

8. Rudnerian Calculations

The last section outlined the natural affinity between the Rudnerian view of science and the traditional methodology of significance testing. Both are entirely focused on categorical judgements and do not assign any significant role to credences in scientific practice.

Somewhat curiously, however, the Rudnerian approach to theory choice itself hinges essentially, if implicitly, on credences. It needs them for its calculations

---

[8] For more on this critique, see Papineau 2018; Bird 2021.
[9] From a philosophical point of view, there is little to recommend 'frequentist' as a description of Neyman–Pearson significance testing. The distinguishing feature of that methodology is that it relies solely on objective probabilities, not that it favours a frequency interpretation of those objective probabilities over alternative propensity or chance interpretations. Still, this usage is now so well-established that there seems little point in resisting it.

about how high or low to set evidential thresholds. In this section we would like to digress briefly to explain this. The point is not crucial to our overall argument, but will be worth rehearsing, if only because it adds to the case that there is no cogent way of avoiding credences in science.

Rudnerians argue that we need to assess the possible costs and benefits of different evidential thresholds. However, Rudnerians are rarely specific about the calculations that this might involve. To make things definite, let us assume, in line with much of the Rudnerian literature, that the evidential thresholds in question are p-values. For example, to stick with the conventional options, should we set the significance level at 5% or 1%? The former will give us a 5% and the latter a 1% chance of false positives (taking neonicotinoids to be toxic when they aren't, for example). Then there is also the issue of false negatives. To figure out the chances of these we will need to estimate the *power* of the tests at issue. Suppose for the sake of the argument that we are able to do this. This will then tell us how much higher the chance of false negatives (not taking neonicotinoids to be benign when they are) will be with a 1% than a 5% p-value.

Let us thus suppose that we have so determined the relative chances of false positives and negatives for both 5% and 1% p-values. The crucial Runderian idea is then that we need to add in *evaluations* of the costs of false positive and negatives to get from these chances to a decision about evidential thresholds. As the Rudnerians see it, the ecologists will be more worried about false negatives than the agriculturalists, and vice versa about false negatives, and so the ecologists will tend to favour the 5% significance level.

Well, the ecologists might be *more inclined* to the higher p-value than the agriculturalists. But will they actually opt for it? If they are rational, that will have to depend on a further factor, namely their *credence* in N—what probability do they attach neonicotinoids being toxic?[10] After all, even the ecologists will presumably attach *some* value to maintaining food production. And so they won't want to threaten to reduce food production without some reasonable expectation of compensating benefit. But, for all that's been said so far, that could well be precisely the upshot of opting for the 5% significance level. To make the point vivid, suppose the ecologists only had a one in a million credence that neonicotinoids are toxic. Then it would be very odd for them to opt for the 5% rather than 1% significance level. That would multiply the chance of stopping spraying and

---

[10] Should this be the prior or posterior credence in N? We leave it to the Rudnerians to tell us. After all, it's their calculation and they certainly need some credence for N to carry it out. They're probably better off invoking their *prior* credence in N—the idea would thus be to choose the significance level before gathering evidence for or against N. Alternatively, they could in principle get the evidence, update their credence in N on that basis, and then use that *posterior* credence to choose the significance level for deciding whether or not to accept N—it seems particularly weird, though, to design a significance test to take you from the evidence to a categorical judgement when you have already used that same evidence to update your degree of belief.

reducing crop yield fivefold, even though such a cessation is almost certain to be pointless.

All right, we could imagine extreme ecological values—no amount of wheat is worth even one bee's life—that would lead ecologists to favour the higher p-value even if their credence in neonicotinoid toxicity was miniscule. But the general point should be clear. If you attach any value to food production, you already need a credence about the toxicity of nicotinoids to assess whether the increased threat to food production posed by the 5% p-value is worth it.[11]

9. Higher-Order Judgements

As we said, Bayesians like Richard Jeffrey point out that science can avoid all the Rudnerian contortions straightforwardly enough by simply avoiding categorical judgements and sticking to quantitative probabilities. A long-standing objection to Jeffrey, however, is that categorical judgements will inevitably re-emerge at a higher level. When the scientists say that some hypothesis *P is probable to such-and-such a degree*, won't that itself be a categorical judgement about a probability, and so won't it reintroduce all the issues of inductive risk?

This anti-Bayesian objection strikes us as muddling up subjective and objective probabilities.

We take *subjective probabilities* (or *credences*, or *degrees of belief*) to characterise the strength of a person's commitment to a proposition, the extent to which they expect it to be true. If your subjective probability that Trump will win the 2024 presidential election is 0.4, say, then you expect it to that degree, as manifested in the choices you make, for example by hedging proportionately more against his losing than his winning. (And we take *rational* subjective probabilities/credences/ degrees of belief to be the extent to which a person *ought* to expect the relevant proposition to be true.)

By contrast, we take *objective probabilities* to characterise the tendency for a certain kind of outcome to happen in a certain kind of situation, such as the

---

[11] Here's a table to make the issue clear. False negatives in the left-hand column, false positives in the right. Let us suppose, for the sake of the argument, that opting for a 1% p-value rather than 5% will reduce the power of the test from 90% to 80%. Environmentalists will be more worried about the left-hand false negatives than the agriculturalists, and vice versa for false positives. But are the environmentalists worried enough to make it worth choosing 5% and so substantially increasing the chance of a false positive and reducing crop yields unnecessarily? This must depend, not just on how much they are concerned about the bees, but also on their overall credence for Toxic versus Benign.

	Toxic	Benign
5% p-value	Pr(Allow) = 0.10	Pr(Ban) = 0.05
1% p-value	Pr(Allow) = 0.20	Pr(Ban) = 0.01

tendency for fair coins to come down heads when tossed, or for radium atoms to decay within 1,000 years. The nature of objective probabilities is much debated, but they are clearly distinct from subjective degrees of belief. Radium atoms would still have a certain objective tendency to decay within 1,000 years even if no humans or any other creatures had ever thought about the matter.

Now, objective probabilities are part of the *subject* matter of many scientific claims. For example, we might be interested in:

(M) the objective probability of lifetime breast cancer for women with the BRCA1 gene is over 50%.

On the other hand, subjective probabilities are a matter of *attitudes to* scientific claims. Someone believes a scientific claim to a certain degree—25%, 50%, 90%, whatever. The scientific claim in question might have objective probability as part of its subject matter, as in M, or it might not.

Rational subjective probabilities are not entirely unrelated to objective ones. There is an obvious and natural connection between them—David Lewis called it the 'Principal Principle' (1980). This connection applies in situations which do involve objective probabilities, like coin tosses and radium atoms. The connection is then that, to the extent we know the objective probability of some outcome, we ought to set our subjective degree of belief equal to that objective probability. For example, if you think some coin's landing on heads has an objective probability of ½, then you ought to have a credence of ½ in its landing on heads.

However, this kind of constraint on subjective probabilities has no obvious application to the kind of case we have been considering, where we attach a subjective probability, not to the claim that a coin will land on heads, or that a radioactive atom will decay, but rather to a general scientific hypothesis about the toxicity of insecticides, say, or the risks associated with the BRCA1 gene. Even if objective probabilities enter into subject matter of such hypotheses, as in the latter example, this doesn't at all mean that the hypothesis itself will have an objective probability. It is not as it there is some physical mechanism that systematically ensures that a certain proportion of such scientific hypotheses are true, in the way that the physics of coin tossing is responsible for a certain proportion of coins landing on heads. So we don't here have the same kind of constraint on credences that arises with coin tosses and the like. Our credences might be directed at claims *about* objective probabilities, but those credences won't themselves be *reflections* of any objective probabilities. Rather those credences will simply express the confidence that scientists have in the relevant hypotheses, given their background knowledge and the existing evidence.

The objection mounted against Jeffrey was that, even if credences of the form *P are probable to such-and-such a degree* and do not commit categorically to non-probabilistic facts, they at least involve a categorical judgement about a probability, and this in itself is enough to reintroduce all the issues of inductive risk. We can now see what is wrong with this objection. Someone who expresses a certain degree of belief in a hypothesis isn't thereby making a categorical claim about some objective probability. There are just expressing their confidence in the hypothesis, and that's it.

The point is clearest when the hypothesis at issue isn't itself about objective probabilities. If I say I am 80% confident that *mad cow disease is caused by prions*, I am not ascribing a probability to anything, categorically or otherwise. The sole proposition in play is that the non-probabilistic *mad cow disease is caused by prions*, and the 80% only characterises my attitude *to* that proposition, not any probabilistic aspect *in* my subject matter. The same point applies even when we're dealing with a hypothesis about objective probabilities. I am 80% confident that M—the objective probability of lifetime breast cancer for women with the BRCA1 gene is over 0.5. Again, the 80% only characterises my attitude *to* M. I don't make any judgement, categorical or otherwise, *about* anything being 80% probable.[12]

In short, when we say scientists should express credences, not make categorical judgements, we don't mean that scientists should make categorical judgements about probabilities. Rather they should simply express their confidence that P is true, grounded in their background knowledge and their evidence. Nothing in that involves any categorial higher-order judgements about anything being probable.[13]

10. The Problem of the Priors

The last section argued that we should understand scientists' credences in hypotheses, not as judgements about objective probabilities, but simply as expressions of their informed expectations that the hypotheses in question are true. However, this now returns us to the 'problem of the priors' mentioned earlier. For Bayesians, rational credences are a function of the empirical evidence and prior credences. A rational subject will start with some initial degree of belief in a hypothesis H. When evidence E is found, they will increase (or decrease) their degree of belief in H to the extent H implies a higher (lower) objective probability for E than the alternative hypotheses. While this updating procedure hinges in part on objective

---

[12] Sometimes credences about objective probabilities will be expressed in the form of 'confidence intervals'. For example, we might have a 95% degree of belief that the mean of the underlying probability distribution lies within a certain distance of some observed statistic. (Note, however, that if confidence intervals are so understood as expressing degrees of belief, they too will depend on prior subjective probabilities, and cannot, as is widely but erroneously supposed, simply be read off from the statistic and the knowledge that, say, it is normally distributed.)

[13] We find it surprising that this response to the higher-order objection to Jeffrey is not more widely accepted. Perhaps part of the reason is that Jeffrey's own comments about the issue at the end of his original paper are unclear and confusing. We would attribute this to his maintaining, along with other Bayesians of the time, that objective probabilities should be entirely eliminated in favour of subjective probabilities. (We would like to thank Richard Bradley for advice about Jeffrey's early views.)

probabilities, it also depends on the prior degree of belief the subject starts with. And it is unclear what basis this prior credence could have, beyond the subject's initial intuition, hunches or prejudices.

On the face of things, this seems a serious worry about Jeffrey's Bayesian line. Do we really want to stop scientists making definite statements and instead restrict them to statements like 'Given my evidentially ungrounded initial assessment of H's credibility, plus the evidence that has now come in, I'd say that the odds in favour of H are 75%'? Not only would the initial credences seem to lack any support, but they are also likely to vary between scientists, with the result that they scientific community will end up sending mixed messages in cases where clear guidance is needed.

Still, what is the alternative? When the evidence leaves room for reasonable experts to disagree, is it really better for them to speak with one categorical voice, just because that opinion supports some policy that they or some favoured section of the population would choose?

Fortunately, this dilemma is not inescapable. In our view, the so-called 'problem of the priors' is overblown. Even if scientists were to turn away from Runderian dogmatism, and embrace the Bayesian alternative, there is no reason why their expression of credences should not command general respect.

An initial point is that the prior credences of scientists won't just be random guesses, but informed by their education and experience, even if not by newly generated empirical evidence. Worries about prior credences are often amplified by citing archaic medical procedures that rested on nothing but the folklore of senior physicians. Traditional medical science, however, is arguably a poor basis for condemning all prior scientific judgement. As a general rule, educated scientists in empirically grounded disciplines will have a reasonable sense of which hypotheses are likely to be true, and of which should be discounted until they are backed up by substantial supporting evidence.

A second point relates to the way credences change once empirical evidence is uncovered. New evidence will tend to wash out differences in priors and push credences in the same direction. For example, when Barry Marshall and Robin Warren first proposed that bacteria were largely responsible for stomach ulcers, most medical scientists thought the idea absurd (Marshall and Warren 1983). However this scepticism was soon enough eliminated as more evidence came in. In truth, there's no reason why the Bayesian judgements of informed scientists shouldn't display reassuring conformity even in cases where they start off with significantly divergent priors.

In particular, this consensus will often amount to de facto certainty. Back in the 1950s, at the time of Rudner's original article, the implicit presupposition was that a claim was only firmly established if it was *logically entailed* by the evidence. Anything short of that meant that the issue remained underdetermined. This is the rationale for Rudner's insistence on unavoidable 'inductive risk'. Categorial commitment always incurs a real chance of error. More recent philosophy of science, however, has recognised that the demand for logically compelling evidence sets the bar far too high. Yes, the empirical evidence is never logically compelling, but it is often sufficient to rule out all but the most fanciful alternatives. And in those cases even Bayesians will agree that scientists can issue unqualified endorsement of hypotheses. One can imagine fanciful alternatives to the claims that *COVID-19 is caused by a virus*, that *smoking causes cancer*, or indeed that *bacteria affect stomach ulcers*—perhaps an evil demon wants to deceive us, perhaps an alliance of world governments is conspiring against us—but, if these are the only alternatives consistent with the evidence, Bayesians will view the probability of the orthodox theories as indefinitely close to one, and expect scientists to endorse them accordingly.

So the 'problem of the priors' does not inevitably condemn Bayesian scientists to destructive disagreement. For a start, they will be able fully to endorse scientific claims in those many cases where the evidence leaves no real room for doubt. Second, even when the evidence is less than conclusive, it will tend to push their degrees of belief in the same direction. And, finally, even when the evidence has not yet produced consensus, the credences of scientists will generally express educated judgements rather than arbitrary guesses.

11. Against Categorical Beliefs

In our view, as we have said, the whole debate about inductive risk starts in the wrong place. The real issue is not the best way to arrive at categorical beliefs in the face of such risks, but whether we need categorical beliefs at all. All the problems of inductive risk only arise if we aspire to categorical beliefs.

So what would go wrong if we didn't bother with categorical beliefs at all and dealt only in credences? While a great deal of mainstream epistemology is concerned with the formation of categorical beliefs, it contains surprisingly little discussion of the need for them.[14]

One thought commonly aired, though not necessarily developed in detail, is that probabilistic calculations quickly become too complicated for ordinary humans, and so we need categorical beliefs to render our everyday decision-making tractable (see for example Sturgeon 2020).

Maybe so. Still, devices designed to simplify decision-making would seem inappropriate for scientifically informed policy making, even if they are necessary for ordinary citizens in everyday contexts. When we are dealing with serious issues

---

[14] Within mainstream epistemology, the debate about 'pragmatic encroachment' covers many of the points we have been discussing. Unfortunately it would take us too far afield to engage with this debate here.

which do not need to be decided in a hurry, what reason is there not to figure out which options will really maximise expected utility given the evidentially supported credences? As a rule we can expect policy makers to be perfectly capable of doing calculations and weighing up costs and benefits properly. Communicating scientific results via categorical beliefs rather than credences will only create room for policy makers to arrive at sub-optimal decisions. Maybe the cost of categorical beliefs is inevitable in many everyday contexts, but that is no argument for importing them into serious policy-based decision-making.

A rather different rationale for categorical beliefs is offered by our King's College London colleague David Owens. He observes that, given the way humans are psychologically constituted, we can't engage emotionally with the world without forming categorical beliefs. You can't *blame* somebody for their bad behaviour, or be *angry* with them, or *pleased* at their success, or *happy* because Spurs won... if you only think it's *likely* these things happened. We need fully to believe before we can react emotionally. And since our emotional engagement with the world is central to our lives, argues Owens, we cannot do without categorical beliefs (Owens 2017).

We do not want to take issue with Owens' argument. But, once more, we do not regard it as relevant to serious decision-making. Let us grant that we need categorical beliefs to form emotional reactions. That is no argument for thinking that policy makers need them when evaluating the pros and cons of different options. If optimal decisions are best arrived at on the basis of informed credences, basing them on categorical beliefs instead can only degrade decision-making.[15]

12. Racing to the Bottom

In much of the literature on Rudner's proposal, the dangers of *manipulation* and *wilful misunderstanding* are standardly cited as reasons for scientists making categorical pronouncements rather than communicating credences. Rudner's supporters argue that vested interests and anti-science factions will latch onto any hint of scientific uncertainty or doubt to undermine the scientific case for sound policies. They will declare that the evidence is equivocal, that there is no scientific proof of insecticide harms, vaccine safety, global warming, and so on (Douglas 2000, 2009; John 2015a).

---

[15] This point perhaps deserves further discussion. What if the options at issue themselves involve *acting out of an emotion*? For example, Clayton Littlejohn (2020) has argued that criminal courts cannot convict without being in a position to *blame*, and so in effect face a choice between freeing the defendant and *convicting-on-categorical-belief*. Our response is that in such cases categorical beliefs will feature among the *ends* aimed at, but that the choice of *means* to those ends should still be guided by rational credences. Littlejohn (personal communication) does not disagree.

The solution, according to Kevin Elliott (2011) is for scientists to adopt a 'no passing the buck principle'. It's no good their passing on their credences to the politicians and policy makers and then leaving it to them to make rational decisions. Their failure to make categorical pronouncements will only be exploited by bad actors. Instead the scientists should grasp the nettle themselves, identify the right course of action, and promulgate as definitively established those claims that support it.[16]

Perhaps this is indeed the pass we have come to in the modern world of fake news and social media bubbles. Maybe there is no alternative to pre-empting debate in the way Elliot urges. If that is so, however, a great deal will have been lost. Let us conclude by pointing out the downsides of Elliot's recommendation, in order to emphasise the importance of seeking another way.

We will do well to remember that the bad actors in our story are not necessarily evil. In the first instance, all that distinguishes them from the consensus of right-thinking scientists is that they have different priorities. They might place more weight on their immediate household incomes than the long-term future of the planet. Or perhaps they prioritise generation of profit over environmental impact. In themselves these divergences are not pathological. They are the kinds of differences in values and interests that we can expect among reasonable people, and there seems no immediate reason why democratic processes should not be able to accommodate and resolve them.

Rudner's supporters might respond that it is not these diverging priorities themselves that are now being invoked to justify categorical scientific pronouncements. Rather it is the consequent sin of manipulating the facts, of seizing on uncertainty to make it seem as if there is no evidence that difficult measures are needed. We entirely agree that this is a sin that should be condemned. Democratic processes will not be able to find the right way of resolving conflicting priorities if they are beset by misinformation about the relevant facts.

Still, is the solution for the scientists to get their manipulation in first?[17] To our mind, the cure seems just as bad as the problem. At the bottom, Elliot's 'no passing the buck' principle responds to the threat of fake news by urging we manipulate the facts in the opposite direction. Rather than honestly communicating the evidential situation, Elliot and other Rudnerians want scientists to present partially evidenced claims as conclusively proven, whenever that will support the policies that right-thinking citizens favour.

The real danger with this strategy is that it is likely to set us off on a race to the bottom. Perhaps it will have commendable upshots in particular cases, ensuring

---

[16] Douglas similarly holds that scientists have a moral responsibility to promulgate value-based categorial judgements (Douglas 2000, 563).
[17] 'Let's get our retaliation in first' urged captain Willie John McBride in his team talk before the first test of the 1974 Lions rugby tour of South Africa.

that we arrive at policies that would be democratically favoured in the absence of propaganda from powerful bad actors. Still, it threatens a deeper cost. If the Rudnerian strategy comes to be adopted as standard practice, it is all too likely to undermine trust in science. Those who are unsure of the scientists' preferred policies will have every reason to suspect that the scientists are overegging the evidence.

It has been put to us that this kind of mistrust would be misplaced if the scientists were guided, not by their own values, but by those of the democratic majority, in the way suggested earlier. But this misses the point. The whole rationale for 'not passing the buck' is to outmanoeuvre those who do not share the majority's priorities. Evidential uncertainties are to be concealed from them, in the interests of bolstering the case for favoured policies. They are to be told that things are beyond doubt, when they aren't. Once this becomes standard practice, the danger is that the intended audience will turn against science. Why should they credit official scientific pronouncements, if they are designed precisely to outwit them? 'We've had enough of experts' would seem an entirely rational response.

Democratic decisions need to be informed by accurate scientific information, including information about partial credences. Policy makers can then take into account differences in values and interests and use the science to see how best to resolve these. But our institutions will struggle to do this if all sides are wielding propaganda to support their preferred policies. In the end, allowing evaluations to shape scientific decisions only threatens to devalue the coin of science.

References

Alexandrova, A. (2018) "Can the science of well-being be objective?." *British Journal for the Philosophy of Science* 69, 421–45.

Anderson, E. (2002) "Situated knowledge and the interplay of value judgments and evidence in scientific inquiry," in *In the Scope of Logic, Methodology and Philosophy of Science*, vol. 2, (eds.), Gärdenfors, P., Kijania-Placek, K., and Woléński, J., 497–517. Dordrecht: Springer.

Baker, M. (2016) "1,500 Scientists lift the lid on reproducibility." *Nature* 26, 452–4.

Betz, G. (2013) "In defence of the value free ideal." *European Journal for Philosophy of Science* 3, 207–20.

Bird, A. (2021) "Understanding the replication crisis as a base rate fallacy." *British Journal for the Philosophy of Science* 72, 965–93.

Douglas, H. (2000) "Inductive risk and values in science." *Philosophy of Science* 67, 559–79.

Douglas, H. (2009) *Science, Policy, and the Value-Free Ideal*. Pittsburg: University of Pittsburgh Press.

Elliott, K. (2011) *Is a Little Pollution Good for You?* New York: Oxford University Press.

Frank, D. M. (2017) "Making uncertainties explicit: The Jeffreyan value-free ideal and its limits," in *Exploring Inductive Risk: Case Studies of Values in Science*. New York: Oxford University Press.

Hempel, C. (1965) "Science and human values," in *Aspects of Scientific Explanation and Other Essays in the Philosophy of Science*, (eds.), Elliott, K. and Richards, T., 81–96. New York: The Free Press.

Jeffrey, R. C. (1956) "Valuation and acceptance of scientific hypotheses." *Philosophy of Science* 23, 237–46.

John, S. (2015a) "Inductive risk and the contexts of communication." *Synthese* 192, 79–96.

John, S. (2015b) "The example of the IPCC does not vindicate the value free ideal: A reply to Gregor Betz." *European Journal for Philosophy of Science* 5, 1–13.

Kirchin, S. (ed.) (2013) *Thick Concepts*. Oxford: Oxford University Press.

Kitcher, P. (2001) *Science, Truth, and Democracy*. New York: Oxford University Press.

Levi, I. (1967) *Gambling with Truth*. New York: Knopf.

Lewis, D. (1980) "A subjectivist's guide to objective chance," in *Studies in Inductive Logic and Probability*, vol. II, (ed.), Jeffrey, R., 263–93. University of California Press.

Littlejohn, C. (2020) "Truth, knowledge, and the standard of proof in criminal law." *Synthese* 197, 5253–86.

Marshall, B. and Warren, J. (1983) "Unidentified curved bacilli on gastric epithelium in active chronic gastritis." *The Lancet* 321, 1273–5.

Open Science Collaboration. (2015) "Estimating the reproducibility of psychological science." *Science* 349(6251).

Owens, D. (2017) *Normativity and Control*. Oxford: Oxford University Press.

Papineau, D. (2018) "Thomas Bayes and the crisis in science." *Times Literary Supplement website*, available at https://www.davidpapineau.co.uk/uploads/1/8/5/5/18551740/bayestlspapineau.pdf.

Rudner, R. (1953) "The scientist qua scientist makes value judgements." *Philosophy of Science* 20, 1–6.

Steele, K. (2012) "The scientist qua policy advisor makes value judgments." *Philosophy of Science* 79, 893–904.

Sturgeon, S. (2020) *The Rational Mind*. Oxford: Oxford University Press.

Wilholt, T. (2009) "Bias and values in scientific research." *Studies in History and Philosophy of Science Part A* 40, 92–101.

Winsberg, E. (2012) "Values and uncertainties in the predictions of global climate models." *Kennedy Institute of Ethics Journal* 22, 111–37.

8
From the Right to Science as an Epistemic-Cultural Human Right to the Right to Expertise

Michela Massimi
University of Edinburgh

> There is no human activity from which every form of intellectual participation can be excluded: *homo faber* cannot be separated from *homo sapiens*.
>
> Antonio Gramsci[1]

1. The Right to Science

Among the rights listed in the United Nations (UN) Declaration of Human Rights (UNDHR) in 1948, there is the so-called 'Right to Science':

Art. 27 (1) Everyone has the right freely to participate in the cultural life of the community, to enjoy the arts and to *share in scientific advancement and its benefits* (emphasis added).

The International Covenant for Economic, Social and Cultural Rights (ICESCR) in 1966 reiterated with Art 15 (1)(b) the Right 'to Enjoy the Benefits of Scientific Progress and its applications' (REBSP). In 2009 the United Nations Educational, Scientific and Cultural Organization (UNESCO) convened a meeting to discuss limits in the implementation of REBSP (UNESCO 2009). In 2020 the United Nations published a General Comment n. 25 to clarify the normative content of REBSP and its wider relations to economic, social, and cultural rights (Committee on Economic, Social and Cultural Rights 2020).

The emphasis of the General Comment n. 25 on ensuring an equal and non-discriminatory access to the REBSP for the most vulnerable people in society is

---

[1] (Gramsci 1971), 9.

Michela Massimi, *From the Right to Science as an Epistemic-Cultural Human Right to the Right to Expertise*
In: *Expertise: Philosophical Perspectives*. Edited by: Mirko Farina, Andrea Lavazza, and Duncan Pritchard,
Oxford University Press. © Michela Massimi 2024. DOI: 10.1093/oso/9780198877301.003.0008

somewhat at odds with the rather specific (albeit fairly standard) operative notion of 'science' underpinning the General Comment n. 25 (ibid.), 1:

> 5. Thus, science, which encompasses natural and social sciences, refers both to a process following a certain methodology ('doing science') and to the results of this process (knowledge and applications). Although protection and promotion as a cultural right may be claimed for other forms of knowledge, knowledge should be considered as science only if it is based on critical inquiry and is open to falsifiability and testability. Knowledge which is based solely on tradition, revelation or authority, without the possible contrast with reason and experience, or which is immune to any falsifiability or intersubjective verification, cannot be considered science.

The word 'falsifiability' repeated twice in this passage is worth highlighting. It refers to a well-defined view articulated by the philosopher of science Karl Popper back in the first half of last century as the hallmark of science and the demarcation criterion between science and pseudoscience. For Popper science is defined by the ability to run severe tests that can falsify hypotheses, or conjectures as he called them.

The underlying picture of the UN General Comment n. 25 seems then to be the following. On the one side, there are the *scientists* 'doing science' according to 'a certain methodology', 'critical inquiry', 'falsifiability and testability'. On the other side, there is the general public under the proxy *Everyone* (including the most vulnerable and disadvantaged people), who are meant to enjoy the benefits resulting from scientists' critical inquiry. Couched in this form, the Right to Science seems identifiable with the right to do scientific research (defined by testability and falsifiability as main criteria) by *freely* pursuing particular research programmes in a way that benefits democratic societies. Most importantly, the General Comment stresses the demarcation between scientists' 'doing science' and other forms of knowledge (including traditional knowledge) that might fall under the remit of cultural rights and are protected in ICESCR Art 15 (1)(a)[2]—but 'cannot be considered science'.

Here is a puzzle at the heart of the Right to Science so understood. How is scientists' 'doing science' meant to align with everyone's *entitlement* to benefit from the outcomes of scientific research pursued by scientists? How to understand this prima facie unlikely alliance of scientists' *freedom of research* and everyone's *entitlement* to benefit from it? My question goes right to the heart of the problem of how to understand the nature of *scientific knowledge*, *who* produces it, and *who*, accordingly, should get to benefit from its advancements. Philosophy of science enters at this crucial juncture of the discussion on the normative foundations of the Right to Science.

---

[2] On cultural rights and ICESCR Art 15(1)(a), see General Comment n. 21 (Committee on Economic, Social and Cultural Rights E/C.12/GC/21 2009).

2. Sharing in Scientific Advancement versus Enjoying the Benefits of Scientific Progress

2.1 The Argument from Membership versus the Argument from Activity

Let me return for a moment to the two different formulations of the Right to Science to be found in the 1948 UNDHR and the 1966 ICESCR. The former speaks of the right to *share in scientific advancement and its benefits*. The latter is phrased as the right to *enjoy the benefits of scientific progress and its applications* (REBSP). In what follows, I take scientific progress and scientific advancement as interchangeable terms and I shall not discuss philosophical views about progress.[3] I focus instead on the difference between the language of *sharing in* and *enjoying the benefits of*. What hangs in the change of formulation from 1948 to 1966? Something philosophically important is signalled by this terminological change, I maintain. Consider the following contrast class of examples about *sharing in*:

- Martha and Anna share in the grandparents' inheritance.
- Three companies share in an investment for renewable energy sources.
- Humankind shares in the deep seabed and its mineral resources.

In daily parlance, *to share in* suggests some kind of apportionment of resources and goods among several parties. Thus, one possible way of reading the 1948 formulation is along the lines of a non-exclusive use of science as a public good whereby *everyone* has a public use right which does not exclude others from exercising their own public use right. In the third previous example, different nation States can for example request to the International Seabed Authority to exercise public use rights to the ocean floor as a non-exclusive public good for e.g. deep-sea mining.

There are two main problems with this possible reading. First, the very ability to exercise a public use right presupposes having infrastructures, financial capital, and human resources. Countries with low-income economies are not on a level playing field compared with countries with high-income economies when it comes to non-exclusive public use rights of either the deep-sea bed, or scientific advancements. Second, the underlying assumption of scientific advancements qua 'public goods' would need to be justified and substantiated. Are scientific advancements public goods in the economists' sense of non-exclusive and non-rival goods? Or are they common goods in the legal Latin sense of *res communis* i.e. for communal/public use but not publicly owned? Or are they participatory public

---

[3] For a discussion on moral progress and scientific progress, see (Kitcher forthcoming).

goods (see Besson 2023, building on Réaume 1998) in analogy with e.g. friendship which implies the ability of two agents to share a good that is inherently participatory?

I will not pursue these interesting questions in what follows because they are tangential to my focus here. Suffice to say that 'doing science' is not typically a two-way relation—with the public and the scientists as relata—unless one conceives of a broad enough class to which scientists and the public are both *members* of, and 'doing science' is in turn conceived of as a relation superimposed on this class. I argue in what follows that there are reasons for scepticism about the very idea of 'membership', and even more so for the prospects of identifying a class that could function as the unit of analysis for such a relation to hold upon (see Section 2.2).

Let us turn now to the 1966 ICESCR formulation in terms of *enjoying the benefits of scientific progress* and let us consider a possible contrast class of examples in daily parlance:

- Lucy and Mary are enjoying the benefits of their hard work.
- Sophie and Mia are enjoying the benefits of the gym subscription.
- I am enjoying the benefit of a democratic voting system.

To *enjoy the benefits of* something does not suggest the *apportionment* of a good or commodity among parties, but instead some kind of acquired *entitlement* to something in virtue of either an activity (e.g. hard work) or a membership (e.g. gym membership; membership of eligible voters). Hence one can envisage two possible kinds of arguments behind the Right to Science so formulated, which I call the argument *from activity* and the argument *for membership,* and run respectively as follows (Table 8.1):

Table 8.1 Comparing two possible arguments for the ICESCR 1966 formulation of the Right to Science: the argument from activity versus the argument from membership

Argument *from activity*	Argument *from membership*
1. To enjoy the benefits of some X is to stand in a particular relation to X.	1. To enjoy the benefits of some X is to gain membership access to X.
2. The particular relation to X involves some *activity* (e.g. hard work) that contributes to X, or brings about X.	2. Membership access to X can be gained in various (transactional or non-transactional) ways, e.g. paying fees to the gym, or civil protests for the voting system.
3. Therefore, to enjoy the benefit of some X is to do some *activity* that contributes to X, or brings about X.	3. Therefore, to enjoy the benefit of some X is to engage in (transactional or non-transactional) ways to gain membership access to X.

In the third previous example, I enjoy the benefit of a democratic voting system by gaining membership of the community of eligible voters when, back in 1928, the suffragette movement led by Emmeline Pankhurst fought hard to have women's right to vote legally recognised in the UK. In the rest of this paper, I concentrate on this ICESCR 1966 Art 15 (1)(b) formulation of the Right to Science (abbreviated as REBSP henceforth).

What grounds REBSP? Should REBSP be interpreted as a right which holds by *standing in a particular relation* that involves some kind of *activity* that contributes to scientific progress? Or, should it be understood along the lines of the argument from *membership*, namely, as an acquired entitlement similar to enjoying the benefits of a democratic voting system? The current formulation of REBSP does not help us see clearly through this distinction. As a result, it is not clear whether the pronoun *Everyone* in Art. 15(1)(b) should be interpreted as referring to a universal right-holder whose right has been acquired by doing some *activity* that contributed to science; or, by gaining some kind of *membership* of the relevant community reclaiming legitimate access to enjoying the benefits of scientific progress (in a sense that remains to be clarified).

One problem here is that if enjoying the benefits of some X is understood as being acquired *via membership*, a precondition for it is to gain membership. But what does it take to gain membership of the relevant community? How should the relevant community be understood in this context? What if membership is refused or denied? It seems that unless one already belongs to the relevant community, one cannot enjoy the benefits associated with membership of that community as per argument *from membership*. I cannot participate in the voting system and enjoy the benefits of a democratic voting system unless I am already a recognised member of the community of eligible voters. But as the story of the UK suffragette movement shows, entrenched power structures of exclusion often prevent people from becoming members of the relevant community who can enjoy the benefits of X.

When it comes to REBSP, it is unclear how the relevant community should even be defined. It cannot be just the *scientists* themselves who are engaged in scientific research, because REBSP is a universal human right that applies to *everyone*. For the same reason, the relevant community cannot be identified either with a selected group of economically high-income countries who can benefit from scientific progress by manufacturing new technological tools and commodifying scientific innovations. How to ensure that people in low-income countries do similarly enjoy benefits of scientific progress if not by recognising that they too are *somehow* members of the relevant global community? How to better qualify this community? Is *everyone* a scientist?

The argument *from activity* eschews these problems. For it does not presuppose some non-better-defined community of membership, but it assumes that *doing some activity* that contributes to scientific progress is all that is required for enjoying the right. In what follows, I build on this insight and unpack what I take to be the philosophical-normative foundations of the REBSP. I shall argue that REBSP should be understood, first and foremost, as an *epistemic-cultural* human right: namely, a right which applies to *everyone* in virtue of a range of *activities* which are an integral part of the *situated practices* of countless multicultural *epistemic communities*. Key to my philosophical analysis is that the right-holder proxy pronoun "*everyone*" refers to individual human beings in virtue of belonging to one or more among a multitude of *epistemic communities*: i.e. communities that produce scientific knowledge via their *situated practices* and associated activities thereof.

2.2 Epistemic Communities and Their Situated Practices. Or, Why REBSP Is not a Group Right

One might reply at this point that universal human rights are rights whose duty-bearing institutions are nation States who sign up and ratify in domestic law documents such as ICESCR. Therefore, *membership of* a State is a pre-requisite for the enjoyment of human rights, including REBSP. In response, I do not doubt that membership of a State is a legal pre-requisite for the enjoyment of any human right. However, it is not this kind of membership—qua *legal* membership of a nation State—that is my topic here and the target of my aforementioned discontent with the argument from membership, but rather a different kind of membership: *membership of epistemic communities*. This in turn raises two further questions:

(1) What is an epistemic community? How to define it vis-à-vis other kinds of communities, i.e. cultural communities, geo-political communities, linguistic communities, demographical, and other varieties of social communities?
(2) Do epistemic communities (as I understand the notion) count as right-holders so that for example the REBSP might be regarded as a 'group right'?

In reply to (2), epistemic communities, as I use the term, cannot be right-holders because they are not a 'conglomerate collectivity' in the word of (French 1984, 5–18), namely they are not a collection of individuals with a well-defined unity and identity, by contrast with, say, cultural communities, linguistic communities, religious communities, or geopolitical communities. In reply to (1), an epistemic community, as I understand it, shares a historically and culturally situated scientific practice, which is how I define a 'scientific perspective' in (Massimi 2022, 5–6),[4]

---

[4] '*Scientific perspective (sp)*: A scientific perspective *sp* is the actual—historically and culturally situated—scientific practice of a real scientific community at a given historical time. Scientific practice should here be understood to include: (i) the body of *scientific knowledge claims* advanced; (ii) the

building on an important tradition about the situated knowledge thesis championed by feminist philosophers of science, e.g. (Wylie 2003; Haraway 1988; Hartsock 1998). Let me briefly expand on these two points.

Any scientific practice is always situated in that the experimental, theoretical, and technological tools to *reliably* advance claims of knowledge are always idiosyncratic to the community which shares *that* practice at *that* historical point in time and in *that* culture. However, the situated knowledge of any epistemic community should be regarded neither as 'siloed', nor as membership-defining for the community. I stressed in (Massimi 2022, 337), how 'the view that scientific knowledge is *defined by* the specific historical-geographical-cultural *membership* of particular epistemic communities' is a remnant of what I call 'Kuhnian communitarianism'. Thomas Kuhn articulated a view of scientific knowledge production linked to *membership* of a scientific paradigm (e.g. Ptolemaic astronomy, phlogiston theory, etc.). In this respect, Kuhn's scientific paradigms are at a distance from what I call 'scientific perspectives'.

In *Perspectival Realism*, I articulated a dynamical and fluid view of how scientific knowledge forms and evolves through a multitude of scientific perspectives, and I argued that it is possible to track the evolution of claims of knowledge about particular phenomena in nature via

> historical lineages [that] span and ramify beyond geographical, national, and sociocultural boundaries. They have a history, evolve, and branch out rather than statically demarcate well-defined territories, scientific homelands or shared memberships (ibid., 343)

In practical terms, this philosophical move implies the recognition and reinstatement of a number of *epistemic communities* whose varieties of situated knowledges are usually 'severed' from scientific narratives about *who* produces knowledge, and *who* accordingly should be benefitting from it, including e.g. know-how about kelp-making being important for glass manufacture and for the subsequent emergence of e.g. cathode rays searches in the late 19th century (Massimi 2022, 307ff).

Ultimately, the view of scientific knowledge that I defend in *Perspectival Realism* is cosmopolitan (Massimi 2022, ch. 11) in emphasising how different epistemic communities have travelled and traded their tools and techniques. It transcends specific national, geo-political, and cultural boundaries. It allows one to position the Chinese Han geomancers along the same historical lineage of *interlacing perspectives* featuring Norse sailors, Amalfi mariners, and William

---

experimental, theoretical, and technological resources available to *reliably* make those scientific knowledge claims; and (iii) second-order (methodological-epistemic) principles that can *justify* the *reliability* of the scientific knowledge claims so advanced'. Massimi (2022), 5–6.

Gilbert in Elizabethan England (among many others) in the production of a series of revised and improved knowledge claims about specific phenomena (i.e. what we currently call the Earth's magnetic field and its declination).

Returning one more time to point (2) above, it becomes clear then why epistemic communities (as I understand the notion) are not organisational or institutional groups with a governance structure like universities, banks, churches, or with a well-defined unity and identity (e.g. the Scottish-Gaelic speaking community of the West Coast of Scotland; or the Ladin-speaking community of South Tyrol in the Italian Alps; and so forth).

Epistemic communities are *porous*, *itinerant*, and *granular*. They are porous in that an individual can be part of more than one epistemic community at the same time, depending on their training, skills, research interests, and activities. For example, Ebenezer Everett was a professionally trained glass-blower working for J.J. Thomson at the Cavendish Lab to produce high-quality cathode rays as well as being trained as a chemist—see (Navarro 2012, 51). The epistemic communities of e.g. the glass blowers and of the chemists are not insular, disjoint, or isolated communities.

Epistemic communities are itinerant: their knowledge claims and their underpinning tools and methods and practices travel over time and across geographical regions and cultures. Through trades, travels, and encounters, varieties of knowledges get transmitted, re-interpreted, and re-used. Being porous and itinerant means that individuals in an epistemic community might often belong to different ethnic, linguistic, religious, and geopolitical groups. The epistemic community of beekeepers—see (Massimi 2022, ch. 8)—spans Yucatán beekeepers, Scottish heather honey producers, and my grandfather Eligio in Italy collecting *millefiori* honey, among other examples. Their situated practices might bear family resemblances in how apiaries are built and where are typically located, but also critical differences in what kind of local vegetation the local species of bees feed on, and the varieties and methods of honey production. Yet to be a beekeeper is to have the epistemic upper hand on a number of phenomena—from pollination peak to the seasonality of honey production—that do not fall under the remit of other epistemic communities such as botanists, entomologists, or plant morphologists.

Epistemic communities are granular: despite sharing a situated practice that delivers knowledge of distinctive phenomena, the beekeeping of my grandfather Eligio and the surrounding community in Lazio is different from the beekeeping of say ancient Romans, or Mayans, or nineteenth-century Scottish heather honey producers, or contemporary Yucatán beekeepers. Epistemic communities are granular in containing a multitude of *differentiated groups* whose knowledge is distinctively *situated for* different purposes.

I draw here on (Massimi 2022, ch. 11, 341ff) in distinguishing between two notions of situatedness: *situated in* and *situated for*. What qualifies a community as *epistemic* is their practice being *situated in* a historical-cultural context has the experimental, theoretical, and technological resources to *reliably* make claims of knowledge about *some phenomena* which prove modally robust—i.e. phenomena that can be identified and re-identified across scientific perspectives. But practices are also *situated for* a particular purpose. The intertwined notions of *situated in* and *situated for* cater to different questions:

- A practice being *situated in* a historical-cultural context highlights the particular experimental-theoretical-conceptual tools and methodological-epistemic principles that underpin scientific knowledge production (as per my definition of a scientific perspective, ft. 4).
- A practice being *situated for* a specific purpose stresses instead the epistemic needs and purposes to which the knowledge so produced is put to use in that community, or better in the differentiated group within a much larger—porous, itinerant, and granular—epistemic community.

Incidentally, it is at the level of *situated for* that often practices become ways of worldmaking, namely ways in which particular groups within wider epistemic communities entangle their knowledge of particular phenomena with sets of metaphysical beliefs and values. For example, evidence from anthropological archaeology about Mayan pre-Columbian beekeeping suggests that it was associated with 'two complementary productive industries: *balché* (honey wine with hallucinogenic properties), and metallurgical production, through the use of beewax to make casting models' (Paris, Peraza Lope, Masson, et al. 2018, 1). For pre-Columbian Mayan beekeepers, honey was a 'key ingredient in the preparation of *balché*, a fermented beverage used in ceremonial intoxication, purging and as offerings to deities' (ibid., 5) in a way that it was not for my grandfather's local community in Lazio, Italy, second half of 20th century, where beekeeping was mostly functional to local consumption of honey in milk and production of propolis sold in local apothecaries, often still run by monks in local abbeys (e.g. Abbazia di Farfa).

Hence, the granularity of any epistemic community is a way of recognising this plurality of *situated-for* knowledges that are often identity defining for particular groups (including Indigenous people and local communities, IPLC), and could be linked to cultural rights for individuals belonging to groups (on the model of Kymlicka's 'group-differentiated rights'—see Kymlicka 1995). To better understand the complex relationship between REBSP and cultural rights more widely, it is then useful to acknowledge the difference between *situated in* and *situated for* when it comes to epistemic communities, differentiated groups thereof, and their respective practices. For the REBSP the relevant notion is that of *situated in*, whereas *situated-for* is mainly relevant to cultural rights.

Zooming in on the defining feature of epistemic communities and their situated-in practices, there is a range of *activities* (experimental, theoretical, technological in a broad enough sense to include artisanal and orally transmitted ones) that any epistemic community might be engaged with to *justifiably* produce *reliable* claims of knowledge about *phenomena*—as per (ii)–(iii) in my definition of a scientific perspective in footnote 4. Crucially, those reliable claims of knowledge concern *modally robust phenomena*—i.e. phenomena that can be identified and re-identified by different communities using different pieces of evidence within the inferential boundaries of their perspectives (see Massimi 2022, ch. 6 for an analysis).

The emphasis on modally robust phenomena is important to understand how I use the term *activities* in what follows, and to distinguish it from similar uses to be found in the pragmatist literature (compare e.g. Mitchell 2023 on affordances), most notably in Hasok Chang's 2012, 15–16 use of *epistemic activities* qua 'a more or less coherent set of mental or physical operations that are intended to contribute to the production or improvement of knowledge in a particular way' and typically have an 'inherent aim' and one or more 'external functions' (Chang 2022, 35–6). Chang's notion of epistemic activities can be regarded as a kind of *internalism* in assessing their success in terms of operational coherence (see Chang 2022, 40–47). Mine, by contrast, is a form of *externalism* in that activities denote an array of *skilful performances*[5] that latch onto stable events in nature and lead to the *identification* of *modally robust phenomena*.[6]

Disentangling activities from coherent sets of operations and linking them to the identification of phenomena bypasses a series of thorny problems concerning the relation between varieties of local, traditional, artisanal knowledge, on the one hand, and scientific knowledge, on the other hand, as I discuss in more detail here below. If scientific knowledge is ultimately knowledge of modally robust phenomena (and open-ended groupings thereof via natural kinds understood as sortal concepts),[7]

---

[5] A word of caution on this terminology. The term 'skilful performance' has been used in the wake of the STS tradition which I discuss in Section 3.1 mostly in the context of management and organisation studies (MOS) and industry studies to unpack 'the *resource-based view* (RBV) . . . [i.e.] that a firm's sustained competitive advantage in markets is a function of the specific resources it possesses, combines, and applies in its performance'—see 'Introduction', 5 of (Sandberg, Rouleau, et al. 2017). This literature is at some distance from the remit, scope, and concerns that motivates my paper and its underpinning philosophical analysis, much as we share a common ancestor in Collins and Evans's STS work. However, there are some interesting analogies and disanalogies between what I call in Section 4 *inter-perspectival expertise* (building on my work on perspectival realism in philosophy of science) and what for example, (Nicolini, Mørk, and et al. 2017) define as 'trans-situated' expertise qua 'a rhizome, a disorderly aggregate of uneven nodes, local roots, partial connections between nodes, shoots that become new rhizomes, and dead ends that lead nowhere' (29), which Nicolini et al. illustrate with the example of how the respective competences of cardiologists, heart surgeons, and anesthesiologists are needed to perform a radical innovation such as Transcatheter Aortic Valve Implantation (TAVI).

[6] In Massimi (2022), ch 6, 207 I have offered an analysis of phenomena as follows: 'Phenomena are stable events indexed to a particular domain of inquiry, and modally robust across a variety of perspectival data-to-phenomena inferences' and I made a distinction between the stability qua law-likeness of events in nature and the modal robustness of phenomena as a secondary quality.

[7] In Massimi (2022) chs.7–10, I defend an inferentialist view of natural kinds which I call Natural Kinds with a Human Face (NKHF). I see natural kinds as sortal concepts (ch. 9) for '(i) historically identified and open-ended groupings of modally robust phenomena, (ii) each displaying lawlike dependencies among relevant features, (iii) that enable truth-conducive conditionals-supporting inferences over time' (Massimi 2022, 249).

then one can begin to appreciate why individuals can reclaim as their own the entitlement to enjoy the benefits of scientific progress in virtue not of *membership* but in virtue of a range of skilful performances which are part of their situated-in practices (or scientific perspectives, as I call them).

The phenomena-first ontology delivered by perspectival realism hooks scientific knowledge to phenomena. In so doing, it reinstates to their rightful role a range of epistemic communities that have been traditionally severed[8] from canons about scientific knowledge production, maybe because they do not have the requisite 'systems of knowledge' in terms of what philosophers of science have often called essential properties, causal dispositions, or powers. (Does anyone recall the old adage that if one wants to know what gold *really* is, one should ask the physicist for the atomic number, rather than a jeweller or an assayer?)[9]

The far-reaching consequence of this philosophical move is that the normative foundations of REBSP become very much continuous with the foundations of cultural rights (both rights are protected in Art 15(1)(b) and (a) respectively), as a number of legal scholars have long advocated for—see (Bishop 2010), (Shaheed and Mazibrada 2021), (Bideault 2021).

The price to pay for this interpretive move is to do away with the elitist term 'science' still operative in the General Comment n. 25. The term 'science' is too often used as a tacit proxy for a well-established canonical tradition that begins in Europe with the so-called Scientific Revolution. If REBSP is to be understood as universal human right, the broader notion of 'scientific knowledge' lends itself more naturally to fulfil the task and to ground REBSP as an epistemic-cultural human right. Scientific knowledge, as I understand the term, is social reliable knowledge of modally robust phenomena (Massimi 2022), 347, whose historical lineages can be traced across different epistemic communities that have 'methodologically intersected' and 'historically interlaced'. This philosophical stance is a powerful ally for interpreting the REBSP as an epistemic-cultural human right, as I argue in what follows.

3. From REBSP as an Epistemic-Cultural Right to the Right to Expertise

3.1 The Problem of Extension of Expertise. Lessons from Anthropology and Science Studies

Key to my aforementioned argument from activity is the idea that 'to enjoy the benefit of some X is to do some *activity* that contributes to X, or brings about X.' If

---

[8] See Massimi (2022, 349–362) for the two epistemic injustices of epistemic severing and epistemic trademarking.

[9] One of the first philosophers who challenged this narrative and drew new attention to the important works of jewellers and assayers was (Hacking 1991).

one takes X to stand for 'scientific knowledge' (rather than the narrower term 'science'), the next step consists in asking *what kind of* activities might be legitimately regarded as contributing to, or bringing about scientific knowledge, where remember how I defined activities back in Section 2.2 qua *skilful performances* latching onto stable events in nature and leading to the *identification* of *modally robust phenomena*. Let us consider some examples.

A mechanical engineer runs simulations for aircraft's resistance to wind and strain. A technician calibrates a measuring instrument in a lab. A crofter harvests seaweed. A beekeeper collects honey. These are all examples of *activities*. What is common to them all is that *skills* are required to perform well in each case. The engineer needs be skilful in devising the simulation in a way that closely captures real storm conditions. The technician requires skills in identifying and rectifying possible miscalibrations of the measuring instrument. The crofter displays skills in distinguishing among varieties of seaweeds, their suitability for kelp manufacturing, and the best season for harvesting. A beekeeper needs be skilful in positioning apiaries in the vicinity of blossoming plants and monitoring the api-botanical cycle which is key to honey production.

Each of these skills come in turn in *degrees*. *Expertise*, I contend, comes in degrees in relation to the *skilful performances* associated with the *identification* of particular *modally robust phenomena*, which in these examples, respectively, are aircraft's *resistance* that the engineer is trying to model; *calibration* for the measuring instrument; kelp *manufacture* (production of ashes of seaweed rich in potash and soda); honey *production*. The emphasis on *skilful performances* follows a tradition that has highlighted varieties of 'experience-based expertise' vs. 'certified expertise' (Collins and Evans 2002); or what Boyer (Boyer 2008, 40) calls the 'experiential-performative' pole of 'skilled knowing and doing' by contrast with the 'social-institutional' pole; or, what is sometimes still referred to as 'lay expertise' vis-à-vis 'epistemic expertise', see (Turnhout, Tuinstra, and Halfmann 2019).[10] Let me single out some key aspects of this tradition of studies that matter for my discussion here.

The literature on expertise has flourished in the so-called 'Third Wave of Science Studies – Studies of Expertise and Experience (SEE)' (Collins and Evans 2002), but also in the anthropology and ethnography of experts—see (Mitchell 2002; Jeffery and Heslop 2020). For example, the anthropologist (Boyer 2008, 38) defines expertise as 'a relation of epistemic jurisdiction' where the salient question is the following: 'On what basis does the representative of one culture of expertise

---

[10] As Turnhout et al. (2019, 184) point out, 'Lay expertise is often defined as the opposite of scientific knowledge because it is considered to be: contextual and localised rather than universal; culturally embedded rather than objective;... practice and experience based rather than based upon methodological principles'. And Martini (Martini 2019, 120) adds that 'To be an expert is not to possess a more or less fixed number of traits, but to stand in a relation with someone else, namely, a layperson. In this sense, the epistemology of expertise can only be social epistemology, because expertise can only exist in a relational (i.e. social) context'.

(the anthropologist) claim legitimate analytical jurisdiction over the members of another culture of expertise and how is this claim enacted?' (ibid., 41). Hence, the perennial risk of what Boyer calls 'interjurisdictional rivalry and epistemophagy' (ibid., 43), i.e. the tendency to consume and incorporate other epistemic jurisdictions when cultures of expertise come in contact.

Collins and Evans (2002) have urged us to abandon the oxymoron of 'lay expertise' and to address the Problem of Extension ('How far should participation in technical decision-making extend'?). Drawing on Wynne's famous case study concerning the relevance of local knowledge by the Cumbrian sheep farmers (and institutional neglect thereof) in the aftermath of the Chernobyl nuclear disaster—see (Wynne 1992; Wynne 1993)—Collins and Evans have made a plea for a more holistic portrait of expertise that extends beyond what they call the 'core set's expertise' to include other pockets of uncertified expertise via social interactions. They call the latter 'interactional expertise' and distinguish it from 'contributory expertise', which is taken as 'continuous with the core set's expertise' (Collins and Evans 2002, 252).

Boyer's concept of 'epistemic jurisdiction' is helpful to delineate the more general and thorny dialectic often at play between scientific knowledge and varieties of local knowledge and to avoid the perennial risk of epistemophagy lurking around the corner. In this respect a few caveats are in order:

1. I declare my positionality as a white non-English native speaker philosopher of science in drawing these considerations about the nature of varieties of expertise cutting across the boundaries of 'experience-based' and 'certified expertise'.
2. I recognise as such the epistemic limits of my positionality and the need to tread carefully when discussing a broader notion of expertise which ought to treat varieties of local knowledge (especially indigenous knowledge) as legitimate ways of knowing *in their own right*, independently of and regardless of what they might (or might not) contribute to Western ways of conceptualising what 'scientific knowledge' is.
3. Accordingly, discussions about intercultural exchanges should be dealt with carefully and not couched in terms of overlapping, merging, integrating perspectives, or by considering different ways of knowing as interchangeable and feeding in some ahistorical Science (with the capital S). Doing so would only reaffirm historically unbalanced power structures over varieties of local knowledges—hence the risks of 'scientisation' of lay expertise and epistemophagy.

It is with caveats 1–3 in mind that I have presented my notion of activities qua skilful performances as linked to stable events and to the identification of modally robust phenomena, rather than linked to 'systems of knowledge'. The emphasis on the latter tends to raise inevitable questions about how and why discussions about local knowledge are often presented and couched from either a Western scientific point of view, or from the assumption that the two are somehow interchangeable at the risk of (willingly or unwittingly) reinstating Western forms of scientific authority (*whose* system of knowledge? By *whose* lights?).

By contrast, placing emphasis on how situated practices and their skilful performances latch onto stable events in nature is a way of putting a marker on what really matters here: namely, understanding how reliable knowledge production is effected by myriad historically and culturally situated communities over time. This is reliable knowledge production that different epistemic communities—beekeepers, cosmologists, plant morphologists, crofters, and so on—can each reclaim as *their own*.

A critic might reply at this point that my expanded notion of scientific knowledge qua reliable knowledge is too broad to be of any use. After all, there is a reason why the term 'science' is used to indicate particular varieties of reliable knowledge, my critic might insist: namely, those that have the hallmarks of critical inquiry and falsifiability. My envisaged critic might formulate the objection under the name of the Problem of Extension of Expertise (PoEE):

PoEE: 'What kind of skilful performances count (or do not count) as expertise'?

My critic might be (justifiably) worried that a generous account of what counts as reliable knowledge production and expertise like the one I am proposing may risk blurring important demarcation lines between bona fide scientific knowledge and all sorts of simply false beliefs, spurious connections, and bogus claims. The worry is real, especially in our time when social media is spreading scientific misinformation, and poses a threat to the public and democratic institutions.

Thus, to live up to the normative foundations of REBSP qua an epistemic-cultural human right, it seems that one must *both* acknowledge that truly global knowledge production requires, first of all, an answer to the PoEE that is not built on traditional dichotomies—i.e. scientific expertise vs. lay expertise; or, core vs. non-core; or scientific knowledge vs. local knowledge—*and* at the same time one must also shelter scientific-knowledge-qua-reliable-knowledge from the perennial danger of misinformation, false theories, and crackpot ideas. How to steer clear from both?

3.2 When Perspectives Intersect and Interlace. Inter-Perspectival Expertise

My answer to both PoEE and the demarcation problem takes the name of *inter-perspectival expertise*, a notion that I lay out in what follows building on my work on perspectival realism (Massimi 2022). I am going beyond Collins and Evans in suggesting that what is needed is a more refined notion of expertise that is akin to what Collins and Evans call 'contributory expertise' in being continuous with core expertise, and therefore more substantive than what they call 'interactional expertise'. I brand such notion of expertise as *inter-perspectival* in recognising that is it the product of a multitude of historically and culturally situated practices of *epistemic communities*. What is inter-perspectival expertise? And how does it work as an answer to both PoEE and the demarcation problem?

Back to my perspectival realism (Massimi 2022, 5), where I offered a definition of 'scientific perspective' that is practice-centred and takes on board the situated knowledge thesis without taking any perspective as more foundational or core than any others. From a strictly epistemic (as opposed to anthropological or sociological) point of view, the title of 'scientific perspective' (see footnote 4) pertains to a variety of situated-in practices that have the experimental, technological and theoretical tools to *reliably* advance claims of knowledge and are able to *justify* such reliability too. This is already a stringent enough criterion to rule out practices that either do not advance claims of knowledge or, if they do, they lack the resources to do so *reliably* and *justifiably* so, in response to the worry about the demarcation problem.

But coming to PoEE, a critic might insist that my definition of expertise is overreaching, and one might question the extent to which e.g. the activities of the crofter in kelp manufacturing or of the beekeeper in producing honey—each embedded into their respective situated-in and situated-for practices—count as 'scientific perspectives' at all. In reply, I use the term 'scientific' in the same strategic and provocative way in which Sandra Harding uses the term 'science' to refer to the practice of Micronesian navigators.[11] But how to negotiate among 'epistemic jurisdictions' so as to avoid Western-centric 'scientisation of lay expertise', on the one hand, and the risk of diluting 'certified expertise', on the other hand? This is where inter-perspectival expertise comes handy as a notion, I contend.

Inter-perspectival expertise is expertise understood as skilful performance embedded in a nexus of *intersecting* and *interlacing* of scientific perspectives. Scientific perspectives *methodologically intersect* when they 'can be brought to bear on one another to refine the *reliability* of particular claims of knowledge'

---

[11] To be sure, the term "science" is not what indigenous cultures use to refer to their knowledge systems. Indeed, Galileo, Newton, and Boyle were "natural philosophers" to their contemporaries and to later generations. It was not until the early nineteenth century that the term "science" was introduced by William Whewell.... So it might seem like one more piece of Eurocentric appropriation to refer to indigenous knowledge as sciences, as I will do here. Yet I do so for strategic reasons. I intend to level the epistemological playing field so that we can begin to understand the costs to us and to indigenous cultures of conceptualizing indigenous knowledge only as myth, magic, and superstition, or only as a residue of tradition that should be replaced by modern Western sciences' rationality and technological expertise. (Harding 2015, 81)

(Massimi 2022, 10). They *historically interlace* when they encounter other perspectives and trade 'with one another some of their tools, instruments and techniques over time... [so as to] track the evolution of knowledge concerning particular phenomena in what I call a "historical lineage"' (ibid.).

In chapter 8 of the book I argued that e.g. when it comes to knowledge of the phenomenon 'pollination' for particular plants (e.g. a melliferous plant called *Gymnopodium floribundum* and very common in Mexico), it is necessary to *intersect* the knowledge of botanists and pollination ecologists with that of melissopalynologists and local beekeepers. The latter have the epistemic upper hand in knowledge of the peak season for the plant which is key to the honey production. Scientific knowledge of particular phenomena (e.g. pollination of the *Gymnopodium*) is produced by bringing to bear the perspective of the melissopalynologists on that of the plant morphologists, by asking whether the pollen under-representation of *Gymnopodium* in the honey of the region can be explained by the reproductive biology of the plant. And, in turn, the perspective of the beekeepers can be brought to bear on that of the plant morphologists by fine-graining the analysis of pollination to the timing of the nectar peak and related pollination peak which the beekeepers know best.

Claims of knowledge get refined and made more reliable through the *synchronic intersecting of perspectives* ranging from beekeepers' local knowledge of the api-botanical cycle to plant morphologists' knowledge of the reproductive biology of the plant. This is an example of the *intersectional* variety of inter-perspectival expertise. Its role is to improve the *reliability* of claims of knowledge by either bringing new data to the phenomena under study; or better triangulating between the data and the phenomena inferred; or, by questioning some of the methodological-epistemic principles used to justify the reliability of the claim within a perspective.

The process of methodologically intersecting respects the situated-in knowledge of each perspective. It does not fall prey of any epistemophagy by overlapping, merging, integrating perspectives. Nor does it undermine core expertise within each perspective because it does not downplay the botanist's knowledge, nor does it merge the beekeeper's knowledge with the knowledge of the plant morphologist. It does not assess or evaluate the validity of one epistemic jurisdiction in terms of another one, reiterating patterns of epistemic exploitation and Western authority. The beekeepers' expertise has its own epistemic jurisdiction, which is distinct from the plant morphologist's expertise and associated epistemic jurisdiction. Yet perspectival pluralism—read through the lenses of this multicultural kaleidoscope of situated-in practices—does not see them as siloed, insular, and disjoint.

Inter-perspectival expertise of the *intersectional variety* accrues by respecting the situated knowledge of each relevant community vis-à-vis the phenomena under study (e.g. pollination), and simply bring them to bear on each other so as to improve the reliability of knowledge claims concerning the phenomena. The central idea is that knowledge of a phenomenon like the pollination of a particular plant *requires* this plurality of perspectives and cannot be delegated or sanctioned by any one singular epistemic community.

Another variety of inter-perspectival expertise is the *interlacing* variety. Here the expertise concerns mostly know-how related to particular tools, instruments, material cultures that were traded among situated communities and their practices *diachronically* over time. In my book, I give the example of how an object for divination (e.g. the Han-Sung dry and wet compass) became an object for nautical use (e.g. the mariner's compass) to track the evolution of knowledge concerning a phenomenon such as the Earth's magnetic field. The notion of historically interlacing perspectives does not anachronistically 'scientise' the local knowledge of Han-Sung geomancers. The use of dry and wet compasses was *situated in* the practice of the Han-Sung geomancers, and was *situated for* a particular purpose, namely divination—see (Massimi 2022, 344–8).

At the same time, inter-perspectival expertise of the interlacing type originates when different perspectives come in contact and trade their tools, instruments, ideas, and techniques within a well-defined historical genealogy of material cultures. Reconstructing how those tools changed role and function over time and across different communities becomes a way of tracing the evolution of knowledge concerning particular phenomena. Inter-perspectival expertise of the interlacing variety takes then a diachronic long vista on the historical evolution of human knowledge by contrast with the synchronic outlook of inter-perspectival expertise of the intersecting variety. For example, Hebridean crofters' kelp-manufacturing is part of a historical lineage where techniques for potash and soda production became eventually industrially produced with the Leblanc process in 1791 and later Solvay process in the 1860s.

In both the case of intersecting and interlacing perspectives, as I stressed in (Massimi 2022, 341), it is paramount that these intercultural exchanges among epistemic jurisdictions do not become an opportunity for further epistemic injustices, epistemic extractivism, and exploitation. In the book I give four normative pointers to fend off this always lurking risk:

(1) respect the historical and cultural situatedness of each perspective;
(2) mutually and reciprocally agreed upon norms and methods of knowledge production sharing;
(3) transparent mechanisms and pathways without blending and obfuscating the contribution of each community;
(4) prevent exploitative systems of appropriation resulting in the commodification of knowledge at the exclusive socioeconomic benefit of one community.

When these pointers are violated, epistemic severing and epistemic trademarking ensue as distinctive varieties of epistemic injustices, I argued in (Massimi 2022, 349–63).

To conclude, inter-perspectival expertise of both the intersectional and interlacing variety bypasses dichotomies such as core vs. non-core expertise, scientific knowledge vs. local knowledge. According to perspectival realism, *all* knowledge is always situated and perspectival. This does not gainsay the all-important question of how to demarcate genuine reliable knowledge from misinformation and conspiracy theories. All that is needed as a deterrent against the latter is an analysis of the historical interlacing and methodological intersecting of perspectives that does the heavy lifting in constraining (without policing) the boundaries of *reliable knowledge production* concerning modally robust phenomena across perspectives.

3.3 Toward a Right to Expertise. Concluding Remarks

In this chapter I have addressed some foundational questions about the Right to Science. I have offered an argument from activity as the best way of understanding the philosophical-normative foundations of REBSP and of addressing the lingering elitism implicit in the term 'science' still operative in the UN General Comment n. 25. I have advocated for replacing 'science' with the broader notion of 'scientific knowledge' and for articulating a generous view of what counts as 'reliable knowledge production' that does not trademark local knowledge as alien to, or peripheral to it.

If these steps are taken, then the REBSP can be regarded as an epistemic-cultural human right: i.e. a right concerning scientific knowledge (*episteme* in ancient Greek as opposed to *doxa* qua opinion or bogus knowledge) qua part of wider cultural practices. Such a shift has the potential to overcome the inevitable gap that seems to exist in the current normative foundations of REBSP between the *few scientists* pursuing freely their research, on the one hand, and *everyone* else as the generic proxy right-holder of the benefits resulting from scientists's research, on the other hand.

I have argued that the philosophical-normative foundations of REBSP qua an epistemic-cultural right must be sought in an analysis of expertise that does not repeat and perpetuate distinctions between scientific knowledge and local knowledge, core expertise and peripheral one, certified vs experience-based. Latching on the literature in STS and anthropology, I offered a view of inter-perspectival expertise via intersecting and interlacing scientific perspectives.

This definition is broad enough to encompass a wide range of skilful performances—from seaweed harvesting to calibrating instruments. I have argued that such a view of expertise shares with the pragmatist tradition an emphasis on 'activities' and is at home in perspectival realism, which takes all knowledge as ultimately local, situated, and perspectival. The challenge of demarcating genuine scientific knowledge from the perennial risk of misinformation and flat falsehoods lies within the internal dynamics of how perspectives methodologically intersect and historically interlace to deliver reliable knowledge of modally robust phenomena over time.

Therefore, for the REBSP to hold as an epistemic-cultural right, one needs to expand the framework in which discussions of REBSP have taken place so far and supplement it with a Right to Expertise (RtE) that belongs to individuals insofar as they are part of one or more situated epistemic communities:

> (RtE) Every situated epistemic community has a *right to expertise* in a relevant epistemic jurisdiction so long as expertise is understood in terms of skilful performances embedded in a nexus of intersecting and interlacing scientific perspectives.

The Right to Expertise so understood gives situated epistemic communities their due. It does not epistemically sever and trademark their local knowledge. Nor does it assimilate or 'scientise' their local knowledge. Most importantly, it has the potential of re-aligning the REBSP qua epistemic-cultural right with the Convention on Biological Diversity and UN General Comment n. 21 that have long legally recognised the importance of indigenous knowledge and local expertise in matters concerning biodiversity and cultural rights. It is time for the Right to Science to catch up with this legal trend. One way of doing so is by dissecting the very notion of 'science' tacitly at work behind REBSP as I have done in this chapter.

Acknowledgements

I am grateful to Mirko Farina and Duncan Pritchard for inviting me to contribute to this volume. An earlier draft was presented at a workshop on the Right to Science at the University of Edinburgh and I am grateful to Samantha Besson for detailed and constructive comments, and to all participants for their helpful questions. Many thanks to Laura Jeffery for stimulating discussions and references on the ethnography of expertise; and to Eduardo Schenberg for reading and commenting on an earlier draft and thought-provoking conversations on Mazatec knowledge. Any error remains entirely my own responsibility.

References

Besson, S. (2023) "The 'Human Right to Science' *qua* right to participate in science." *The International Journal of Human Rights*, DOI: 10.1080/13642987.2023.2251897.

Bideault, M. (2021) "Considering the right to enjoy the benefits of scientific progress and its applications as a cultural right," in *The Right to Science. Then and Now*, (eds.), H. Porsdam and S. Porsdam Mann, 140–9. Cambridge: Cambridge University Press.

Bishop, L. (2010) "The right to science and culture." *Wisconsin Law Review* 121, 141.

Boyer, D. (2008) 'Thinking through the anthropology of experts." *Anthropology in Action* 15, 38–46.

Chang, H. (2012) *Is Water H₂O? Evidence, Realism, and Pluralism.* Dordrecht: Springer.

Chang, H. (2022) *Realism for Realistic People. A new pragmatist philosophy of science.* Cambridge: Cambridge University Press.

Collins, H. M. and R. Evans. (2002) "The third wave of science studies: Studies of expertise and experience." *Social Studies of Science* 32, 235–96.

Committee on Economic, Social and Cultural Rights E/C.12/GC/21. (2009) "General Comment n. 21, Right of Everyone to Take Part in Cultural Life."

Committee on Economic, Social and Cultural Rights, E/C.12/GC/25. (2020) "General Comment No. 25 (2020) on Science and Economic, Social and Cultural Rights (Article 15 (1) (b), (2), (3) and (4) of the International Covenant on Economic, Social and Cultural Rights)." United Nations, available at https://docstore.ohchr.org/SelfServices/FilesHandler.ashx?enc=4slQ6QSmlBEDzFEovLCuW1a0Szab0oXTdImnsJZZVQdxONLLLJiul8wRmVtR5Kxx73i0Uz0k13FeZiqChAWHKFuBqp%2b4-RaxfUzqSAfyZYAR%2fq7sqC7AHRa48PPRRALHB.

French, P. A. (1984) *Collective and Corporate Responsibility.* New York: Columbia University Press.

Gramsci, A. (1971) *Selections from the Prison Notebooks,* (eds.), Q. Hoare and G. Nowell Smith. New York: International Publishers.

Hacking, I. (1991) "A tradition of natural kinds." *Philosophical Studies* 61, 109–26.

Haraway, D. (1988) "Situated knowledges: The science question in feminism and the privilege of partial perspective." *Feminist Studies* 14, 575–99.

Harding, S. (2015) *Objectivity and Diversity: Another Logic of Scientific Research.* Chicago: University of Chicago Press.

Hartsock, N. C. M. (1998) "The feminist standpoint revisited," in N. C. M. Hartsock *The Feminist Standpoint Revisited and Other Essays,* 227–48. Boulder, Colorado: Westview Press.

Jeffery, L. and L. Heslop. (2020) "Roadwork: Expertise at work building roads in the Maldives." *Journal of the Royal Anthropological Institute* 26, 284–31.

Kitcher, P. (forthcoming) *Descartes Lectures.*

Kymlicka, W. (1995) *Multicultural Citizenship.* Oxford: Clarendon Press.

Martini, C. (2019) "The epistemology of expertise," in *The Routledge Handbook of Social Epistemology,* (ed.), M. Fricker, 122–15. London: Routledge.

Massimi, M. (2022) *Perspectival Realism.* NY: Oxford University Press.

Mitchell, S. (2023) "The Bearable Thinness of Being: A Pragmatist Metaphysics of Affordances". In *The Pragmatist Challenge. Pragmatist Metaphysics for Philosophy of Science*, edited by H.K. Andersen and S. Mitchell. New York: Oxford University Press.

Mitchell, T. (2002) *Rule of Experts: Egypt, Techno-Politics, Modernity*. Berkeley: University of California Press.

Navarro, J. (2012) *A History of the Electron. J.J. and G.P. Thomson*. Cambridge: Cambridge University Press.

Nicolini, D., B. E. Mørk, et al. (2017) 'Expertise as trans-situated: The case of TAVI," in *Skillful Performance: Enacting Capabilities, Knowledge, Competence, and Expertise in Organizations*, (eds.), J. Sandberg, L. Rouleau, et. al. Oxford: Oxford University Press.

Paris, E. H., C. Peraza Lope, M. A. Masson, et al. (2018) "The organization of stingless beekeeping (meliponiculture) at Mayapán, Yucatan, Mexico." *Journal of Anthropological Archaeology* 52, 1–22.

Réaume, D. (1998) "Individuals, groups, and rights to public goods." *University of Toronto Law Review* 38, 1–27.

Sandberg, J., L. Rouleau, et al., (eds.) (2017) *Skillful Performance: Enacting Capabilities, Knowledge, Competence, and Expertise in Organizations*. Oxford: Oxford University Press.

Shaheed, F. and A. Mazibrada. (2021) "On the right to science as a cultural human right," in *The Right to Science. Then and Now*, (eds.), H. Porsdam and S. Porsdam Mann. Cambridge: Cambridge University Press.

Turnhout, E., W. Tuinstra, and W. Halfmann. (2019) *Environmental Expertise. Connecting Science, Policy and Society*. Cambridge: Cambridge University Press.

UNESCO. (2009) 'The Right to Enjoy the Benefits of Scientific Progress and Its Applications—UNESCO Digital Library." 2009, available at https://unesdoc.unesco.org/ark:/48223/pf0000185558.

Wylie, A. (2003) 'Why standpoint matters," in *Science and Other Cultures: Issues in Philosophies of Science and Technology*, (eds.), R. Figueroa and S. Harding, 26–48. New York: Routledge.

Wynne, B. (1992) "Public understanding of science research: New horizon or hall of mirrors?." *Public Understanding of Science* 1, 37–43.

Wynne, B. (1993) "Public uptake of science: A case for institutional reflexivity." *Public Understanding of Science* 2, 321–37.

9
Studies of Expertise and Experience

Demarcating and Defending the Role of Science in Democracy

Harry Collins
School of Social Sciences, Cardiff University

Robert Evans
School of Social Sciences, Cardiff University

1. Introduction

Recent years have seen unprecedented polarisation in the valuation within democratic societies of the role and status of science, with the beliefs of US citizens about climate change providing the iconic example: in the spring of 2022, 78 per cent of Democrat supporters viewed climate change as a 'major threat', compared to only 23 per cent of Republicans (Tyson et al. 2023). What is remarkable about this divide is that it was recorded one year after the Intergovernmental Panel on Climate Change (IPCC) concluded it is now 'unequivocal that human influence has warmed the atmosphere, ocean and land', 'that the scale of recent changes [is] unprecedented' and that 'human-induced climate change is already affecting many weather and climate extremes in every region across the globe' (IPCC 2023, paras. A1, A2, A3). In addition to the obvious problems this causes for policymakers, the populist appeal of scepticism about climate science also raises some uncomfortable questions for social scientists. In particular, how should constructivist fields like science and technology studies (STS) respond to the continued support for politicians such as Donald Trump whose policies will, if we trust the scientists, do nothing except make an extremely grave situation worse. Is it possible for fields such as STS to argue that all knowledge, including science, is socially constructed *and* that there are reasons and circumstances in which the right thing to do is give more weight and status to scientific knowledge?

In this chapter, we argue that it is possible to retain the social constructivist perspective that informs STS and, at the same time, find reasons for choosing science over other forms of knowledge. The chapter is structured as follows. First, we set out the fundamentals of our approach. We define expertise as the outcome of successful socialisation into expert groups and hence as including both the acquisition of tacit knowledge and the internalisation of community norms and values. Next, we examine the consequences of this view for the role of science in society, stressing its fallibility when judged in narrowly epistemic terms but also that science is not politics by other means. The practical consequences of these ideas are then explored through three examples: vaccination, climate change, and the COVID-19 pandemic.

2. Studies of Expertise and Experience

Understanding expertise as a social practice requires a combination of philosophy and sociology. The philosophical foundations are found in Ludwig Wittgenstein's ideas of a form of life and a language game (Wittgenstein 1953) and their subsequent exposition by Peter Winch (Winch 1958). Wittgenstein shows how the meanings of words and concepts can only be understood through their use in the course of practical actions, whilst Winch makes the connection with the regulation of these actions through shared expectations about their use and meaning. As Winch puts it:

> The concepts we have settle for us the form of the experience we have of the world... when we speak of the world we are speaking of what we in fact mean by the expression 'the world': there is no way of getting outside the concepts in terms of which we think of the world... The world *is* for us what is presented through those concepts. (Winch 1958, 15 emphasis in original)

Sociology sees the same relationship between ideas and actions but starts with the other side of the coin. Being a member of a society means sharing a form of life with others, with this being demonstrated in the correct deployment of the actions that instantiate the ideas belonging to that group's language game (Bloor 1973; Collins 1992). Sociology also gives a name to the process through which this 'knowing how to go on' is learnt—socialisation—and to the mysterious quality that is acquired as this happens—tacit knowledge (Collins 2010; Collins and Kusch 1998).

The revolution in the social studies of science and technology that occurred during the early 1970s—what we have called the Second Wave of Science Studies in order to distinguish it from the preceding 'First Wave' that uncritically accepted the epistemic superiority of science—was the realisation that the language and practice of science could be understood in exactly the same way as those of any other social group (Bloor 1991; Collins 1992; Knorr-Cetina and Mulkay 1983; Latour and Woolgar 1979). STS scholars showed that the resolution to classic philosophical problems like the under-determination of theory and the problems of induction and falsification were to be found not in the logical structure of arguments but in the social agreements made between scientists about who to trust and how to interpret the available data (Shapin 2007).

Seen this way, science is no longer justified by its closer correspondence to, or alignment with, an external reality. Instead, what power and authority science has accrues from the political and other institutions whose interests it is now seen as serving. Given the historic, contemporary, and ongoing examples of technical and scientific knowledge being used to exclude the marginal and enable the powerful (Arksey 1998; Epstein 1996; Fischer 2000; Irwin and Wynne 2003; Ottinger 2013), this critical perspective has much to commend it. Demonstrating how scientific knowledge is socially constructed so that those who have been wrongly excluded may have their voices heard remains an important task. What the recent past has made clear, however, is that simply opening the door to 'new voices' is not enough. A more sophisticated position is needed that allows for both the inclusion and exclusion of new voices.

2.1 Third Wave of Science Studies: Tacit Knowledge and Formative Aspirations of Science

The earliest articulation of this alternative position was published as the 'Third Wave of Science Studies' (Collins and Evans 2002). This set out the foundations of a more complex typology of expertises, later published as the 'Periodic Table of Expertises' (Collins and Evans 2007), and drew what would become an important distinction between the normative aspirations of the scientific community and those that characterise other institutions (Collins et al. 2020; Collins and Evans 2017).

The typology of expertise, and the role played by tacit knowledge in its acquisition and distribution, can be thought of as the 'technical' part of SEE. Unlike the extant literature of the time, SEE takes a realist approach to expertise (Carr 2010; Evans and Collins 2007). In contrast, most of STS took, and continues to take, a relational approach that focuses on the attribution of expert status and legitimacy. Research in this latter tradition is typically concerned with the political and other social interests at play in claims to expertise and the boundary work (Gieryn 1983, 1999) that takes place as each party seeks to define the domain of legitimate expertise to include their allies and exclude their critics. In many cases, the critique that emerges is of an unjustified technocracy in which the problem framing and assumed expertise of the scientific community is used to silence the equally expert, but less prestigious, voices of indigenous peoples, local communities, and victims of industrial or medical neglect (for a recent overview see Felt et al. 2017).

What is often left unremarked in these case studies is that their impact depends on the reader sharing the sense of epistemic injustice (Fricker 2007) felt by the excluded groups. In other words, they only work as critique because the excluded group is not chosen at random: their experiences mean they have a real and relevant expertise that is not available in the scientific community. If this case can't be made, then the sense of epistemic injustice and critique evaporates. After all, there is no reason to think that (say) shop workers, *qua* shop workers, have any specialist expertise on issues such as the pollution in fenceline communities, the design of clinical trials or radioactive fallout in rural hill-farming areas, so why should we be concerned that their views on these issues were not sought?

At one level, then, the Third Wave paper was about the reasoning that goes into these kinds of fieldwork decisions and how they reveal that even STS researchers know there is more to expertise than the attribution of expert status. The initial classification of specialist expertise that came out of these reflections included just three types of expertise.

1. *No Expertise*: that is the degree of expertise with which the fieldworker sets out; it is insufficient to conduct a sociological analysis or do quasi-participatory fieldwork.
2. *Interactional Expertise*: this means enough expertise to interact interestingly with participants and carry out a sociological analysis.
3. *Contributory Expertise*: this means enough expertise to contribute to the science of the field being analysed.
   (Collins and Evans 2002, 254 emphasis in original)

There are two important features of this typology. The first is the category of interactional expertise and the recognition it gives to language use as a distinct kind of expertise. This idea has turned out to be surprisingly important, generating a new research method called the Imitation Game, recasting sociological fieldwork in terms of language acquisition rather than embodied experience, and providing a way of understanding what is necessary for a complex division of labour to succeed (Caudill et al. 2019; Collins et al. 2016; for more on the idea of interactional expertise and the Imitation Game see Collins and Evans 2014, 2015). For now, however, it is enough to note that, with sufficient immersion in the relevant community, it is possible to learn the language that describes the practices, beliefs, and values of that group well enough to participate in that discourse as if one were a native member (Collins 2011).

The second feature, and the one that makes the approach a realist one, is the emphasis on social fluency as a marker of expertise (Evans 2008). This reflects the importance of tacit knowledge for the system of classification. Sometimes defined as 'know-how', and contrasted with the 'know-that' of the explicit knowledge, tacit knowledge is the taken-for-granted knowledge that is routinely used in social interactions but whose existence is only revealed when these go awry. As tacit knowledge can only be acquired via socialisation, it is the practical acquisition of this group-specific tacit knowledge that differentiates the expertise of those with extensive in-group interactions from the knowledge of those who remain outside the community, no matter how much the latter may read or study other sources of explicit knowledge (Collins 2018; Collins and Evans 2007).

This basic axiom—that expertise is the outcome of successful socialisation—leads to a more complex classification of expertise based on the different ways in which it is possible to interact with social groups. In the case of interactional and contributory expertise—the highest levels of specialist expertise—prolonged, in-person socialisation is the only way to develop the necessary tacit knowledge. This comes with an obvious opportunity cost, however. The complexity of modern societies makes sustained participation in anything but a small number of social groups impossible, creating a distribution of specialist expertises that reflects the differential access to occupational, leisure, and lifestyle groups within those societies (Collins 2013). Because no-one can be an expert in everything, it follows that the members of any society are constantly dealing with experts in domains they do not fully understand. This, in turn, gives rise to a set of meta-expertises—expertises about expertise—that are needed to make decisions about the quality of the expertise of others.

These different types of expertise form the basis of a Periodic Table of Expertises (Table 9.1), the main elements of which can be summarised as follows.

Table 9.1 Periodic Table of Expertises

UBIQUITOUS EXPERTISES					
DISPOSITIONS			Interactive Ability		
				Reflective Ability	
SPECIALIST	*UBIQUITOUS TACIT KNOWLEDGE*			*SPECIALIST TACIT KNOWLEDGE*	
EXPERTISES	Beer-mat Knowledge	Popular Understanding	Primary Source Knowledge	Interactional Expertise	Contributory Expertise
				Polimorphic	
					Mimeomorphic
META-	*EXTERNAL (Transmuted expertises)*		*INTERNAL (Non-transmuted expertises)*		
EXPERTISES	Ubiquitous Discrimination	Local Discrimination	Technical Connoissuership	Downward Discrimination	Referred Expertise
META-CRITERIA	Credentials		Experience		Track-Record

Source: (Collins and Evans, 2007).

- **Ubiquitous expertises**: these are the expertises developed through primary socialisation and which are shared by all members of a society. Ubiquitous expertises include such things as natural language speaking, general social norms and values, and an understanding of common social institutions. Because ubiquitous expertises are, by definition, widespread they are often not seen as expertises at all.
- **Specialist expertises**: the right-hand end of this row is contributory expertise, which is the category that corresponds most closely to the everyday meaning of expertise: a contributory expert is someone who, through extensive participation in the domain, is able to speak and act in ways that their peers recognise as being correct. Specialist expertises are typically acquired through secondary socialisation and can be relatively common (e.g. driving a car or riding a bicycle) or highly esoteric (e.g. a scientific specialism or an unusual hobby). Because expertise is always the product of successful socialisation, differences in the status accorded to different domains of specialist expertise are matters of sociology not epistemology. Moving left along the specialist expertise row, the levels of expertise get progressively lower, with the sharp discontinuity occurring between interactional expertise, which still requires extended socialisation and so includes the specialist tacit knowledge of the group, and primary source knowledge, which does not include any social interaction with the specialist group and so only includes the general tacit knowledge associated with the ubiquitous expertises. Moving still further left, the sources used become increasingly unsophisticated and generic, leading to progressively weaker and more limited forms of expertise.
- **Meta expertises**: as with specialist expertises, there is a distinction between expertises that are limited to the tacit knowledge associated with ubiquitous expertise, including everyday heuristics about appearance, demeanour, or interests, and those that require some familiarity with the domain in question. Given the opportunity costs of acquiring specialist expertise, this row of the table is where the expertises that most people use most of the time are found.

2.2 Third Wave of Science Studies: Norms and Values

Because SEE characterises all expertise in terms of socialisation and the acquisition of tacit knowledge, it does not distinguish between science and other domains of knowledge on the basis of their epistemology. Nevertheless, the problem this approach was intended to solve—how to defend the value of science—requires some way of distinguishing between different claims to expertise and different domains of expertise. What we called the 'problem of extension' (Collins and

Evans 2002) was intended to draw a contrast with the 'problem of legitimacy' that had been, and to a large extent remains, the focus of Wave Two STS.

> To save misunderstanding, let us admit immediately that the practical politics of technical decision-making still most often turn on the Problem of Legitimacy; the most pressing work is usually to try to curtail the tendency for experts with formal qualifications to make ex-cathedra judgements curtained with secrecy. Nevertheless, our problem is not this one. Our problem is academic: it is to find a clear rationale for the expansion of expertise. But a satisfying justification for expansion has to show, in a natural way, where the limits are. Perhaps this is not today's practical problem, but with no clear limits to the widening of the base of decision-making it might be tomorrow's. (Collins and Evans 2002, 237)

The question posed by the problem of extension is, therefore, how to demarcate science from other forms of expertise whilst remaining consistent with the social constructivist approach of Wave Two. The answer follows directly from the decision to treat expertise as the cultural and collective property of a social group: just as it is possible to distinguish between social groups based on their norms and values, so it is possible to distinguish science from other social groups on the basis of the norms and values that shape its form of life.

Some clarification is needed here about the way in which the role of norms and values is to be understood. Earlier work in the sociology and philosophy of science attempted to use the 'norms of science'—Merton's CUDOS norms (Communism, Universality, Disinterestedness, and Organised Scepticism) are the most well-known (Merton 1973)—as rigid demarcation criteria. This effort is widely seen as having failed because so-called 'counter norms' (Mitroff 1974) were also identified, suggesting that there were no distinctive norms after all. SEE takes a different approach. Rather than seeing norms as providing a rigid structure that must be adhered to at all times, we see norms as providing the formative aspirations (Collins 2019; Collins and Kusch 1998) that set the overall expectations for members of that group and against which subsequent actions are held accountable. Seen this way, deviations from group norms may require special justification but they do not refute the relevance or status of the over-arching norms.

Instead, our view is that the CUDOS norms identified by Merton, combined with the traditional concerns of the philosophy of science literature, such as the value placed on empirical data, attempts to falsify or corroborate theoretical claims, and a number of sociological characteristics that we have derived from the SEE approach, provide a coherent set of formative aspirations for the scientific community (Collins and Evans 2017). The complete list is summarised in Table 9.2.

Table 9.2 Formative aspirations of science

Differences from Wave One	As described in Wave Two	Introduced by Wave Three
Shallow-rooted (i.e. moral foundation) not deep-rooted (i.e. not correspondence truth) Expertise not truth Approach and norms not facts Best decisions not right decisions	*Observation* *Corroboration/ replicability* *Falsification* *Universalism* *Disinterestedness* *Openness to criticism*	Honesty and integrity Locus of legitimate Interpretation Clarity Individualism Continuity Open-endedness Generality Value of expertise

Source: adapted from (Collins and Evans, 2017, 55).

We do not claim that all these values are unique to science, though their integration into one set almost certainly is. The idea is that these aspirations are necessary for a scientific community to be 'scientific' and that their salience for the scientific community provides a way distinguishing it from other social groups.

2.3 Science Is not Politics: Technical and Political Phases

Distinguishing science from other communities is just the start, however. Solving the problem of extension also requires a proposal for how the expert and non-expert groups should be enabled to contribute to technological decision-making in the public domain. Within SEE, this is done by distinguishing between the technical and political phases of technological decision-making and the role of experts and lay people in each (Collins et al. 2010; Collins and Evans 2002).

To begin with, we should note that the technical and political phases are not chronologically ordered. Instead, the term 'phase' is used the physical science sense of 'phase of matter' in which the 'same' substance can take different forms (solid, liquid, gas) under different combinations of temperature and pressure (e.g. ice, water, steam). The idea here is that, depending on what is being demanded, the solution might require a technical or political judgement. Another way of thinking about it, is to see the technical phase as the 'filling' in a sandwich, with the political phase both 'upstream', setting and framing the questions to which expert work is directed, and 'downstream', responding to and acting upon the information that is provided (Collins and Evans 2017; Evans and Plows 2007).

Regardless of how we visualise the relationship between technical and political phases, we need to describe the ways in which they are different from each other.

Table 9.3 Technical and political phases

		Phase	
		Political	Technical
Nature of	Politics	Extrinsic	Intrinsic
	Rights	Stakeholder	Meritocratic
	Representation	By survey	By action
	Delegation	By proxy	Impossible

Source: (Collins and Evans, 2002, 262).

In the 'Third Wave' paper, these differences were set out in terms four dimensions—politics, rights, representation, and delegation—as shown in Table 9.3.

The technical phase is the expert-led part of technological decision-making and is characterised by an adherence to scientific norms and high-levels of expertise. Participants in the technical phase are interactional or contributory experts in relevant domains of practice who are willing to work in ways that reflect the values listed in the right-hand column. For example, political considerations should be eliminated from judgements insofar as that is possible; thanks to Wave Two we know that some irreducible, intrinsic politics will remain, but as noted previously, it is the aspiration that matters. Likewise, when it comes to the right to participate, this is determined meritocratically by the relevance and level of expertise, and not by financial or other interest. The final two rows follow from this emphasis on experiential expertise: expert communities have to actively represent themselves as they embody their expertise and attempting to delegate this task to non-experts makes no sense.

The left-hand column shows how these same concerns play out very differently if a political judgement is required. Now extrinsic politics are obviously and legitimately relevant as it would be perfectly appropriate for a political decision to be taken on the basis that it is what 'the public' wanted. Likewise, the right to participate in the political phase is not conferred by having expertise but by having a stake. In the most inclusive definition, this might mean being a citizen in the jurisdiction; alternatively, distinctions might be made between different kinds and levels of stakeholder. As with the technical phase, the final two rows follow from these considerations. Stakeholders can be represented by others, either in the form of surveys or votes that aggregate preferences, or by the appointment of a human proxy who then represents their interests.

Putting these different elements together gives a basis for solving the problem of extension whilst also retaining the insights from the Second Wave of Science Studies. The problem of legitimacy results from too much weight being given to the technical phase and to scientific experts in particular. An important part of the

Table 9.4 Minimal default position

	Wave of STS	Can policy-makers deny legitimacy of expert advice?	Can policy-makers refuse to act on expert advice?
Technocracy	Wave 1	No	No
Technological Populism	Wave 2	Yes	Yes
Elective Modernism	Wave 3	No	Yes

solution is to open up the technical phase to a wider range of experts so that technical judgements reflect a wider and more appropriate range of expertises (cf. Harding 2006; Longino 1990). The problem of extension arises, and is resolved, by recognising that the technical phase requires expert participants and follows different norms to the political phase. As such, the technical phase cannot be opened to everyone as non-experts lack the specialist tacit knowledge needed to take part. This does not mean that there is no place for non-expert citizens in the decision-making system, however. Instead, their participation must be organised and encouraged through the political phase, which needs to ensure it is truly open, democratic and, where necessary, willing to challenge any advice and recommendations that might emerge from the technical phase (Collins et al. 2010; Collins and Evans 2017; Weinel 2007). The three possible outcomes are summarised in Table 9.4, with the final row—elective modernism—being the position advocated by SEE:

SEE thus navigates between the science-led technocracy of Wave One and the unfettered technological populism of Wave Two by arguing that expertise should provide a limit on how policy choices are justified, not how they are chosen. Where there is a strong scientific consensus, then this should be acknowledged and any decision to ignore this made clear. Likewise, where no such consensus exists, the risks of putting all the policy eggs in a single basket should be similarly emphasised. In each case, the choice remains the responsibility of the elected leaders—their only obligation is to explain it honestly and clearly.

3. Science in Society

Thus far, we have set out the ways in which SEE has engaged with debates or ideas drawn from the philosophy of science. In this section, we turn to SEE's contribution to topics that are usually found within political philosophy: the nature of democracy and the institutions and practices that are needed for it to survive. In doing so, we shift the focus from the specialist expertises that define the multiple and overlapping social groups that make up any modern society to the more

STUDIES OF EXPERTISE AND EXPERIENCE   157

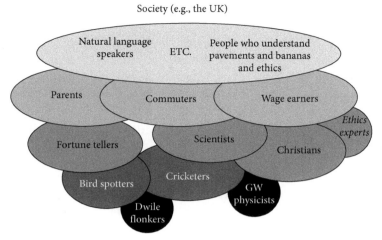

Figure 9.1 Fractal model of society
Source: Collins, 2019, p. 9.

general ubiquitous expertises that are shared by all the members of that society. The relationship between ubiquitous and specialist expertises is summarised in Figure 9.1, which depicts the fractal-like nature of a society made up of different social groups.

The fractal metaphor is intended to capture the idea that each group is characterised by an expertise, and that this expertise has the same structure regardless of how widely it is distributed or how highly it is valued. At the top of the diagram, there is the 'top-level' fractal that represents the society as a whole. For the purpose of this discussion, we can think of the 'top-level' as being a nation-state, with the ubiquitous expertises referring to the shared national culture and traditions that are available to all its citizens. These ubiquitous expertises would include things like being able to speak the native language, recognise what counts good manners, display everyday knowledge of popular sports and culture, and describe the role of different institutions. Below this are the various groups, occupations and organisations to which members of that society might belong, some nested inside each other by virtue of being more specialist versions of a larger activity. In general terms, the ubiquitous expertises in the top-level fractal will 'filter down' and provide the context within which other, more specialist, expertises are exercised, but there is also scope for the relationship to work in the other direction. For example, a general concern about animal rights in the wider society might be reflected in gradual decline in their use for scientific research, but increasing scientific understanding of the suffering caused by experimentation and the development of alternatives might flow 'up' to the wider society, thereby increasing the volume and legitimacy of those concerns.

If the expertise possessed by any community or domain of practice is always the same kind of 'stuff', then we can characterise that expertise as having two different dimensions or faces (Collins et al. 2020). On the one hand, there is the 'organic' face that represents the fundamental formative intentions that define that particular social group and set the normative standards expected of its members. In the case of science, for example, the organic face would include the values set out in Table 9.4. Likewise, for the top-level fractal in a democratic society, we might expect to see commitments to free and fair elections, the rule of law, and the equal treatment of all citizens.

This organic face is complemented by an 'enumerative' face, that represents the different ways in which the possibilities offered by the organic face are expressed. In the case of science, the enumerative face represents the array of different disciplines and methods that are found across the scientific community. These different ways of 'being scientific' are possible due to the open-ended nature of organic face, which gives the enumerative face its dynamic and innovative character and, over time, creates the possibility of changes in the organic face.

3.1 Checks and Balances

In the case of democratic societies, the organic face of the top-level fractal should include a commitment to pluralism, by which we mean a recognition that the views and experiences of minorities matter and that the winners of any election have an obligation to govern for the entire society. In practice, all elected governments will enact policies that tend to favour the groups and sectors that support them, but there is a limit to how far this tendency can be indulged before democratic government descends into a tyranny of the majority. The potential for this kind of negative outcome has long been recognised and accounts for the use of institutional checks and balances such as written constitutions, multiple elected chambers, and different levels of government as protections against the excessive centralisation of power. These institutions of government are then further constrained by an independent judiciary and an independent media.

Where SEE goes beyond the standard accounts found in political philosophy is that it includes science within these institutional checks and balances. As explained previously, science cannot, and should not, directly determine policy choices but it can, and should, limit the ways in which those choices are explained and justified. Seen this way, science often becomes a target for populist leaders because it constrains their ability to act with impunity. Assuming STS scholars want to defend democracy, we argue that there are two ways in which they, and other social scientists committed to a social constructivist view of science, can argue for the importance of an independent science within the system of checks and balances.

The first reason is that it is surely better that policy decisions are informed by advice from people who know what they are talking about. What SEE adds to the debate is the ability to operationalise this claim: the technical phase restricts the scope of technical advice to propositional questions posed by democratic institutions and the ideas of interactional and contributory expertise set out what is required to 'know what you are talking about'. In addition, the focus on values provides a basis for preferring scientific advice over that generated in other ways, whilst the emphasis on socialisation leaves open the possibility of including experience-based expertise alongside traditional science.

The argument for giving special weight to science is that expertise developed via the norms and values set out in Table 9.4 is morally superior (i.e. just plain 'better') to knowledge based on other values. For example, it is simply better if the knowledge used to inform policy makers about the natural world is generated in settings that are open to empirical testing and not in settings where the authority of ancient texts or charismatic leaders means such testing is either not needed, not necessary, or not permitted. Similarly, it is simply better that knowledge used to inform policy is generated in settings where the source of a knowledge-claim is not taken as a direct indicator of its veracity rather than one where ascribed characteristics form the basis of belief systems. To be clear, we are not saying that science is perfect: there is clearly much to do around improving the way norms of universality, disinterestedness, and organised scepticism work in practice but the very fact that most, if not all, scientists would agree with these concerns only testifies to the constitutive role these values play as formative aspirations (Oreskes 2021).

Using these same values also provides a way of determining the role that should be given to other groups. The decision rule is simple: where the community in question has experience-based expertise that is relevant and can demonstrate that it is willing to work within the norms and values of science then either its representatives or others with the interactional expertise needed to represent the group should be included in the technical phase; where this willingness cannot be demonstrated, the community should be excluded. Using these sociological criteria, it is possible to distinguish between the actions of patient or community groups seeking to represent their experiences in the best way they can, and whose voices should be heard, and the actions of religious groups, fringe science communities, and 'think tanks' sponsored by corporations.

The second argument for giving special weight to science draws on the fractal model (Figure 9.1). As noted earlier, there is a two-way relationship between the values and norms that inform the ubiquitous expertises of the top-level fractal and those that are found in the more specialist, lower-level fractals. Norms and values flow 'down' from the higher levels but also filter back 'up' the fractal layers through their use in these diverse settings. In the case of science, the core moral value is that of aspiring to know the truth and, by striving towards 'the truth', science resists the erosion of the difference between truth and lies in society as a whole, thereby influencing the national culture and its institutions and making a descent into authoritarianism less likely.

4. Examples: Vaccines, COVID, Climate Change

Having set out the main ideas that inform SEE—the Periodic Table of Expertises, the distinction between technical and political phases, the fractal model, sociological demarcation criteria for science, and the value of scientific values for society as a whole—we now illustrate their application by considering three examples: vaccination, climate change, and the response to the COVID-19 pandemic.

4.1 Vaccinations

The controversy around the MMR vaccine that followed the publication of Andrew Wakefield's now discredited paper prompted much of our early thinking on these issues (Wakefield et al. 1998). The controversy centred on a press conference in which Wakefield claimed there was a link between the measles, mumps, and rubella (MMR) vaccine and the development of autism. The text of the now-retracted paper, which is based on a sample of just twelve children, shows that they 'did not prove an association' but the very fact that the possibility was considered, coupled with Wakefield's subsequent insistence that the risk was real, led to a substantial public debate about the safety of the combined MMR vaccine.

Much has been said about this controversy. For example, it is clear that the media's use of 'balanced' story formats, in which the claims of Wakefield and the parent groups that supported him were given equal space and attention to the unanimous assessment of epidemiologists and other medical professionals that there was no evidence that the vaccination posed any risk, played a significant role in promoting and prolonging the controversy (Boyce 2007; Hargreaves et al. 2003). Here, however, we are concerned with the role of STS and whether there was anything within our discipline that might have helped parents make good choices about whether to vaccinate their child. We argue there are at least three lines of argument that STS scholars could have made without compromising their intellectual integrity or giving up a commitment social justice.

First, technical connoisseurship based on many years of studying science ought to have provided the skills needed to see that a research paper with a sample size of twelve, and whose conclusions include the phrase 'did not prove', is not a sound basis for overthrowing a public health policy supported by epidemiological evidence stretching back many years and from many countries.

Second, contributory expertise in the realm of constructing logical arguments should enable the critical awareness necessary to know that the truism that you can't prove a negative is too blunt an instrument for serious academic work. Whilst it is logically possible that the MMR vaccine does, via some yet-to-be-understood pathway have this specific adverse side effect on some, as yet unspecified sub-population whose risk factors cannot (yet) be described, the same claim could be made about literally anything. Whilst resorting to the precautionary principle might appear to put a veneer of respectability onto the argument, it does no intellectual work unless some vaguely plausible evidence for the risk and the mechanism by which it might be realised are provided. The contrast with the cases championed elsewhere in the STS literature seems clear: farmworkers protesting about the failure to ban the herbicide 2,4,5,T were able to show that the safety precautions everyone agreed were needed for 'safe use' were not possible within the actual work settings they encountered (Irwin 1995).

Finally, and to put this in its wider context, all these arguments require some background judgements about the mainstream medical epidemiology to which Wakefield's claims represent a challenge. As far as we know, there has been no suggestion that the epidemiological research routinely cited in support of the MMR vaccine falls below the expected standards, though it would, no doubt, be possible to construct the standard, sceptical argument that it reflects interests and concerns of nation states keen to control their populations and multi-national corporations seeking to maximise their profits. Even if this were to be done, however, then barring a faith-based insistence that the underdog is always right, it is hard to see how any of the evidence marshalled against the MMR vaccine was powerful enough to warrant the change in policy its advocates demanded.

4.2 Climate Change

Climate change poses a different kind of challenge for STS. Here being on the 'right' side of the argument—i.e. the one that is easily framed in terms of social and environmental justice—requires accepting that the mainstream science is sufficiently trustworthy that major policy decision can, and should, be based on it. It is possible to show, as Oreskes and Conway (2010) have done, how corporations with vested interests in the fossil fuel industries have worked to undermine the credibility of climate change science by actively constructing a 'sceptical' public discourse in ways that seem unfair and unreasonable. But, as with any case study that fits this 'problem of legitimacy' framework, the power of the argument depends on accepting the injustice, which in this case means accepting that the mainstream science is 'right'. This poses a problem for STS as it believes IPCC science is just as socially constructed as claims made by the sceptics who seek to undermine it. On what, basis, therefore can STS scholars be upset or concerned when world leaders like President Trump reject climate change science? Or, to put it another way, what did the STS community think when Greta Thunberg told the US Congress in September 2019 that 'I don't want you to listen to me, I want you to listen to the scientists'.

To the extent that these questions were debated during the Trump presidency, it seems those committed to a Wave Two version of STS had surprisingly little to say in of support of Thunberg's central claim that the science mattered. (Collins et al. 2017; Jasanoff and Simmet 2017; Lynch 2017; Marres 2018; Sismondo 2017b, 2017a) Instead, the argument was made that symmetrical STS studies do not take sides but reveal the

> construction of more-or-less stable sociotechnical orders, which bring together local achievements and supporters, scientific arguments, techniques, technologies, capital and ideologies... [to show how epistemic authority] is established and drawn upon, even when the result is criticized.   (Sismondo, 2017a, 589)

What is lacking, however, is any willingness to say that some of these sociotechnical orders are morally preferable to others. It is this that SEE brings to the argument, recognising the logical indefeasibility of the social constructivist claims but also arguing that some claims are better than others because of the moral aspirations and related methodologies that support them. Again, some judgement is needed but it seems reasonable to argue that the contributory expertise of the STS community should include sufficient awareness of the research done on, by, and with the IPCC to justify the conclusion that, in this case at least, the socio-technical order it represents is morally preferable to that of its industry-sponsored critics.

4.3 COVID-19 Pandemic

Like climate change, the effects of the COVID-19 pandemic were global. Unlike climate change, however, affluent countries were at least as likely to be badly affected as more disadvantaged nations, with some of the highest per capita death rates occurring in the richest and most developed nations in the world. How then, are we to understand the range of different outcomes and the potential contribution of STS to debates about preparedness for future pandemics and other disasters?

One interesting suggestion comes from the Comparative COVID Response project, which analyses the approaches taken in twenty-three different countries and identifies 'three broad and dramatically different patterns [control, consensus and chaos], connecting policies and outcomes' (Jasanoff et al. 2021, 10). Of particular importance for SEE is the suggestion that these different outcomes are not to be explained by standard geopolitical, economic, or demographic characteristics. Instead, the crucial variable is the 'social compact' within each of these countries, which includes questions such as:

> What are the fundamental obligations of the state to its citizens? How is authority to make decisions delegated and to whom? What are the rights, obligations, and proper roles of citizens? To justify decisions that constrain the polity, what forms of public reasoning, including kinds of argumentation and evidence, are required? Since citizens will never completely agree on the nature of the good or the allocation of power and resources, how does the polity reach binding settlements to achieve justice without irreparably fracturing the social order?
> (Jasanoff et al. 2021, 14)

The resonance of these ideas with the distinction between the organic and enumerative faces of a national culture is clear, as is the challenge they pose to STS. For example, how does the tendency to argue for more public debate about and with the experts themselves, such that the assumptions, histories, and limitations of the epidemiological models used are made (more) visible (Pearce 2020; Saltelli et al. 2020), square with the sense that problems of 'chaos countries' could be attributed to a surfeit of debate rather its absence? Or, to put it the other way, what characterised the more successful consensus countries was not the abundance of public debate but the public's trust in their democratic institutions, including those providing expert advice. Interestingly, the consensus countries, which include Sweden, France, and Australia, also show that many different policy outcomes were compatible with the scientific advice that was available at any given time. The argument is, therefore, not that science determines policy but that trust in scientific expertise allows policy makers to act in ways that fit their local context whilst retaining the support of their citizens. In contrast, chaos countries lacked this trust and so struggled to find any coherent and stable framework around which to unite as a nation.

In response, SEE suggests that there are several ways in which democratic societies might look to restore or retain the trust in expert institutions that the 'consensus countries' show is a necessary part of any effective response to public health emergencies. These include the acknowledgement of the different phases of technological decision making through which it is possible to define and justify a specific role for expert advice giving institutions, recognition of the diverse sources of scientific and experience-based expertise that might be relevant, greater reflexive awareness about how the questions asked of expert advice are framed, and wider commitments to programmes of civic education such that the importance of a social compact within which the value of science is recognised alongside its role as an institutional check and balance on government power (Collins et al. 2022; Evans 2022).

5. Summary

This chapter has set out the main elements of the studies of expertise and experience (SEE) and their application to problems of technological decision-making in democratic societies. The founding principle of SEE is that expertise is characterised by the acquisition of tacit knowledge through a period of extended socialisation within the relevant domain of practice. Depending on the degree and type of interaction, different kinds of expertise can be developed. These include the ubiquitous expertises that characterise a national culture, specialist expertises that correspond to demographic, lifestyle, or occupational groups, and the meta-expertises by which judgements are made about the specialist expertise possessed by others.

The idea of expertise as the property of social groups also gives rise to a more general fractal model of society within which these layers and types of expertise are distributed. Within this model, it is important to distinguish between the organic face of particular social group, which includes the formative aspirations that define its normative commitments, and the enumerative face that represents the different ways in which these aspirations can be enacted. This, in turn, provides a way to distinguish science from other social institutions: its distinctive commitment to the truth, and the values that support this quest, not only provide robust sociological demarcation criteria they also explain why scientific institutions should be valued within the system of checks and balances found within pluralist democracies.

These arguments were then applied to a number of knowledge-based controversies in order to show how they enable a critical, but symmetrical, analysis of policy advice that is able to argue for more inclusive institutions, but also draw boundaries such that the idea and role of expertise is not lost in a mono-tonic call for ever more democratisation. In this way, SEE remains true to the social constructivist origins of STS whilst also providing a platform from which contemporary problems of democratic societies can be addressed.

References

Arksey, H. (1998) *RSI and the Experts: The Construction of Medical Knowledge*. London: UCL Press.

Bloor, D. (1973) "Wittgenstein and Mannheim on the sociology of mathematics." *Studies in History and Philosophy of Science Part A*, 4(2), 173–91, available at https://doi.org/10.1016/0039-3681(73)90003-4.

Bloor, D. (1991) *Knowledge and Social Imagery* (2nd edn). Chicago: The University of Chicago Press.

Boyce, T. (2007) *Health, Risk and News: The MMR Vaccine and the Media*. New York: Peter Lang.

Carr, E. S. (2010) "Enactments of expertise." *Annual Review of Anthropology* 39(1), 17–32, available at https://doi.org/10.1146/annurev.anthro.012809.104948.

Caudill, D. S., Conley, S. N., Gorman, M. E., and Weinel, M. (2019) *The Third Wave in Science and Technology Studies: Future Research Directions on Expertise and Experience*. Cham, Switzerland: Palgrave Macmillan, available at https://dx.doi.org/10.1007/978-3-030-14335-0.

Collins, H. M. (1992) *Changing Order: Replication and Induction in Scientific Practice* (2nd edn (1st edn 1985)). Chicago: The University of Chicago Press.

Collins, H. M. (2010) *Tacit and Explicit Knowledge*. Chicago: The University of Chicago Press.

Collins, H. M. (2011) "Language and practice." *Social Studies of Science* 41(2), 271–300, available at https://doi.org/10.1177/0306312711399665.

Collins, H. M. (2013) "Three dimensions of expertise." *Phenomenology and the Cognitive Sciences* 12(2), 253–73, available at https://doi.org/10.1007/s11097-011-9203-5.

Collins, H. M. (2018) *Artifictional Intelligence: Against Humanity's Surrender to Computers*. Medford, MA: Polity Press.

Collins, H. M. (2019) *Forms of Life: The Method and Meaning of Sociology*. Cambridge, MA: The MIT Press.

Collins, H. M. and Evans, R. (2002) "The Third Wave of science studies: Studies of expertise and experience." *Social Studies of Science* 32(2), 235–96, available at https://doi.org/10.1177/0306312702032002003.

Collins, H. M. and Evans, R. (2007) *Rethinking Expertise*. Chicago: The University of Chicago Press, available at http://dx.doi.org/10.7208/chicago/9780226113623.001.0001.

Collins, H. M. and Evans, R. (2014) "Quantifying the tacit: The imitation game and social fluency." *Sociology* 48(1), 3–19, available at https://doi.org/10.1177/0038038512455735.

Collins, H. M. and Evans, R. (2015) "Expertise revisited, Part I—Interactional expertise." *Studies in History and Philosophy of Science Part A* 54, 113–23 available at https://doi.org/10.1016/j.shpsa.2015.07.004.

Collins, H. M. and Evans, R. (2017) *Why Democracies Need Science*. Cambridge, UK: Polity Press, available at https://www.wiley.com/en-gb/Why+Democracies+Need+Science-p-9781509509607.

Collins, H. M., Evans, R., Durant, D., and Weinel, M. (2020) *Experts and the Will of the People: Society, Populism and Science*. Cham, Switzerland: Palgrave Pivot, available at https://doi.org/10.1007/978-3-030-26,983-8.

Collins, H. M., Evans, R., Innes, M., Kennedy, E. B., et al. (2022) *The face-to-face principle: Science, trust, democracy and the internet*. Cardiff: Cardiff University Press, available at https://doi.org/10.18573/book7.

Collins, H. M., Evans, R., and Weinel, M. (2016) "Expertise revisited, Part II: Contributory expertise." *Studies in History and Philosophy of Science Part A* 56, 103–10, available at https://doi.org/10.1016/j.shpsa.2015.07.003.

Collins, H. M., Evans, R., and Weinel, M. (2017) "STS as science or politics?" *Social Studies of Science* 47(4), 580–86, available at https://doi.org/10.1177/0306312717710131.

Collins, H. M. and Kusch, M. (1998) *The Shape of Actions: What Humans and Machines Can Do*. Cambridge, MA: MIT Press.

Collins, H. M., Weinel, M., and Evans, R. (2010) "The politics and policy of the Third Wave: New technologies and society." *Critical Policy Studies* 4(2), 185–201, available at https://doi.org/10.1080/19460171.2010.490642.

Epstein, S. (1996) *Impure Science: AIDS, Activism, and the Politics of Knowledge*. Berkeley, CA: University of California Press.

Evans, R. (2008) "The sociology of expertise: The distribution of social fluency." *Sociology Compass* 2(1), 281–98, available at https://doi.org/10.1111/j.1751-9020.2007.00062.x.

Evans, R. (2022) "SAGE advice and political decision-making: 'Following the science' in times of epistemic uncertainty." *Social Studies of Science* 52(1), 53–78, available at https://doi.org/10.1177/03063127211062586.

Evans, R. and Collins, H. M. (2007) "Expertise: From attribute to attribution and back again," in *Handbook of Science and Technology Studies*, (3rd edn), (eds.), E. J. Hackett, O. Amsterdamska, M. Lynch, and J. Wajcman, 609–30. Cambridge, MA: MIT Press.

Evans, R. and Plows, A. (2007) "Listening without prejudice?: Re-Discovering the value of the disinterested citizen." *Social Studies of Science* 37(6), 827–53, available at https://doi.org/10.1177/0306312707076602.

Felt, U., Fouche, R., Miller, C., and Smith-Doerr, L. (eds.) (2017) *The Handbook of Science and Technology Studies* (4th edn). Cambridge, MA: The MIT Press.

Fischer, F. (2000) *Citizens, Experts, and the Environment: The Politics of Local Knowledge*. Durham, NC: Duke University Press.

Fricker, M. (2007) *Epistemic Injustice: Power and the Ethics of Knowing*. Oxford: Oxford University Press.

Gieryn, T. F. (1983) "Boundary-Work and the demarcation of science from non-science: Strains and interests in professional ideologies of scientists." *American Sociological Review* 48(6), 781, available at https://doi.org/10.2307/2095325.

Gieryn, T. F. (1999) *Cultural Boundaries of Science: Credibility on the Line*. Chicago: The University of Chicago Press.

Harding, S. G. (2006) *Science and Social Inequality: Feminist and Postcolonial Issues*. Urbana: University of Illinois Press.

Hargreaves, I., Lewis, J., and Speers, T. (2003) *Towards a Better Map: Science, the Public and the Media*, 64. Swindon, UK: Economic and Social Research Council

(ESRC), available at http://www.esrc.ac.uk/_images/towards_a_better_map_tcm8-13558.pdf.

Intergovernmental Panel on Climate Change (IPCC) (2023) "Summary for policymakers," in *Climate Change 2021: The Physical Science Basis. Contribution of Working Group I to the Sixth Assessment Report of the Intergovernmental Panel on Climate Change*, 3–32. Cambridge, UK: Cambridge University Press, available at https://doi.org/10.1017/9781009157896.001.

Irwin, A. (1995) *Citizen Science: A Study of People, Expertise, and Sustainable Development*. London: Routledge.

Irwin, A. and Wynne, B. (eds.) (2003) *Misunderstanding Science? The Public Reconstruction of Science and Technology* (1st paperback edn). Cambridge, UK: Cambridge University Press.

Jasanoff, S., Hilgartner, S., James, W., and White, L. (2021) *Learning from COVID-19: A 23-Nation Comparative Study of COVID-19 Response, with Lessons for the Future of Public Health* (Comparative Covid Response Project, 30) [Synthesis]. Cornell: Cornell University, available at https://compcore.cornell.edu/.

Jasanoff, S. and Simmet, H. R. (2017) "No funeral bells: Public reason in a 'post-truth' age." *Social Studies of Science* 47(5), 751–70, available at https://doi.org/10.1177/0306312717731936.

Knorr-Cetina, K. and Mulkay, M. (eds.) (1983) *Science Observed: Perspectives on the Social Study of Science*. London: Sage Publications.

Latour, B. and Woolgar, S. (1979) *Laboratory Life: The Social Construction of Scientific Facts*. Beverly Hills: Sage Publications.

Longino, H. E. (1990) *Science as Social Knowledge: Values and Objectivity in Scientific Inquiry*. Princeton, NJ: Princeton University Press.

Lynch, M. (2017) "STS, symmetry and post-truth." *Social Studies of Science* 47(4), 593–9, available at https://doi.org/10.1177/0306312717720308.

Marres, N. (2018) "Why we can't have our facts back." *Engaging Science, Technology, and Society* 4(0), 423–43, available at https://doi.org/10.17351/ests2018.188.

Merton, R. K. (1973) *The Sociology of Science: Theoretical and Empirical Investigations*. Chicago, The University of Chicago Press.

Mitroff, I. I. (1974) "Norms and counter-norms in a select group of the Apollo moon scientists: A case study of the ambivalence of scientists." *American Sociological Review* 39(4), 579–95, available at https://doi.org/10.2307/2094423.

Oreskes, N. (2021) *Why Trust Science?* Princeton, NJ: Princeton University Press.

Oreskes, N. and Conway, E. M. (2010) *Merchants of Doubt: How a Handful of Scientists Obscured the Truth on Issues from Tobacco Smoke to Global Warming* (1st US edn). New York: Bloomsbury Press.

Ottinger, G. (2013) *Refining Expertise: How Responsible Engineers Subvert Environmental Justice Challenges*. New York: New York University Press.

Pearce, W. (2020) "Trouble in the trough: How uncertainties were downplayed in the UK's science advice on Covid-19." *Humanities and Social Sciences Communications* 7(1), available at https://doi.org/10.1057/s41599-020-00612-w.

Saltelli, A., Bammer, G., Bruno, I., Charters, E., et al. (2020) "Five ways to ensure that models serve society: A manifesto." *Nature* 582(7813), available at https://doi.org/10.1038/d41586-020-01812-9.

Shapin, S. (2007) *A Social History Of Truth: Civility and Science in Seventeenth-Century England* (4th print). Chicago: The University of Chicago Press.

Sismondo, S. (2017a) "Casting a wider net: A reply to Collins, Evans and Weinel." *Social Studies of Science* 47(4), 587–92, available at https://doi.org/10.1177/0306312717721410.

Sismondo, S. (2017b) "Post-truth?" *Social Studies of Science* 47(1), 3–6, available at https://doi.org/10.1177/0306312717692076.

Tyson, A., Funk, C., and Kennedy, B. (2023, April 18) "What the data says about Americans' views of climate change." *Pew Research Center*, available at https://www.pewresearch.org/short-reads/2023/04/18/for-earth-day-key-facts-about-americans-views-of-climate-change-and-renewable-energy/.

Wakefield, A. J., Murch, S. H., Anthony, A., Linnell, J., et al. (1998) "RETRACTED: Ileal-Lymphoid-Nodular hyperplasia, non-specific colitis, and pervasive developmental disorder in children." *The Lancet* 351(9103), 637–41, available at https://doi.org/10.1016/S0140-6736(97)11096-0.

Weinel, M. (2007) "Primary source knowledge and technical decision-making: Mbeki and the AZT debate." *Studies in History and Philosophy of Science Part A* 38(4), 748–60, available at https://doi.org/10.1016/j.shpsa.2007.09.010.

Winch, P. (1958) *The Idea of a Social Science and its Relation to Philosophy*. London: Routledge & Kegan Paul.

Wittgenstein, L. (1953) *Philosophical Investigations* (trans., G. E. M. Anscombe). Oxford: Blackwell.

PART 5
EXPERTISE AND VIRTUE

10
Humility for Experts

Linda Zagzebski
University of Oklahoma

1. Authorities and Experts

We live in a society suspicious of authority, and for good reason. People in authority have abused their position so many times that trust between them and their subjects has almost vanished, and many of us believe that nobody deserves to be treated as an authority, even in limited intellectual domains. In place of intellectual authorities, we have "experts." Experts have specialized knowledge and are consulted for their opinion in their area of expertise. They have no authority to dictate a belief or action, and they may not even have a reasonable expectation that their opinion will be accepted or that their advice will be followed. They might not even have a reasonable expectation that they will be recognized as an expert.[1] In contrast, an intellectual authority adopts an epistemic contract with her subjects that gives both sides responsibilities. The person with epistemic authority takes on the responsibility to be right and to convey what she believes to be the truth to her subjects clearly, and the subjects know that they are expected to accept her opinion and to act on her recommendations. There is no such contract between the expert and the non-expert. That can lead to confusion and conflict because sometimes experts think of themselves as mere experts when the subjects take them to be authorities, or the expert may expect to be treated as an authority when the non-experts have no intention to do so. What's more, some people may treat an expert's testimony as authoritative whereas others treat it only as advice, advice that they might think has been defeated by other experts or their own judgment. Still others judge that the putative expert is not an expert. All of that happened countless times during the COVID pandemic.

In *Epistemic Authority* (2012), I defended the existence of epistemic authority from the standpoint of the subject, arguing that consistent epistemic self-trust rationally commits us to have epistemic trust in others, and some of those others satisfy the conditions for authority modeled on Joseph Raz's (1988) justification for authority in the political domain. Using that model, I argued that another

---

[1] When you read a sentence in a newspaper column that begins "Experts say...", do you believe that whoever they are, they are experts? I doubt it unless you already have a high degree of trust in the columnist.

person is an epistemic authority for me when I conscientiously judge that I am more likely to get a true belief and avoid a false one if I believe what the authority tells me than if I try to figure out what to believe on my own. In such cases it is my own rational self-trust that leads me to see that I ought to accept the authority's testimony (Zagzebski (2012), chs 5–6). That reason can be defeated, e.g., in cases of competing authorities, and in the future, I might withdraw my judgment that the authority meets the conditions for authority. But as long as I continue to conscientiously judge that a person has successfully met the requirements for justifiable authority, I am committed to believing on his or her authority by the exercise of my own rational self-governance.

Following Raz's analysis of the nature of authority in general, an epistemic authority gives me a pre-emptive reason to believe what she testifies. That is what makes an authority an authority rather than a mere expert. A pre-emptive reason is one that replaces my other reasons for adopting a belief p and is not simply added to them. In contrast, an expert only gives me a reason for p that I can add to my total stock of reasons to believe p—perhaps a strong reason, but not a pre-emptive one. To act on authority is to act on a pre-emptive reason. For instance, if I stop at a traffic light because that is what the law says, I am acting on authority. If I stop because I believe that all things considered, it is safer to do so, and the fact that the law says to stop merely strengthens the case for stopping, I am not acting on authority. I am treating the law as advice, not authority. Some people treat their physicians as epistemic authorities and act on their directives pre-emptively. Others treat their physicians as medical advisors. The same point applies to lawyers and financial advisors.

In *Epistemic Authority* I approached the rationality of belief on authority from the perspective of the individual subject. In this paper I will investigate the other side of the authority-subject relation: the position of the person who holds intellectual authority. This relationship is especially complicated when there are multiple potential subjects, perhaps reaching into the millions. Some of those potential subjects see the authority as authoritative in the sense I have described; many others see him as a mere expert whose opinion they can take or leave as they see fit; still others do not view him as an expert at all, much less an authority. This can result in anger, recriminations, increased belief polarization, and even violence among the non-experts. We the non-experts do not handle the expertise of the experts well. The experts often do not handle it well either. Speaking as an expert is fraught with moral dangers. It is easy to be an expert. It is not easy for the expert to properly relate to the non-expert.

2. Trust and Distrust in Experts

Trust is the critical element in the relationship between experts and non-experts, but it is widely claimed that there has been a serious decline in public trust in scientific and medical experts. That perception hides an important finding. A recent article on trust in science and scientific institutions (Oreskes and Conway 2022) indicates that trust in science is strongly polarized along political lines. Most Americans trust scientific institutions and believe that scientific research is beneficial.[2] The crisis in trust is mostly among conservatives, who are much more likely than the public at large to distrust scientific findings because of the way science has been used to support government action. Oreskes and Conway trace this trend in the US back to the Reagan administration when scientific evidence of environmental crises was used to justify a governmental response. Recently, we have seen the same reaction to the use of science in supporting government action to combat COVID.

It is easy to claim that conservatives oppose science—a position that does not deserve the dignity of a response, but there is a big difference between opposing science as a practice, and opposing the use of science to support the implementation of policies that are perceived to be harmful for selfish or unselfish reasons. I seriously doubt that many people of any political persuasion oppose scientific research, but lack of trust in scientific experts is another matter. Trust in a person is quite different from trust in a field of knowledge. Listeners think that they perceive character traits when they hear an expert speak. When they do so, of course, they are often prone to jumping to conclusions. Do we really perceive arrogance in a person, or dishonesty, or self-serving bias? What we see is behavior that indicates negative traits like arrogance, or selfish ambition, or dishonesty, or positive traits like sincerity, or fairness, or wisdom. It is relatively well-known that we are not good at attributing traits of character to other persons,[3] but the perception of a trait affects the health of the social fabric even if the person does not actually have the trait. That suggests to me that the expert must be attuned to the way her behavior is perceived.

What should we say about the lack of trust in scientific and medical experts among an important segment of the population? Oreskes and Conway write:

[T]he distinctly partisan pattern of American distrust in science suggests that its origins are likely to lie more in political beliefs and commitments than in anything that scientists themselves have done or failed to do. To be sure, poor

---

[2] A 2018 poll by Research!America found that more than 70 percent of Americans believe that government investments in science and technology are worth the cost to the public. The American Academy of Arts and Sciences (2018) analyzed the poll and other data and found that most Americans believe that science is beneficial, trust scientists to tell the truth, and support an active role for science and scientists in shaping public policy.

[3] Gilbert Harman (1999) and John Doris (2002) made suspicion of our ability to attribute traits of character to others and even the existence of character traits a matter of debate in the first decade of the present century. For a good overview of the debate, see Christian Miller (2014).

> communication by scientists does not help their cause, but—absent other factors—missteps by scientists would likely generate skepticism across the political spectrum rather than in one part of it. (111)

I have quite a different reaction to the evidence. There is no reason to think that everyone would react the same way to "missteps" by the experts. If conservatives are more likely than others to be distrustful of scientific experts, is it entirely due to mistakes by the conservatives? In fact, if there is an identifiable segment of the population that is suspicious of experts, that is all the more reason why the experts should make a special attempt to attend to the way their public assertions are perceived by a part of the lay public. This point also applies to racial groups because race is another predictor of trust or distrust in scientific and medical experts. Native American communities have been historically alienated from expert institutions and report low trust in scientific experts. Like many political conservatives, they wish to keep government out of their lives as much as possible. Cultivating their trust requires awareness of the reasons for their distrust, and sensitivity to their preexisting attitudes; likewise for political conservatives.[4]

Lack of trust is a problem for both sides. It is a problem for those who do not trust, and it is a problem for those who are not trusted. We live in a society that prefers to look for structural solutions to societal problems. Rarely do we find public discussion of the place of personal traits in generating and solving those problems. But the perception of personal traits in experts explains a lot about our divisions over the authority of experts and the differing responses to their public testimony. We are right to hesitate to call someone dishonest or arrogant or power-hungry, but I think it is important for the expert to be able to identify behaviors that can be interpreted as expressive of those traits. We are always told that those in a position of authority must avoid the *appearance* of a conflict of interest because the appearance of being self-serving can be as detrimental to their ability to perform their social function as it would be if they were in fact self-serving. For the same reason, I think that they should avoid the appearance of dishonesty, selfish ambition, arrogance, insincerity, disrespect, and any other trait that undermines trust in their advice.

3. Responsibilities of the Expert

I assume that the responsibilities of authorities and experts are the same, differing only in degree. If people take your word as a pre-emptive reason for them

---

[4] Bennett (2022) discusses studies that show a significant difference in level of trust in the government's response to the COVID-19 pandemic among races, with Indian/Alaskan Natives reporting lower trust compared to Whites, Blacks, and Asians. American Indian or Alaskan Native participants also had the highest rates of vaccine refusal.

to accept what you say, you must be especially careful to adhere to your responsibilities religiously, but even if they treat you only as an advisor and you inadvertently mislead them, the consequences can be serious, especially if your expertise is in an area of human health or well-being.

Given that trust is the key element in the relationship between experts and non-experts, it follows that the primary responsibility of an expert is to be deserving of trust and to act in a way that is perceived to be trustworthy.[5] That leads to a number of specific responsibilities for the expert. I assume that the most basic one is the responsibility to be as accurate as possible in their testimony and to display their accuracy in their behavior. That means admitting inaccuracy when it occurs. Some experts are hesitant to admit mistakes for fear that the admission will undermine their credibility, but I believe that the truth is the opposite. Inaccuracies are almost always discovered eventually, and if not forthrightly admitted by the expert, their trustworthiness sinks. What people call "flip-flopping" is reputedly confusing, but surely there is nothing wrong with a change in position if that is what the evidence warrants and the grounds for the change in judgment are clearly explained to the public. At the beginning of the COVID pandemic when its means of transmission was unknown, medical experts advised frequent hand washing and the use of sanitizers on hands and even grocery bags. When it was discovered that its transmission was primarily through the air, they quietly stopped urging surface sterilization, but they did not clearly explain that there was a change and why, and consequently, many people spent unnecessary time and effort at sterilizing the surface of every item they touched.

It would not be necessary for experts to change recommendations frequently if they admitted to a lack of confidence in their recommendations in the first place. Uncertain advice is often a more accurate expression of the expert's knowledge than firm recommendations, but sometimes experts believe that they need to display more confidence than they have. Studies by Gaertig and Simmons (2018) show that this is a mistake. Their studies indicate that people prefer confident advisors, but they do not dislike uncertain advice. Participants approved of advisors who expressed uncertainty by giving a range of outcomes with numerical probabilities, or saying that one outcome was more likely than another. When faced with an explicit choice, the participants were more likely to choose an advisor who provided uncertain advice over one who was certain. Gaertig and colleagues conclude that people prefer advisors who are confident of the level of knowledge they possess, but people dislike false certainty.

---

[5] Following Onora O'Neill (2018), some writers distinguish trustworthiness in the sense of what makes trust well-placed from trustworthiness as a virtue of trustworthy people. Someone could be trustworthy even though it would not be appropriate for a person in certain circumstances to trust them. This distinction shows the importance for the expert to not only have the virtues that make them responsible experts, but to attend to whether their speech and actions make trust in them by people in different racial or political groups well-placed.

Few things ruin trustworthiness more than lying to the non-experts. That rules out the "noble lie," which Plato made famous in the *Republic* (414b–415c). Plato proposes that people should be taught a myth to get them to accept the ideal city's class structure. They are to be told that although all members of the republic are brothers, while they were down within the earth being molded for their life above, God fashioned gold in those who were to rule, silver in their helpers, and iron and brass in the farmers and artisans. If people can be made to believe this story, they will be motivated to care for the city and to perform their function in it. The tale is a lie, but it is told with good intentions.

I have no idea whether anyone could get away with the noble lie in Plato's imaginary republic, but they have not gotten away with it in our society, and the attempt has had dangerous consequences. Perhaps the most well-known noble lie in recent years was the proclamation by health experts early in the COVID pandemic that wearing masks would not limit the transmission of the virus and that people would not know how to wear them anyway. The motivation was to preserve masks for health care workers, whose need was greater than the need of the general public. This was a serious mistake, as Zeynep Tufekci (2020) argued in an op-ed piece in the *New York Times*. The move quickly backfired and caused greater subsequent harm than the harm they were trying to prevent.[6]

Another statement that was in part a noble lie was Dr. Fauci's change in the target range for herd immunity. In a telephone interview with Donald McNeil, who at the time was a reporter for the *New York Times*, Fauci said that emerging studies suggested a higher target for herd immunity, but he added a revealing comment:

> When polls said only about half of all Americans would take a vaccine, I was saying herd immunity would take 70 to 75 percent. Then, when newer surveys said 60 percent or more would take it, I thought, "I can nudge this up a bit," so I went to 80, 85.[7]

People do not like being lied to, but even worse is being treated in an arrogant and condescending manner. People thought, "How stupid do they think we are? If masks don't work, why are the health care workers trying to get more?" And later: "Why does Fauci think that the goal of herd immunity has changed in such a short time?" The public is much more knowledgeable than in the past. Gone are the days when a patient would do whatever her doctor ordered without question, often not knowing what medications she was taking, much less their risks. We

---

[6] Fauci admitted that he told a lie a few months later. He said that if he had endorsed mask wearing, mass panic would have ensued, leading to a shortage of surgical masks among health care workers who needed them more. Reported by Powell and Prasad (2021).

[7] Related by Powell and Prasad (2021).

know that there is rampant misinformation, but there is also a great deal of accurate and helpful information that allows ordinary lay people to make autonomous judgments about their own health and wellbeing. They want and deserve to be treated with respect by the experts. It is easy for an expert to show respect for a non-expert who is eager to accept the expert's opinion, but the test of whether the expert is truly respectful comes when the non-expert expresses skepticism and asks for a reply to objections and contrary views given by other experts.[8] When experts disdain contrary views and belittle those who doubt their pronouncements, they are perceived as arrogant by part of the population who then fail to benefit from the expert's advice, possibly even resulting in death.[9] Breakdowns in trust have far-reaching consequences, affecting the perceived veracity of the expert in important future communications with the lay public.

The need for the experts to be perceived as sincerely attempting to convey the truth to non-experts also gives experts the responsibility to display their awareness of contrary hypotheses and to treat them seriously. If they are convinced that a contrary hypothesis is false and possibly dangerous, it is important that they respond calmly and with careful attention to the details of the contrary view. Otherwise, non-experts will suspect that they have motives other than accurate testimony. The reaction of Anthony Fauci and Francis Collins to the Great Barrington Declaration was, in my opinion, an example of a high-handed failure to meet this responsibility. The Great Barrington Declaration was written by Dr. Jay Bhattacharya, professor at Stanford University Medical School, Dr. Sunetra Gupta, professor at Oxford University, and Dr. Martin Kulldorff, professor of medicine at Harvard University. In their words:

> The Declaration was written from a global public health and humanitarian perspective, with special concerns about how the current COVID-19 strategies are forcing our children, the working class, and the poor to carry the heaviest burden. The response to the pandemic in many countries around the world, focused on lockdowns, contact tracing, and isolation, imposes enormous unnecessary health costs on people. In the long run, it will lead to higher COVID and non-COVID mortality than the focused protection plan we call for in the Declaration.[10]

---

[8] This can happen when the expert thinks of himself as an authority rather than a mere expert. An authority does not need to display the evidence supporting his directives.

[9] A report by Wallace et al. (2022) from the National Bureau of Economic Research showed substantially higher death rates for Republicans when compared to Democrats in Ohio and Florida, with most of the difference occurring in the period after vaccines were widely available. This is evidence that people died because of their lack of trust in the vaccines. The death rates for Native Americans are also much higher than in the general population. See Akes and Reber 2021 for detailed comparisons.

[10] See Great Barrington Declaration (https://gbdeclaration.org) for the statement and signatures.

Dr. Fauci and Francis Collins, who was then the Director of the National Institutes of Health, worked together to "take down" the declaration. Collins wrote to Fauci in an email: "This proposal from the three fringe epidemiologists... seems to be getting a lot of attention—and even a co-signature from Nobel Prize winner Mike Leavitt at Stanford. There needs to be a quick and devastating published take down of its premises." Subsequently, they both did numerous media interviews denouncing the declaration, with Fauci telling ABC it was nonsense, but without addressing the concerns expressed in the declaration.[11]

The noble lie and the suppression of contrary hypotheses are examples of manipulation that fail to respect the autonomy of the lay public. There is evidence that social media elevated certain material about COVID and suppressed other material with governmental collusion. If experts were involved in these actions, they were assuming that they had the right to manipulate personal decisions of the non-experts.[12] Similarly, the lie about masks paternalistically assumed that people cannot be trusted to make judgments for the community good when accurately informed.

Matthew Bennett (2022, Introduction) distinguishes two kinds of trust in experts: epistemic trust and recommendation trust. Epistemic trust concerns believing what the expert says; recommendation trust applies to acting on the expert's advice. Bennett argues that more is needed for recommendation trust than epistemic trust because "novices looking for trustworthy expert recommendations need to establish whether the recommended course of action supports what is important to them and accords with their values." I think that this point is important in understanding the different action responses to COVID recommendations among different political and racial groups. Action recommendations that make value assumptions opposed to the values of certain people are useless at best and disrespectful at worse. To be an effective expert, we need to meet people where they are. If there is a disagreement about values, we must then discuss the values. But it is irresponsible of the expert to promote their own values under the guise of expertise.

The general responsibility that applies here is the responsibility of the expert not to make proclamations beyond their area of expertise. It is tempting to speak outside their field in making policy recommendations that require all-things-considered judgments. At the beginning of the COVID epidemic, Dr. Fauci was justifiably hesitant to make broad recommendations that involved weighing harms to the economy, the education of children, and other aspects of public health such as drug abuse, depression, and suicide against the benefits of combatting COVID. Fauci is not an expert on the economic effects of shut-down

---

[11] Quoted in Finley (2022).

[12] David Zweig released internal Twitter company documents on December 26, 2022 through a Twitter thread. Zweig says that the documents show that "the United States government pressured Twitter and other social media platforms to elevate certain content and suppress other content about COVID-19."

policies on different population groups such as laborers and small business owners, nor is he an expert on the consequences of those policies on children, as he was aware. As an immunologist, he is not an expert on the effects of policies on rates of drug use, depression and suicide, or the social and economic effects of moving workers from office to home. But at some point, he was handed the responsibility to make policy by political leaders. He is not an expert at weighing total benefits and risks. Is anybody? Decisions that should be made in a democratic society by open discussion and debate were handed over to health care experts who were not equipped to make such decisions.

For many years, economists have gone beyond their area of expertise in making policy recommendations. Most economic policies benefit some people and harm others, and that raises questions of justice. But whether a policy is just or unjust is not a part of economic expertise. For example, globalization has had uneven effects on different population groups.[13] Policies to combat inflation affect certain groups of people disproportionately. Economists are experts on economic cause and effect. They are not expert on the morality of bringing about these effects. Like all citizens, they have the right to express their moral viewpoint and policy recommendations, but it needs to be clear that as economists they do not have special moral insight.

A well-known case of experts going beyond their area of expertise and causing a public outcry was the 1996 Herrnstein and Murray book, *The Bell Curve*, which examined data on racial differences in IQ, and they used it to make sweeping public policy recommendations. Because the book could be used to support racist attitudes, it was called pseudo-science, but that was not because of the way they collected the data, but because of the policy recommendations which went well beyond what could be reasonably inferred from the data they had. What is particularly interesting to me about this case is that people thought that if you are not racist, you must claim that there was something wrong with the science in the book, but that was not the problem. The fault was not in data collection; it was in conclusions that required moral judgment that was not part of the authors' expertise as a psychologist and a political scientist. Similarly, economists who make recommendations based on moral and social premises that are not part of economics fail in their responsibility to limit their expert advice to the areas in which they really are expert. It is true that life is all of a piece, and as human beings we need to make judgments that go beyond our area of expertise, and we might want to share our opinions with others. There is nothing wrong with that. Experts have a right as much as everybody else to express their opinions about issues of

---

[13] Normalizing trade relations with China in 2000 had benefits, but also many costs that raise questions of justice. American industrial communities suffered large job losses, and China never made good on its promises to make economic reforms that would make their own economy more just. See Scott Paul's (2022) article in *The Hill* for the point of view of the President of the Alliance for American Manufacturing.

concern to others. But it should be clear when they are speaking as a citizen rather than as an expert.

A responsibility that is especially important in a society that aims to be democratic is the responsibility to try to prevent their expertise from becoming too closely connected with politics. Experts are often beholden to funding institutions that exert pressure on them to take a line that prevents them from fully exercising their responsibilities. Regulatory agencies can be "captured" by private industry professionals or lobbyists to serve their own interests.[14] And even when scientific experts have reached their conclusions responsibly, it is important that their findings are expressed in a way that is accepted by people of all political affiliations. Unfortunately, when Donald Trump was president, he politicized COVID. Once that happened, there was no chance that health recommendations would be accepted across political divisions. That should not have been permitted to happen, and it cost lives. The Trump administration altered or sidelined research that they implied was the point of view of Democrats,[15] but then when only Democrats objected, expert opinion on COVID was perceived as either Democrat or Republican. It is not the fault of the expert if other people politicize their findings, but the more they can distance themselves from partisan politics, the better it will be for all of us.

The responsibilities I have identified are all connected with the basic responsibility of the expert to speak accurately, and to convey their degree and range of expertise and their sincere aim to be truthful to the non-experts. Non-experts must depend upon experts for knowledge and advice that may have significant effects on their lives. If there is a lack of trust, experts cannot perform their role successfully. So, they should try to prevent the use of their expertise to support political agendas that many of their potential subjects oppose for other reasons. They should limit their testimony to their area of expertise. They should always show respect for the autonomy of the non-expert and for experts who disagree with them. They should admit their mistakes. And above all, they should never lie.

4. Humility: The Key Virtue

Humility in the most general sense can be characterized as the virtue of realism about the features of ourselves that give us high status.[16] Traditionally, it was

---

[14] See Andrea Saltelli et al. (2021) for an examination of five cases in Europe in which lobbyists presented themselves as upholders of science and intervened directly in the methodological and ethical aspects of science to further their own agenda.
[15] See Viglione (2020) for reports of cases in which the Trump administration meddled in scientific reports and silenced scientists whose views they rejected.
[16] In what follows I address intellectual humility in a Western context. For a Confucian perspective, see Jin Li (2016). Li argues that the Western values of self-esteem and self-confidence make Westerners

thought to be the virtue of having a lowly opinion of oneself. That view has been rejected for some time because it is hard to see why it would be virtuous to have a false view of oneself, but it is interesting to compare this way of looking at humility with the advice of Aristotle about how to cultivate a virtue. Given his interpretation of virtue as a mean between two extremes—one a vice of excess, and one a vice of deficiency, Aristotle proposes that the way to get to the mean is to aim for the opposite vice. If you are cowardly, aim for foolhardiness. If you are intemperate, aim to be insensitive to pleasure. He assumes that the cowardly person will never become foolhardy, and the intemperate person will never lose the desire for pleasure, but their aims will get them to move in the right direction. It could be proposed in a similar vein that a person with too high an opinion of himself should aim to have a lowly opinion of himself. In doing so, he would move towards a more realistic view of his talents, knowledge, skills, and any other features of which he is especially proud. Aristotle clearly did not think of humility as a virtue,[17] but his strategy in virtue cultivation could be helpful. Nonetheless, it seems to me that humility is not the trait of having a low opinion of oneself.

In one of the earliest works on humility in the contemporary philosophical literature, Robert C. Roberts and Jay Wood (2003, 2007) argue that humility is the virtue that opposes a raft of vices, including arrogance, vanity, conceit, egotism, grandiosity, pretentiousness, impertinence, snobbishness, self-righteousness, domination, selfish ambition, and self-complacency. These vices, they say, help to define humility; they are negatively definitive of it (2007, 236). Their hypothesis is that a single virtue opposes all these vices. They define humility as the virtue of having a low concern for one's status and entitlements. Intellectual humility is the virtue of having a low concern for one's intellectual status and entitlements.

Whitcomb, Battaly, Baehr, and Howard-Snyder (2017) discuss two of the vices identified by Roberts and Wood: vanity and arrogance. Their hypothesis is that humility is the virtue of owning our limitations, and intellectual humility is the virtue of owning our intellectual limitations. With that hypothesis in hand, they make a series of predictions about behavior that can be empirically tested.

I am not interested in giving a precise definition of humility. It seems to me that both the Roberts/Wood definition and the Whitcomb et al. definition are plausible candidates, and both approaches can tell us much about the landscape of traits underlying behaviors in experts that make them responsible or irresponsible in performing their social function. I surmise that the vices opposed to intellectual

---

ambivalent about humility. In contrast, the Confucian conception of humility is geared toward moral self-cultivation in learning. The Confucian view appeals to the constant need to improve, and it emphasizes the trap of conceit. The more one achieves, the greater the need to remain humble.

[17] Aristotle lists magnanimity, or greatness of soul, as one of the virtues (*Nicomachean Ethics* BK IV, chap. 3). The magnanimous person (a man) thinks highly of himself for good reasons. Humility does not make Aristotle's list, and he probably would have thought it very peculiar that anyone would consider it a virtue on any current definition.

humility are special temptations for the expert since expertise gives a person a reputation for special knowledge that can make them susceptible to temptations to power and high status. That can jeopardize their ability to convey their knowledge accurately and sincerely and to meet their responsibilities to non-experts. Experts are just people like everybody else, and their expertise does not make them immune to the character flaws that inflict the rest of us. But there is also some evidence that humility is an attractive trait. If that is true and becomes widely believed, it could make experts more effective and would counterbalance temptations against humility.[18]

The Whitcomb group proposes that the first step in humility is to be aware of our limitations, at least implicitly (516). Intellectual limitations include gaps in knowledge, cognitive mistakes, and intellectual character flaws. They then argue that proper attentiveness to one's intellectual limitations is not enough for intellectual humility. The intellectually humble person will *own* her limitations. Her response to her intellectual limitations includes cognitive, behavioral, motivational, and affective responses. She not only believes that she has specific limitations, she also recognizes that the negative outcomes of her limitations are due to her limitations. She is disposed to admit her limitations to others. She is disposed to care about them and to take them seriously as the context demands. She is disposed to regret her limitations, but not to be hostile towards herself—again, as the context demands (517–18).

Although the Whitcomb group does not discuss experts, I think that the way they characterize intellectual humility can help us see how a single character trait can be directly connected with the expert's ability to successfully exercise her expertise in a public setting and to meet the responsibilities I have outlined. If the expert owns her limitations, she will limit her testimony to her area of expertise. She will admit her mistakes, as the Whitcomb group concludes, and she will not lie. She will respect the non-expert and experts who disagree with her because she will be aware that her limitations can blind her to the point of view of other experts, and her limitations can lead her to exaggerate her superiority to non-experts in certain ways. Owning her limitations probably is not enough to lead her to prevent the use of her expertise for political purposes. As I mentioned, that responsibility is not hers alone. It is mostly the responsibility of political leaders. But the Whitcomb group's final definition of intellectual humility can be helpful for this responsibility also. They propose that humility is "an intellectual virtue just when one is appropriately attentive to, and owns, one's intellectual limitations *because* one is appropriately motivated to pursue epistemic goods, e.g., truth, knowledge, and understanding" (520). The expert is a person whose pursuit of

---

[18] According to Paloma Cantero-Gomez (2019), a self-described expert on leadership and entrepreneurship, highly admired people have some significant traits in common. The first one she mentions is humility, a trait she says makes them very attractive to others.

HUMILITY FOR EXPERTS   183

epistemic goods is shared with others in the community or the society at large. That goal will motivate the expert to attempt to prevent the use of her expertise in any way that blocks a large part of her intended audience from getting the advantages of her expertise. Her role as an expert is intellectual, and it should always be clear that her aims in the testimony she gives non-experts are intellectual aims, not political ones.

Using their account of intellectual humility, the Whitcomb group makes no fewer than 19 predictions about what an intellectually humble person will do, all of which rest upon the single, simple account they have given. Almost all their predictions could be applied to experts and tested in empirical studies.

1. Intellectual humility (IH) increases a person's propensity to admit her intellectual limitations to himself and others.
2. IH reduces both a person's propensity to pretend to know something when she does not, and her confidently answering a question whether or not she knows the answer.
3. IH reduces a person's propensity to blame and explain-away when confronting her own intellectual shortcomings.
4. IH decreases a person's propensity to set unattainable intellectual goals.
5. IH increases a person's propensity to defer to others who don't have her intellectual limitations in situations that call upon those limitations.
6. IH increases a person's concern about her own intellectual mistakes and weaknesses.
7. IH reduces feelings of anxiety and insecurity about one's own intellectual limitations.
8. IH decreases a person's propensity to excessively compare herself to others intellectually.
9. IH reduces the intellectual aspect of the self-serving bias in a person, which is, very roughly, the propensity to attribute to oneself more responsibility for intellectual successes than for intellectual failures.
10. IH increases a person's propensity to revise a cherished belief or reduce confidence in it when she learns of defeaters (i.e., reasons to think her belief is false or reasons to be suspicious of her grounds for it).
11. IH increases a person's propensity to consider alternative ideas, to listen to the views of others, and to spend more time trying to understand someone with whom she disagrees.
12. IH increases a person's propensity to seek help from other sources about intellectual matters.
13. IH increases a person's propensity to hold a belief with the confidence that her evidence merits.
14. IH increases a person's propensity to have a clearer picture of what he knows and justifiably believes and what he neither knows nor justifiably believes.

15. IH reduces a person's propensity to expect or seek recognition and praise for his intellectual strengths and accomplishments.
16. IH reduces a person's propensity to treat intellectual inferiors with disrespect on the basis of his (supposed) intellectual superiority.
17. IH tends to decrease focus on oneself and to increase focus on others.
18. IH increases a person's propensity to accurately estimate his intellectual strengths.
19. IH decreases a person's propensity to be obsessed with his strengths and to boast about them. (521–7).

A comment about #18: they say that although on their view, IH does not consist in accurately estimating one's strengths, it can explain the plausibility of this prediction because if one is property attentive to and owns one's intellectual limitations, then one would also be aware of the limitations of one's strengths, which would require an accurate assessment of those strengths (534). If the Whitcomb group is mistaken about this, we could revert to my more general characterization of humility as the virtue of realism about the features of ourselves that give us high standing. For the expert, that would be realism about his or her intellectual strengths and weaknesses. As I said, I am not aiming for precision in a definition of intellectual humility. Even if the Whitcomb account is not 100% on target, it certainly is close, close enough to give us a reason to surmise that there is a single intellectual virtue underlying a large range of behaviors. They do not discuss experts, but most of these behaviors directly connect to the responsibilities of experts and their success in reaching the goals of their expert testimony. If empirical studies confirm their prediction that these behaviors are consequences of owning one's intellectual limitations, then given that these consequences are directly tied to the responsibilities of experts, we would have good evidence that intellectual humility is a fundamental virtue for the expert.

All of this is very plausible, but experts have sometimes been placed in a special class that makes them exempt from some of their epistemic responsibilities on the grounds that they have more than epistemic goals. Nobody would defend the noble lie unless they thought that the epistemic goals of the expert can be trumped by non-epistemic goals. What if the truth puts people's lives at risk? What if the best way to get people to act in the best interests of the community is make proclamations in an arrogant and simplistic manner? What if making assertions in an area beyond your field of expertise is better than not doing so because there isn't anybody with expertise in that area who is able to influence non-experts? What if people turn away from you as soon as you admit a mistake and end up harming themselves and others?

My first response to this argument is to point out that if an expert puts non-epistemic goals ahead of the epistemic goals of her field of expertise, she is not doing so *as* an expert, but as a person whose ability to evaluate the relative importance of non-epistemic vs. epistemic goals is not part of her expertise. I have offered examples of medical experts and economics experts who make recommendations based on their own interpretation of the general goal of social welfare, an interpretation that is not a component of their expertise. There are people with practical wisdom who are good at weighing risks and benefits of all kinds in decision-making, but the expert qua expert is not one of them. Of course, it is possible that an expert has the degree of practical wisdom needed to make these decisions, but we cannot take for granted that the expert is any better at those decisions than is the ordinary non-expert. In a democratic society, our way of handling this problem is not to hand over decisions about general welfare to scientific, medical, or economic experts, but to settle the matter by open discussion and votes by our elected representatives.

My second response to this argument is that behaviors that violate an expert's responsibilities as an expert are not likely to succeed anyway. In a society with (almost) unrestricted communication, falsehoods are discovered, and since part of the population will continue to believe the expert's falsehoods, the population becomes fractured over trust in the expert. Arrogance and disrespect are also perceived differently by different groups. Some people are quick—even too quick—to perceive disrespect because of their preexisting dislike of the expert, whereas other people are quick to forgive behaviors that they would find disrespectful in another person. That difference also divides the community and causes lack of trust within the community, a very unhealthy situation in a society that relies upon common agreement to make policy decisions.

5. The Courage to Be Humble

It is a challenge for anybody to be intellectually humble, but it is especially challenging for experts because they face greater temptations to protect their ego when they are acting as an expert. But occasionally an expert stands out for her intellectually humble behavior. An example discussed by Jason Baehr (forthcoming) in his book on intellectual virtues and vices is Doris Kearns Goodwin, a Pulitzer-Prize winning presidential biographer, who is known for her meticulous scholarly practices and intellectual carefulness. However, in her 1987 book, *The Fitzgeralds and the Kennedys*, Goodwin failed to properly attribute some source material taken from a work by another historian, Lynne McTaggart. McTaggart contacted her about the mistake, and Goodwin (2002) reports her reaction: "I realized that she was right. Though my footnotes repeatedly cited Ms. McTaggart's work, I failed to provide quotation marks for phrases that I had taken verbatim, having assumed that these phrases, drawn from my notes, were my words. I made the corrections she requested, and the matter was completely laid to rest." Almost fifteen years went by, and then amid a public debate about citations and historical research, the incident with McTaggart resurfaced. Goodwin wrote an essay called "How I Caused That Story" in *Time* magazine recounting her mistake and the overhaul of her research practices it occasioned. She also humbly admitted that she still has limitations and is not protected from future errors:

> [T]here is no guarantee against error. Should one occur, all I can do, as I did 14 years ago, is to correct it as soon as I possibly can, for my own sake and the sake of history. In the end, I am still the same fallible person I was before I made the transition to the computer, and the process of building a lengthy work of history remains a complicated but honorable task.

As Baehr remarks (ch. 1), Goodwin's intellectual conduct and attitudes show her qualities of character as well as her high standards as an historian. She owned her limitations by acknowledging her mistake and making amends when she discovered it. What makes her actions particularly impressive is that she believes her expertise as an historian rests upon her accuracy and meticulous documentation. But she did not shirk from admitting her mistake even when her pride in herself as an historian was threatened. What some people perceived as a scandal was actually an exemplary case of humility that hopefully will give other experts the courage to be intellectually humble when their limitations are discovered. Pretending that these limitations do not exist does not enhance trust in their expertise; it invariably detracts from it.

We are all dependent upon experts. We need accurate information from experts in order to make many choices in our lives, and we rely upon experts to help us function as informed citizens. To successfully perform their social function, experts must be accurate in their testimony, but they must also act in a way that is perceived to be trustworthy. That is harder than being good at their area of expertise because it requires knowing their audience and the rational or irrational impulses of their audience. Whether they are functioning as intellectual authorities or only as intellectual advisors, their power can misfire in many ways. Some of those ways are outside of their control, but I have focused on the ways that are within their control as governors of their character and their actions.

References

Akes, Randall and Sarah Reber. (2021) "American Indians and Alaska Natives are Dying of COVID at Shocking Rates." Brookings Research Report (February 18).

American Academy of Arts and Sciences. (2018) *Perception of Science in America*. Cambridge, Mass: American Academy of Arts and Sciences, available at https://www.amacad.org/publications/perceptions-science-america.

Baehr, Jason. (forthcoming) *Inquiry & Agency: A Theory of Intellectual Virtues and Vices*. NY: Oxford University Press.

Bennett, Mathew. (2022) "Judging expert trustworthiness: The difference between believing and following the science." *Social Epistemology* 36(5). Special issue on Social Indicators of Trust in Experts and Institutions. Guest (eds.), T. Y. Branch and Glrai Origgi, available at https://doi.org/10.1080/02691728.2022.2106459.

Cantero-Gomez, Paloma. (2019) "The Top 10 Characteristics of Highly Admired People," *Forbes* (June 18).

Doris, John. (2002) *Lack of Character: Personality and Moral Behavior*. Cambridge: Cambridge University Press.

Finley, Allysia. (2022) "Public distrust of health officials is Anthony Fauci's legacy." *Wall Street Journal* November 27.

Gaertig, Celia and Joseph P. Simmons. (2018) "Do people inherently dislike uncertain advice?" *Sage* (Scientific Advisory Group for Emergencies UK) 29(4), available at https://doi.org/10.1177/0956797617739369.

Goodwin, Doris Kearns. (2002) "How I Caused that Story," *Time* magazine (January 27). https://content.time.com/time/nation/article/0,8599,197614,00.html.

Harman, Gilbert. (1999) "Moral philosophy meets social psychology: Virtue ethics and the fundamental attribution error." *Proceedings of the Aristotelian Society* 99, 316–31.

Herrnstein, Richard J. and Charles Murray. (1996) *The Bell Curve*. NY: Free Press.

Li, Jin. 2016. "Humility in learning: A Confucian perspective." *Journal of Moral Education* 45(2), 147–65.

Miller, Christian. (2014) *Character and Moral Psychology*. NY: Oxford University Press.

Oreskes, Naomi and Erik M. Conway. (2022) "From anti-government to anti-science: Why conservatives have turned against science." *Daedulus (Journal of the American Academy of Arts and Science)* 151(4), 98–123.

Paul, Scott. (2022) "It's Time To Revoke China's Normalized Trade Relations." *The Hill*, December 2.

Powell, Kerrington and Vinay Prasad. (2021) "The Noble Lies of COVID-19," *Slate*, July 28.

Raz, Joseph. (1988) *The Morality of Freedom*. Oxford: Oxford University Press.

Roberts, Robert C. and Jay Wood. (2003) "Humility and epistemic goods," in *Intellectual Virtue: Perspectives from Ethics and Epistemology*, (eds.), Michael DePaul and Linda Zagzebski. Oxford: Oxford University Press.

Roberts, Robert C. and Jay Wood. (2007) *Intellectual Virtues: An Essay on Regulative Epistemology*. NY: Oxford University Press.

Saltelli, Andrea, Dankel, Dorothy, Di Fiore, Monica, Holland, Nina, et al. (2021) "Science: The endless frontier of regulatory capture." *SSRN* (March 1). https://papers.ssrn.com/sol3/papers.cfm?abstract_id=3795058.

Tufekci, Zeynep. (2020) "Why Telling People They Don't Need Masks Backfired," *NY Times* March 17, available at https://ww.nytimes.com/2020/03/17/opinion/coronavirus-face-masks.html.

Viglione, Giuliana. (2020) "Four ways Trump has meddled in pandemic science—and why it matters." *Nature* (November 3).

Wallace, Jacob, Paul Goldsmith-Pinkham and Jason L. Schwartz. (2022) "Excess Death Rates for Republicans and Democrats During the COVID-19 Pandemic." Bureau of Economic Research Working Paper 30512 (September).

Whitcomb, Dennis, Heather Battaly, Jason Baehr and Daniel Howard-Snyder. (2017) "Intellectual humility: Owning our limitations." *Philosophy and Phenomenological Research* XCIV(3), (May).

Zagzebski, Linda. (2012) *Epistemic Authority: A Theory of Trust, Authority, and Autonomy in Belief.* NY: Oxford University Press.

11
Expertise-in-Action

The Importance of Intellectual and Moral Virtue(s) to Experts' Epistemic Authority

Andrea Lavazza
Andrea Lavazza, Senior Research Fellow,
Centro Universitario Internazionale, Arezzo, Italy and Adjunct Professor
in Neuroethics, University of Milan and University of Pavia, Italy

James Giordano
Pellegrino Center Professor of Neurology and Biochemisty,
and Chief of the Neuroethics Studies Program at
Georgetown University Medical Center, Washington, DC, USA

Mirko Farina
Professor and Head of Human Machine Interaction Lab, Institute for
Digital Economy & Artificial Systems, Xiamen University and
Lomonosov Moscow State University

1. Introduction

As the COVID-19 crisis has demonstrated, subject matter experts have been important in addressing the basis and impact of biomedical, socio-economic, and political engagement. In many respects and in many countries, experts—especially biomedical experts—have acquired, perhaps as never before (Farina and Lavazza 2021a), roles in both (1) the development of public health practices and policies aimed at mitigating the spread of the virus, and (2) advising politicians and economic decision-makers in formulating appropriate frameworks for the containment of COVID-evoked effects (Farina and Lavazza 2021b) on various aspects of society (Pietrini et al. 2022).

Despite ongoing debate among experts about the origins of the pandemic (Lavazza and Farina 2020), and regarding the validity of preventive and/or precautionary measures (e.g., lockdown, mask-wearing, etc.; Farina and Lavazza 2020) experts' knowledge has undoubtedly fortified tactics for dealing with the emergency (Lavazza and Farina 2021; Karimov et al. 2022). And while expert discourse and debate has, in some instances, contributed to differing views and practices among the lay population, it is indubitable that refuting experts' advice has often led to significant worsening of the epidemiological situation (Cairney and Wellstead 2021; Dente et al. 2022).

Denialism of expert-informed decisions seems to be unjustified and should be rationally discouraged (Nichols 2017). With this premise as a starting point, we propose consideration of an important issue of the COVID crisis, which to date, has received a paucity of attention and analysis, yet which is equally referential to public health, safety, and security issues, at large. Namely, that biomedical experts involved in the formulation of public health policies (i.e., as advisors to decision-makers, and/or as decision-makers themselves) have often made choices that have been dictated by criteria of objectivity and neutral rationality, and aimed at optimizing effectiveness, yet seem to be relatively inattentive to ethical principles and certain constitutional rights (Pietrini et al. 2022). For example, in the United States (US) during the period of March–May 2020, eleven of fifty states recommended particular exclusion criteria (e.g., neurologic diseases [8] and/or end stage dementia [3]) for implementing treatment of severe COVID (i.e., requiring ventilatory support) in adults (Piscitello et al. 2020). During the second wave of the pandemic (viz. winter 2020–21) in the UK as well "...people with learning disabilities have been given do not resuscitate orders...in spite of widespread condemnation of the practice last year and an urgent investigation by the care watchdog."[1]

Importantly, these guidelines and recommendations were informed and directed by experts, and thus were valid and valued as being epistemically sound, and—as consistent with the tasking of expertise—were rational, objective, and aimed at maximizing effectiveness and efficiency by pursuing the greatest good for the greatest number of people.[2] In the main, there was relatively large-scale criticism of, and dissent to recommendations that people with the aforementioned pre-existing conditions be excluded from access to ventilators. For example, the Office for Civil Rights at the US Department of Health and Human Services stated that, based on Section 1557 of the Affordable Care Act and Section 504 of the Rehabilitation Act, "persons with disabilities should not be denied medical care on the basis of stereotypes, assessments of quality of life, or judgments about a person's relative 'worth' based on the presence or absence of disabilities or age."[3]

---

[1] https://www.theguardian.com/world/2021/feb/13/new-do-not-resuscitate-orders-imposed-on-covid-19-patients-with-learning-difficulties (last accessed July 2023). Subsequently, the Department of Health and Social Care dubbed those orders as "completely unacceptable."

[2] By experts in these specific situations, we mean individuals with skills and knowledge, who possess a greater epistemic authority over non-experts and by virtue of this authority can make informed decisions and accurate predictions.

[3] HHS Office for Civil Rights in Action (2020). BULLETIN: Civil Rights, HIPAA, and the Coronavirus Disease 2019 (COVID-19), available at https://www.hhs.gov/sites/default/files/ocr-bulletin-3-28-20.pdf. (Last accessed July 2023). https://gizmodo.com/alabama-disavows-plan-to-deny-ventilators-to-Covid-19-p-1842770059 (Last accessed July 2023). Cf. Andrews et al. 2020; Chen and McNamara 2020.

These cases show very clearly that experts can suggest or implement decisions that are perfectly consistent with their specialized expertise but at the same time may go against the sensitivities of certain social groups or even against certain principles that society shares or that the state has established as elements of its legislation. The protection of the disabled or of the fragile or of the disadvantaged falls within this scope. If the only criterion for inclusion/exclusion from scarce care resources is life expectancy, in the context of the epidemic disabled and disadvantaged people may pay twice for their condition. Firstly, by suffering a disadvantage (for which they are not responsible) in their quality of life compared to the rest of the population and—secondly—because of this disadvantage, by being further penalized and not receiving appropriate treatment.

In this sense, biomedical experts (during the pandemic) were undoubtedly among those most involved in technical decisions, which had far-reaching social repercussions, and hence directly affected people lives. To make an analogy, it is plausible that an engineering expert called upon to assess whether it is better to build a bridge or an underground tunnel to connect two locations on either side of a bay would not have to consider any implication. Naturally, it would be assumed that her expertise would fully consider the environmental impact of her work. In the past, however, an engineer called upon to suggest or decide on the question described previously might not have been sufficiently attentive to environmental requirements.

Similarly, with respect to the allocation of scarce care resources during crises, one could argue that a biomedical expert could adopt (based on a legitimate expression of her epistemic authority) a utilitarian or contractualist view and—consequently—her advice or decision should not be necessarily considered as a "failure," at least if such decisions had better aggregate consequences than alternative ones. However, because the outcomes of such decisions often have the potential to be detrimental to—at least—some individuals (in the case of the COVID-19 pandemic we are likely talking about hundreds of thousands of subjects if not millions), we believe that experts should become more sensitive to diverse ethical issues. As the engineering expert is required to be sensitive to concerns (such environmental impact of her work) that could have been easily overlooked in the past, so biomedical experts—when suggesting policies—should be more sensitive to responsibilist virtues.

In the previously mentioned cases, public discord concerning biomedical experts was based upon, and directed toward the espoused stances inattention to moral and ethical principles—and constitutional rights—of individuals (for example, rights affirmed in the US by the Affordable Care Act and the Rehabilitation Act), and not merely society at-large. While these decisions did not necessarily reflect lack or failure of ethico-legal consideration, such examples certainly do reinforce the importance of, and need for expertise in action and articulation, which obligates (1) explicit and specific ethical expertise in informing public health and safety decisions, and (2) relative transparency about the extent, role, and contribution of any such ethical expertise and guidance.

2. The Need for Experts-in-Action

The role of experts and epistemic authority in informing decisions that affect individual and community welfare is obvious. Often, experts are biomedical scientists and/or clinicians, who are recruited to advise policymakers or draft public health policies. In so doing, while there is considerable, if not primary value placed upon epistemic soundness, which often operates within a naturalistic framework, such orientation can favor non-neutral outcomes that burden, or place at risk (or in harm) more vulnerable constituents of a community. Triage-like protocols for decisions about appropriation and allocation of limited goods (e.g., resources, services) while often necessary, and likewise often difficult. The difficulty arises from the obligations and constraints arising when contrasting commutative and distributive justice (Pellegrino 1995).

In part, this apparent ambiguity may be due to the technical language of experts and expertise, which tends to impose neutral values on non-ethically neutral situations/decisions. The issue of neutrality, often in relation to the ideal knowledge of subjects disconnected from the contingencies of real life, has been at the center of debates on expertise for a very long time (Goldman 1978).

Neutral values, while often intentionally resistant to political influence, can also be seen as disconnected from social justice (Sandel 2020). While the objectivity of expert opinion is certainly valuable, if/when accentuated beyond a certain threshold, there can be considerable misunderstandings or rejection with the group/community that relies on expert knowledge and deliberations. But we argue that objectivity need—and should—not be distinct or contrary to sensitivity and responsivity to factors that define (individual and group) vulnerability, need, and some defined balance of commutative and distributive justice.

As Bledsoe et al. (2020) have noted, ... some current proposals, including universal do-not-resuscitate orders, social worth, and life-years, contravene fairness and conflict with ethical principles. They risk clinician moral distress and public distrust, further dividing ... society ... Resource allocation approaches that advocate disadvantaging ... persons, or [other] groups on the basis of diagnosis, perceived social worth ... send a message ... that some lives are valued more than others. This will engender distrust in the medical profession—now, when trust is most needed, as well as into the future." Thus, the adage "no decision about us, without us" is meaningful and applicable. But while individual and group representation is held as a valuable (if not necessary) element when informing decisions about the vulnerable, such engagement of these persons may not always be possible. A contributory and participatory epistemology would at least call for (ethico-legal) expertise as directly relevant to both the enterprise and the individuals and groups affected. But what level or extent of "community" expertise would obtain?

Recently, Landemore (2020) has proposed basing decisional process on an "open mini-public"—a large, jury-like body of selected citizens gathered to define guidelines and policies, in connection with, and as referent to the larger public. The "open mini-public" would not be impermeable to external expertise but would filter and assimilate this input to the communal values desiderata, using methods such as "crowdsourcing" to conjoin a representative collective to inform key elements of the decision-making process.

Open societies are believed to endorse five institutional principles: participatory rights, deliberation, the majoritarian principle, democratic representation, and transparency; yet only some would be consistent with the requisite pacing, competence, and efficiency dictated by emergent circumstances. Indeed, a challenge would be to find balance between ensuring the majoritarian principle, democratic representation, and transparency and the need for contemporary society to nimbly and effectively address crises and emergencies (e.g., such as a pandemic, or other rapidly evolving threat to public health and safety). The complexity of contemporary pluralist societies makes it particularly difficult to define a universally ideal decision-making process that is both adequate and timely to meet the contingencies involved. However, we believe that it is possible that broadening and deepening the epistemic authority of experts could at least partially improve their contribution to informing and developing decisions that more capably incur positive effects upon both individuals and communities at large.

Any such contributory expertise should be sensitive and responsive to diverse needs and preferences. As the often-existential decisions of public health and safety cannot—and we argue should not—be informed and made only on the bases of mere quantitative assessments that are often informed by a welfarist approach. In this light, we propose a version of Thagard Equilibrium—a method for justifying norms by identifying: (1) domain of practices, (2) candidate norms for practices, (3) appropriate goals of practices, with the specific intent of (4) adopting the practices that best accomplish given goals as domain norms (Thagard 2010), which, when taken with a (Rawlsian) *maxi-min* approach (Rawls 1999), would regard the entirety of needs and values of the engaged community to direct both rule and act utility within a duty of care. This cosmopolitan-communitarian orientation is cognizant of overall needs, values, and goods of (all) the communities involved and can be focally applied to address and serve those that are inherent to the individuals and collective(s) at hand (NB: a complete description of these methods is beyond the scope of the this chapter, for further detail, see Lanzilao et al. 2013; Shook and Giordano 2014; Giordano et al. 2015; Shook and Giordano 2020).

In general, an "expert' is regarded (by its community) as a person who possesses excellence in knowledge and ability as gained through experience and/or training and is recognized and identified as such by/within a select group (Funk and Wagnalls Dictionary 1967, 468). An interesting definition of expertise comes from Goldman's functionalist approach. According to Goldman, a subject "S is an expert in domain D if and only if S has the capacity to help others (especially laypersons) solve a variety of problems in D or execute an assortment of tasks in D which the latter would not be able to solve or execute on their own. S can provide such help by imparting to the layperson (or other client) his/her distinctive knowledge or skills" (Goldman 2018). To perform this expert function, the truth condition states that a subject "S is an expert about domain D if and only if S has more true beliefs and fewer false beliefs in propositions pertaining to D than do most people." (Ibid.).It can be argued that the justification for relying on the authority of experts lies in the rationality of the subject who recognizes, on the basis of their own conscientious judgment, that they are more likely to form true beliefs and avoid false beliefs if they believe what the expert tells them instead of trying to work out for themselves what to believe without the help of others (Zagzebski 2012). In this vein, epistemic authority is compatible with autonomy, but epistemic self-reliance is incoherent. It must be acknowledged that others are more reliable than we are if we want to rationally achieve certain epistemic goals. But, as seen in the previous sections, certain epistemic purposes can be achieved by relying on the epistemic authority of others, while having secondary effects of particular significance that impact on moral values and principles.

One may therefore consider supplementing the figure of the "noviced-oriented" expert, i.e., aimed at those who are not experts, with an approach that takes into account the virtues required to be able to respond appropriately to epistemic dependence on them (Croce 2018). While it is rational, as Zagzebski argues, to rely on the epistemic authority of those who are most likely to give us true beliefs, an ideal situation to aim for is one in which the expert is a figure capable of combining epistemic authority with ethical sensitivity, represented by certain specific virtues (Croce 2019). This is to avoid the consequences highlighted, for example, in the field of biomedical expertise.

In light of what have described so far, we offer a definition of what we call "Experts-in-Action;" such that: *experts-in-action are individuals with above average knowledge, achieved in the most objective and repeatable way possible, in a specific field, who use their skills and methods to translate this knowledge into decisions, actions, or suggestions for decisions (or actions) that value a community larger than that of the experts themselves. Such experts should exhibit epistemic virtues and of moral inclinations that allow them to consider the community's shared axiological values and normative principles responsibly and even critically, so that their knowledge can be effectively applied by combining scientific objectivity (or neutrality) with responsibilist virtues.*

3. The Role of Experts-in-Action

The previous definition highlights that experts are, by necessity, people with knowledge and skills, who apply such abilities and attitudes to address and resolve problems for the benefit of the community that engages their participation. Therefore, in practice, experts bring their professional experience and acumen to a shared enterprise of informing and making decisions as relevant to the community involved. Hence, it becomes important to (1) qualify the act of profession as the nexus for attributes that are to be exercised in such constituent activities; (2) specify the endeavor in which the act(s) of profession are to be engaged, and from these, (3) propose those virtues that are fundamental to the apt (and propitious) conduct of profession in the endeavor (O'Neill 2007).

Consider next two examples, which we will discuss in turn to illustrate our proposal a. access to care for disabled people, and b. the construction of a bridge in a highly protected landscape area.

In the former case, the expert in the allocation of scarce medical resources can refer to well-established protocols in the welfarist perspective based on which to maximize a criterion that seems suitable for the situation, for example the life expectancy of patients measured in healthy years (Palacios-González et al. 2022). This may, however, collide with the protection of the disabled or of individuals who have been disadvantaged since childhood and who therefore have a reduced life expectancy without being responsible for it. The expert-in-action should therefore display the ability to broaden their perspective and consider the fate of different people affected by the criteria adopted. The expert-in-action should also need to listen to the stakeholders involved in the choices to be made and the broader cultural and value context in which these choices are made. In this task, the virtues we will list next can help, giving them a sensitivity that is not a bias to their efficiency and technical neutrality, but an enrichment of their ability to intervene effectively and inclusively. Consequently, the expert-in-action could balance the welfarist criterion with certain priorities granted to specific disadvantaged groups or by partially modifying the criterion so that it does not penalize, for example, the disabled.

If an expert is called upon to design the construction of a bridge, she should not only summon the technical expertise to build the bridge so that it is safe, efficient, and not too expensive. Aspects of environmental impact, aesthetics, and how the bridge will be used by the community should also be taken into account. In this context, a negative example could be considered as an instructive failure. The well-known and praised urban planner Robert Moses (Caro 2015), who realized many important public infrastructural works in New York at the beginning of the 20th century "made sure bridges on the parkways connecting New York City to beaches on Long Island were low enough to keep city buses—which would likely be carrying poor minorities—from passing underneath."[4] The motivation would have been his adherence, beyond his indisputable technical expertise, to discriminatory views toward minorities or his desire (perhaps directed by the municipal government of the time) to make those areas more or less exclusively for the benefit of wealthier white citizens.

An expert-in-action called upon to make important decision for her community should therefore consider, in addition to technical and economic considerations, the needs of all potential users, the shared values of the community (e.g., the degree of commitment to environmental sustainability) as well as the impact and costs of her decisions (Valdes-Vasquez and Klotz 2013). In doing this, having cultivated the virtues that we shall detail next can be of significant help.

But here, we feel that some clarifications are in order concerning what functions an expert-in-action should exercise in the policy-making process. It is important to ensure that experts' contributions are not distorted by a lack of consideration of the context in which the experts exercise their epistemic authority. In other words, experts should be cognizant of, and vested to the values, norms, and even preconceptions of the community that they serve.

Yet, we don't want to consider as experts only those who possess certain epistemic and ethical values. All of us indeed need neutral information, for instance about the costs and effects of different policies. However, an allegedly neutral form of scientific reductionism (the idea that there cannot be any relevant truth beyond science), which is often at the roots of many decisions made by experts (as in the COVID-19 pandemic we discussed previously), runs the risk of becoming a metaphysical framework that can trap and negatively influence the formulation and implementation of specific public policies.

In addition, we argue that experts-in-actions are not supposed to replace subject matter experts, nor are they supposed to supplement the advice given by subject matters experts. In a sense, experts-in-action are subject matter experts with a capacity to enter the policy-making process more effectively and responsibly. They are in principle no different from subject matter experts except for the fact that they are fully endowed with a set of epistemic and responsibilist virtues that allow them to apply their knowledge according to a procedural and substantive style that takes fully into account shared values and principles.

This does not mean, however, that the expert-in-actions should unthinkingly endorse the values and principles of a society that systematically discriminated against specific groups (e.g., for racial motives). Experts-in-actions should not

---

[4] A. Valentine (2020) "'The wrong complexion for protection.' How race shaped America's roadways and cities." *Npr*, available at https://www.npr.org/2020/07/05/887386869/how-transportation-racism-shaped-america. This reconstruction, although included in an accurate Pulitzer Prize-winning book, has been challenged by other interpretations of the reasons for the low clearance of the bridges, cf. https://www.washingtonpost.com/politics/2021/11/10/robert-moses-saga-racist-parkway-bridges/.

merely endorse a society's values and principles. On the contrary, they should exercise their capacity to critically reflect on those values and principles.

Furthermore, experts-in-action are experts in their specific subject matter who, having developed a desirable set of skills and virtues (more on which will follow), are better able to contribute to the decision-making process than a subject matter expert who does not have these virtues in their fullest form. In this sense, the qualification of subject matter expert-in-action may be a matter of degree and we acknowledge that it could turn to be hard to discriminate between 'regular' subject matter experts and subject matter experts-in-action.

Indeed, it is implausible to argue that a subject matter expert called upon to participate in the policy making process has none of the virtues we consider in this contribution, or only some of them to a minimum degree. As it emerges from the pandemic-related examples discussed previously, these virtues are relevant to the policy making process and make a subject matter expert an expert-in-action. However, this does not mean that the personal quality of the subject matter expert is necessarily different. Our is not an exercise at ranking virtues of individuals, but rather an attempt at describing their ability toward the construction of fairer and more effective experts when it comes to participating in decisions that have a relevant public impact.

This does not mean endorsing the previously cited "novice-oriented" approach to expertise, which has been criticized in favor of the "research-oriented" one (Croce 2018). However, it does represent an attempt to identify an ideal figure of an expert who is epistemically superior to most of the individuals belonging to the epistemic community she addresses, maintaining Goldman's functionalist and truthful criteria, but is also the bearer of those virtues that make her a subject attentive to the values and moral principles at stake in her decision-making.

This takes us to the issue of ethical expertise. Being an expert-in-action—we believe—means having and applying certain epistemic and responsibilist virtues, but it does not require one to be an expert in ethics. Indeed, there is a tradition in philosophy of doubting that there can be such expertise (Burch 1974; Archard 2011; generally, it is argued that morality cannot be taught, as Socrates claimed, or that morality is giving oneself a rule, as Kant proposed). Experts-in-action need not be moral experts in the full sense, as moral philosophers can be, but should ideally cover both sides of moral expertise (Gesang 2010). On the one hand, they should have moral authority because they are able to reason well in the moral field. This means that they form correct moral judgments more often than most people do or analyze moral situations in all their aspects by framing them more appropriately. On the other hand, they should also be capable of exemplary moral behavior, at least for the domains we have outlined so far. They do not, of course, need to be moral examples for individual conduct, but they are called upon to act well with respect to the collective choices to which their authority and epistemic expertise apply.

There is no need here to assume the strong thesis put forward by Zagzebski that we should rely on the moral authority of the expert based on our conscientious judgment that we are more likely to form a true belief in moral matters and avoid a false belief if we rely on moral authority instead of resorting to our own tentative understanding of what to believe. This is because the ethical preferences of the epistemic community addressed by the expert-in-action should be taken by them as a starting point, albeit potentially also in a critical sense. Yet, it is to be stressed that specialists in ethics who are consequentialists often appear to offer very different moral guidance from ones who are deontologists, and this advice appears different from the ethical advice of specialists in ethics who are virtue ethicists.

The expert-in-action therefore does not need to embrace a specific metaethics. Possessing and cultivating certain epistemic and responsibilist virtues implies, on our view, not a substantive but rather a procedural approach. Procedural, however, does not mean consequentialist. The expert-in-action should have the ability to recognize the situation in which ethical issues arise and take them into account in addition to the technical and objective aspect of their expertise. The expert-in-action should, through their virtues, flag the ethical matters and contribute to addressing them, without having to be an expert in ethics. Moral philosophers, political philosophers, bioethicists, and neuroethicists may be involved in the decision-making process as experts in addressing ethical issues, but we do not think that moral philosophers should be distinguished into experts and experts-in-action. Our proposal applies mainly to subject matter experts from outside the academic disciplines related to ethics.

Ethical issues as well as values and principles to be taken into account by the expert-in-action may entail specialized knowledge that a subject matter expert cannot add to their own, except at the price of a deterioration of their own technical subject matter, be it medicine, economics, or engineering. The expert-in-action should therefore not sacrifice the constant deepening and updating of their specialized knowledge, but only add to it some personal dispositions and attitudes and non-time-consuming procedural skills. Acquiring the virtues we propose thus may entail an effort that is not uniquely cognitive but also affective, and this may imply a kind of personal discipline.

However, as Singer already pointed out, moral philosophers or ethicists in general benefit from their specific training that makes them more competent in argumentation and the detection of invalid inferences. Moreover, their specific experience in moral philosophy gives them a better understanding of moral concepts and the logic of moral arguments. Lastly, scholars in the discipline, if they wish, can think about moral issues full-time, while everyone else needs also deal with other types of issues (Singer 1972).

Becoming an expert-in-action therefore also requires a certain amount of training, but this is of a different kind than that required to become a subject matter expert. Furthermore, there can be no objective certification of the expert-in-action as there is for a subject matter expert in terms of education, academic affiliation, publications, funding obtained, and prizes won. The subject matter expert-in-action is an ideal to which all subject matter experts who wish to contribute effectively to the policy making process should strive, through the incorporation of the epistemic and responsibilist virtues that seem most relevant for this purpose.

Lastly, one might object that there are other and already in-use ways to achieve what experts-in-action are purported to be able to do. For example, a bioethicist or a policy expert can flank the subject matter experts in their decisions. This approach is now standard practice in all major research projects, is certainly not against our proposed argument, and could be surely pursued and further expanded in line with our proposal.

An additional pragmatical requirement could be to demand representatives of the groups affected and potentially harmed by the decision in question to be deciding alongside the experts. This has been recently suggested by Kitcher (2021) (and by us in other works, Farina and Lavazza 2021b).

Both proposals might be deemed as interdisciplinary because they suggest that experts of diverse subject matters (i.e., philosophers, social scientists, and experts by experience) should be directly involved (Lavazza and Farina 2023). We think that these solutions are desirable if compared to the action of a single expert or homogenous group of experts (O'Neill 2012). However, the difficulty of bringing together experts from very different backgrounds or experts and stakeholders in dialog can sometimes lead to suboptimal outcomes.

Therefore, the expert-in-action who has exercised the relevant virtues and has appropriate skills to consider the different interests at stake may have an advantage in practical terms and achieve better results. Naturally, this does not detract from the fact that the involvement of communities or experts of different backgrounds and expertise can also bring benefits when an expert-in-action is involved.

4. Key Virtues of Experts-in-Action

Several virtues can be considered to transform a subject matter expert into an expert-in-action. The virtues are both epistemic and moral (responsibilist). This is required by the fact that from the epistemic point of view, the authority of the expert can be supplemented by a series of skills, attitudes and sensitivities that are not uniquely provided by the technical knowledge of the subject matter expert alone. These are virtues that have been identified and described as being particularly valued in themselves and functional for a better relationship with one's fellow citizens when at stake is the acquisition, sharing and social and political use of knowledge.

From a moral point of view, the virtues proposed in a general sense are not alien to the training of an individual who has gone through all degrees of higher education and is constantly involved in interactions with their peers. What is recommended, however, to the expert-in-action is that they possess them to a high degree and for the specific purpose for which they are suggested. In other words, these responsibilist virtues should be integrated into a coherent and homogeneous whole that gives the expert the aforementioned ability to be an effective participant in the policy making process, combining objective knowledge of their subject matter with an ethically sensitive approach thus resulting in an improved experts' epistemic authority (Anderson and Giordano 2013; Giordano 2012).

Virtue, writ-large, and the various virtues are discussed in abundant literature from the ancients to modernity. Building upon Aristotle, Aquinas (1984) has described the application of intellectual and moral virtues to activity as a right measure of knowledge, insight, and behavior. Pellegrino (1995) expands upon Aristotelian and Thomistic notions of balance to posit that the right balance of knowledge and insight are crucial to the right measure and conduct of action within a given enterprise or practice. This notion of virtue(s)-in-practice is undergirded by MacIntyre's (1988) description of a practice as a "...coherent and complex...socially established cooperative human activity through which good(s) internal to that...activity are realized in...trying to achieve those standards of excellence which are appropriate to, and partially definitive of...that activity...with the result that...conceptions of the ends and goods involved are systematically extended." This definition reinforces that to be authentically practical, ethics must be focal to the endeavor in which it is applied.

To be sure, numerous virtues can be considered valuable in the exercise of expertise, and a comprehensive listing is not intended. Instead, we offer what we believe are those virtues that are fundamental to expert advisement of science, regardless of whether its perspective is technical, ethical, or legal.

4.1 Benevolence

Engaging epistemological capital in domains of human activity should be oriented and strive toward those goods defined by the community that the expert serves (Livnat 2004; King 2005). Defining the relevant stake—and shareholder needs and values establishes the goals and ends of the enterprise in which expertise is sought. This aligns the duties of the expert with those of endeavor, and in so doing, depicts the rule and act utility of the conduct and outcomes of expert agency. The appreciation for, understanding of, and striving toward these goods are therefore essential to authentic expertise.[5]

---

[5] However, it is also critical to define what good[s], and who has decided upon such goods, as good[s] can and often is relative, and what is good for some may be burdensome or harmful to others. Appreciating both the "locality" and the "generality" of good[s] are therefore important in the exercise of expertise. For an overview, see MacIntyre (2007; 1999).

4.2 Fidelity to Trust

In these ways, benevolence (i.e., the desire to articulate defined good[s]) and beneficence (i.e., dedicated actions toward effecting such good[s] in practice) undergirds the trust that a community places in the expert. Given that the expert-in-action is often required to balance relative benefits, burdens, help and harms, the invitation to trust—and fidelity to such trust—obligates insight, judgment, and a viable extent of transparency and communication about benevolent aims, and possible risk(s) (Goold 2001; Brien 1998). Explicit to such trust is the primacy of the best interest(s) of the community served by the expert(s). Thus, the invitation to trust also demands pragmatic non-malevolence (a rational intent to minimize if not prevent any, and all avoidable harms).

4.3 Veracity

Yet, when apprehending the proximate, intermediate, and long-term outcomes of any expert decision or action, there is always the possibility for burden, risk, and harm, either idiosyncratically or systemically. Establishing and sustaining trust requires truthful explication of these benefits, risks, the process(es) utilized to predict and prepare for their occurrence, and the methods and outcomes that could affect the community served. However, apropos Delphic invocation, the expert-in-action must also recognize and acknowledge the limits and limitations both of what is known, and their own capabilities (Williamson 2002). Thus, true veracity obligates the explication of such limits.

4.4 Epistemic Humility

Ergo, veracity demands epistemic/intellectual humility, which can be defined as the "disposition not to adopt epistemically improper higher order epistemic attitudes, and to adopt (in the right way, in the right situations) epistemically proper higher order epistemic attitudes" (Hazlett 2012). As Roberts and Wood (2007) have noted, intellectual humility has both a personal and social character or dimension. At the level of the single cognitive agent, epistemic humility can be a motivating trait or factor that may lead to re-evaluating one's own beliefs in the light of available evidence. So, it can be argued that intellectual humility can allow experts to become better familiarized with current and emerging information (such as findings on mechanisms involved in individual and group cognitive [and ethical] processes; cf. Giordano and Shook 2018: Giordano et al. 2016; Samuelson and Church 2015) Situating this information within a bio-psychosocial paradigm could afford heuristic value when applying expert knowledge to complex, socio-cultural contexts (Giordano 2012; Giordano and Olds 2010; Shook and Giordano 2017). Furthermore, intellectual humility compels experts to seek further perspectives and confirmations of their ideas/proposals. Recently, Christen, Alfano, and Robinson (2014) characterized intellectual humility as:

i. being "sensitive;" that is, open to new ideas and ready to responsibly embrace them;
ii. being "inquisitive;" willing to seek new views and hypothesis driven by curiosity and exploration skills;
iii. being non-pretentiously open to and objectively aware of other cognitive agents' capabilities and limitations, especially in cases wherein there is disagreement.

This third attribute is particularly important as it bridges intellectual humility at a personal level to that which is typically manifested on the social scale. The intellectually humble expert-in-action should realistically consider their own status in the context of the community in which they are solicited to provide professional contribution. This necessitates ongoing self-assessment and self-acknowledgement, as well as recognition of others' capabilities, limitations, and values.

4.5 Pathos

The articulation of the "expert-in-action" occurs through engagement and application of profession toward the condition and/or predicament of others for whom such expertise has been sought and recruited. Fundamental to this enterprise is apprehension and appreciation of the situation of these others. By definition, this is pathos—feelings for others (Funk and Wagnalls Dictionary 1967, 988). At very least, pathos would be required for the expert to acquire both a sense of, and sensibility for the (gravitas and effects of) need(s) of the others for which their expertise is solicited. Pathos is cognate to empathy, which encompasses a diverse array of psychological capacities that are considered integral to the constitution of human beings as social entities. These capacities enable individuals to gain insights into the thoughts and emotions of others, to establish emotional connections, to partake in the exchange of thoughts and emotions, and to exhibit concern for the welfare of others. However, as is well known, empathy is a concept that has recently provoked even negative analyses (Stueber 2019).

4.6 Fortitude

Frank acknowledgement of one's abilities, limitations, and responsibilities when seeking to cooperatively work with others in circumstances of crisis, exigency, and/or uncertainty requires fortitude. Many have opined that fortitude can serve to militate attempts to streamline disagreement, and/or pressures toward a single epistemic point of view—inclusive of one's own (Alfano 2013; Fricker 2003, 2007; Medina 2013). In concert with intellectual humility (as well as benevolence, veracity, and fidelity to trust) fortitude enables the expert to both accept the burden of responsibility inherent to expertise, and to admit to one's incapability, inadequacy, and failure. Fortitude is not fearlessness. On the contrary, it is the ability to "carry on" with the task at hand, in those ways that are consistent with the communally defined good, despite tenuous and often fearful conditions and contingencies.

4.7 Practical Wisdom

Perhaps the nexus of the aforementioned virtues occurs in developing wisdom in and for practical applications. Aristotle defined such practical wisdom (i.e., *phronesis*) as "concerned with human affairs, namely, with what we can deliberate about. For deliberating well, we say, is the characteristic activity of the practically wise person above all.... The person unqualifiedly good at deliberation is the one who tends to aim, in accordance with his calculation, at the best of the goods for a human being that are achievable in action. Nor is practical wisdom concerned only with universals. An understanding of particulars is also required, since it is practical, and action is concerned with particulars" (Aristotle 2014, 1141b). Similarly, MacIntyre (1981) stressed importance of practical experience and derived wisdom to achieve an effective knowledge of human beings and their affairs (cf. O'Neill 2007). More recently, Hacker-Wright (2015) posited that "practical wisdom must take into account very general facts about human beings; these facts shape what counts as good practical reflection, not because human nature is intrinsically normative, but because it is part of the inevitable background against which we understand ourselves."

This *phronetic* approach has been applied to social science (Flyvbjerg et al. 2012) economics (Rindermann 2018), and medicine (Pellegrino and Thomasma 1993; Davis 1997; Kristjánsson 2015; Waring 2000) *Phronesis*—understood as the ability to make the experience of human life and human beings one's own and to understand which elements of situation are the most important from a moral viewpoint, is what allows the respectful and flexible address of the specificities of the endeavor entailed within the particularities of its ecology (Hursthouse and Pettigrove 2018).[6]

---

[6] For a more detailed discussion of ethics-as-human-ecology, see Flanagan(1996).

5. Conclusion: Toward a Synthetic Approach

There has been a debate, about the contrast, and to some extent the possibilities for conflict between duty, utility, and virtue (Drašček et al. 2021). To be sure, duties can conflict, as can aspects of utility (e.g., ends versus means), and this can create vagaries in what virtues are necessary, and the duties and ends toward which such virtues should be applied. Herein we propose a structural and functional approach that is grounded in, defined by, aligned with, and focal to—and in these ways synthetic toward—the ends (i.e., *telos*) of the endeavor. Thus, duty and rule utility provide structure, while act utility and virtue enable functional engagement of expertise in the ecology and context of the enterprise.

Duties and rules can be defined and prescribed (O'Neill 1993); and if virtue is a "final common path" for agents' execution of duty, rule, and acts of defined utility, it may be queried whether and how such virtue can be developed. Certainly, expertise can be acquired through experience and erudition, and we opine that so too can virtue. We offer that these two elements (of experience and education) are not necessarily mutually exclusive, nor necessarily temporally related. Indeed, education can precede experience just as experience can precede education; however, both are necessary.

In this light, we have provided a framework that defines what we propose to be necessary characteristics of the expert-in-action, and an education and training paradigm (mastery rubric) for developing expertise—whether technical, legal, or ethical (Tractenberg et al. 2015; Tractenberg 2013).

Thus, expertise-in-action is a dynamical engagement of duty, rule, and act utility, and intellectual and ethical virtue(s) as decided upon by the community of reference that has solicited the experts, as focal to the ecology, exigencies, and contingencies of the endeavor within and/or by that community. We have asserted, and reiterate more here, that this expertise—and the virtue(s) important for its technical and ethical execution—can and should be inculcated and evaluated as crucial to the sound implementation of informing, guiding, and establishing goods as relevant to a community in need (Anderson and Giordano 2013; Giordano 2013).

In conclusion, we have proposed, and presented those ways, that experts-in-action are vital to inform, advise, and execute decisions about the ways that communities, on a variety of scales, employ tools and techniques to affect the human predicament and condition. We thus claim that the concept of experts-in-action will enable ever more capable engagement of the duties, utilities and agentic responsibilities entailed by particular efforts of human activity in—and across—a variety of scales of execution and effect. It is our hope that our thesis may serve as a prompt not merely for further discourse, but for collaborative progress to establish programs for educating, training, and providing such experts-in-action that will be better prepared to meet the contingencies and exigencies of the 21st century global stage.

References

Alfano, M. (2013) "Identifying and defending the hard core of virtue ethics." *Journal of Philosophical Research* 38, 233–60.

Anderson, M. A. and Giordano J. (2013) "Aequilibrium prudentis: On the necessity for ethics and policy studies in the scientific and technological education of medical professionals." *BMC Medical Education* 19(4).

Andrews, E. E., Ayers, K. B., Brown, K. S., Dunn, D. S., et al. (2020) "No body is expendable: Medical rationing and disability justice during the COVID-19 pandemic." *American Psychologist*, https://doi.org/10.1037/amp0000709.

Aquinas, T. (1984) *A Treatise on the Virtues*, transl. J. A. Oesterle. Notre Dame, IN: University of Notre Dame Press.

Archard, D. (2011) "Why moral philosophers are not and should not be moral experts." *Bioethics* 25(3), 119–27.

Aristotle (2014) *Nicomachean Ethics*, R. Crisp (ed.). Cambridge, UK: Cambridge University Press, 1141b.

Bledsoe, T. A., Jokela, J. A., Deep, N. N., and Sulmasy, S. L. (2020) "Universal do-not-resuscitate orders, social worth, and life-years: Opposing discriminatory approaches to the allocation of resources during the COVID-19 pandemic and other health system catastrophes." *Annals of Internal Medicine*, https://doi.org/10.7326/M20-1862.

Brien, A. (1998) "Professional ethics and the culture of trust." *Journal of Business Ethics* 17(4), 391–409.

Burch, R. W. (1974) "Are there moral experts?." *The Monist* 646–58.

Cairney, P. and Wellstead, A. (2021) "COVID-19: Effective policymaking depends on trust in experts, politicians, and the public." *Policy Design and Practice* 4(1), 1–14.

Caro, R. A. (2015) *The Power Broker: Robert Moses and the Fall of New York*. New York: Random House.

Cf. Valdes-Vasquez, R. and Klotz, L. E. (2013) "Social sustainability considerations during planning and design: Framework of processes for construction projects." *Journal of Construction Engineering and Management* 139(1), 80–9.

Chen, B. and McNamara, D. M. (2020) "Disability discrimination, medical rationing and COVID-19." *Asian Bioethics Review* 12(4), 511–18.

Christen, M., Alfano, M., and Robinson, B. (2014) "The semantic space of intellectual humility," in *Proceedings of the European Conference on Social Intelligence*, (eds.), Herzig, A. and Lorini, E., 40–9.

Croce, M. (2018) "Expert-oriented abilities vs. novice-oriented abilities: An alternative account of epistemic authority." *Episteme* 15(4), 476–98.

Croce, M. (2019) "On what it takes to be an expert." *The Philosophical Quarterly* 69 (274), 1–21.

Dente, M. G., Riccardo, F., Declich, S., Milano, A., Robbiati, et al. (2022) "Strengthening preparedness against global health threats: A paradigm shift based on one health approaches." *One Health*, https://doi.org/10.1016/j.onehlt.2022.100396.

Draščhek, M., Rejc Buhovac, A., and Mesner Andolšek, D. (2021) "Moral pragmatism as a bridge between duty, utility, and virtue in managers' ethical decision-making." *Journal of Business Ethics* 172, 803–19.

Farina, M. and Lavazza, A. (2020) "Lessons from Italy's and Sweden's policies in fighting COVID-19: The contribution of biomedical and social competences." *Frontiers in Public Health* 8, 563397.

Farina, M. and Lavazza, A. (2021a) "The meaning of freedom after Covid-19." *History and Philosophy of the Life Sciences* 43, 3.

Farina, M. and Lavazza, A. (2021b) "Advocating for greater inclusion of marginalized and forgotten populations in COVID19 vaccine rollouts." *International Journal of Public Health*, 66, 1604036.

Flanagan, O. (1996) "Ethics naturalized: Ethics as human ecology," in *Mind and Morals: Essays on Cognitive Science and Ethics*, (eds.), L. May, M. Friedman, and A. Clark, 19–43. Cambridge, MA: MIT Press.

Flyvbjerg, B., Landman T., and Schram, S. (2012) *Real Social Science: Applied Phronesis*, Cambridge, UK: Cambridge University Press.

Fricker, M. (2003) "Epistemic injustice and a role for virtue in the politics of knowing." *Metaphilosophy* 34(1–2), 154–73.

Fricker, M. (2007) *Epistemic Injustice: Power and the Ethics of Knowing*, New York: Oxford University Press.

Funk & Wagnalls *Standard College Dictionary* (1967) Pleasantville, NY: Readers' Digest Publishing.

Gesang, B. (2010) "Are moral philosophers moral experts?." *Bioethics* 24(4), 153–9.

Giordano, J. (2012) "Keeping science and technology education In-STEP with the realities of the world stage: Inculcating responsibility for the power of STEM." *Synesis: A Journal of Science, Technology, Ethics and Policy* 3(1), G1–5.

Giordano, J. and Olds, J. (2010) "On the interfluence of neuroscience, neuroethics, and legal and social issues: the need for (N)ELSI." *AJOB-Neuroscience* 2(2), 13–15.

Giordano, J. and Shook, J. (2018) "Neuroethics: What it is, does—and should do." *Health Care Ethics-USA* 9(2): 15–19.

Giordano, J., Becker, K., and Shook J. (2016) "On the 'neuroscience of ethics'—Approaching the neuroethical literature as a rational discourse on putative neural processes of moral cognition and behavior." *Journal of Neurology and Neuromedicine* 1(6), 32–6.

Giordano, J., Lanzilao, E., Shook, J., and Benedikter, R. (2015) "Guidare la neuroscienza e lo sviluppo della persona nel XXI secolo: Una prospettiva naturalistica e cosmopolita per la neuroetica." *L'Arco di Giano* 80, 147–64.

Goldman, A. I. (1978) "Epistemics: The regulative theory of cognition." *The Journal of Philosophy* 75(10), 509–23.

Goldman, A. I. (2018) "Expertise." *Topoi* 37(1), 3–10.

Goold, S. D. (2001) "Trust and the ethics of health care institutions." *Hastings Center Report* 31(6), 26–33.

Hacker-Wright, J. (2015) "Skill, practical wisdom, and ethical naturalism." *Ethical Theory and Moral Practice* 18, 983–93.

Hazlett, A. (2012) "Higher-order epistemic attitudes and intellectual humility." *Episteme* 9(3), 205–23.

Hursthouse, R. and Pettigrove, G. (2018) "Virtue Ethics," in *The Stanford Encyclopedia of Philosophy*, E. N. Zalta (ed.), https://plato.stanford.edu/archives/win2018/entries/ethics-virtue/.

Karimov, A., Lavazza, A., and Farina, M. (2022) "Epistemic responsibility, rights, and duties during the Covid-19 pandemic." *Social Epistemology*, doi: 10.1080/02691728.2022.2077856.

King, S. B. (2005) *Being Benevolence*. Honolulu, Ha: University of Hawaii Press.

Kitcher, P. (2021) *Moral Progress*. New York: Oxford, UK: Oxford University Press.

Kristjánsson, K. (2015) "Phronesis as an ideal in professional medical ethics: Some preliminary positionings and problematics." *Theoretical Medicine and Bioethics* 36, 299–320.

Landemore, H. (2020) *Open Democracy*. Princeton, NJ: Princeton University Press.

Lanzilao, E., Shook, J., Benedikter, R., and Giordano J. (2013) "Advancing neuroscience on the 21st century world stage: The need for—and proposed structure of—an internationally relevant neuroethics." *Ethics in Biology, Engineering and Medicine*, 4(3), 211–29.

Lavazza, A. and Farina, M. (2020) "The role of experts in the Covid-19 pandemic and the limits of their epistemic authority in democracy." *Frontiers in Public Health* 8, 356.

Lavazza, A. and Farina, M. (2021) "Experts, naturalism, and democracy." *Journal for the Theory of Social Behaviour* 52(2), 279–297.

Lavazza, A. and Farina, M. (2023) "Leveraging autonomous weapon systems: Realism and humanitarianism in modern warfare." *Technology in Society* 102322.

Livnat, Y. (2004) "On the nature of benevolence." *Journal of Social Philosophy* 35(2), 304–17.

MacIntyre, A. (1999) *Dependent Rational Animals: Why Human Beings Need the Virtues*. Chicago: Open Court.

MacIntyre, A. (2007) *After Virtue*, (3rd edn.) Notre Dame, IN: University of Notre Dame Press.

McIntyre, A. (1988) *Whose Justice? Which Rationality?*. Notre Dame, IN: University of Notre Dame Press.

Medina, J. (2013) *The Epistemology of Resistance: Gender and Racial Oppression, Epistemic Injustice, and Resistant Imaginations*. New York, NYC: Oxford University Press.

Nichols, T. (2017) *The Death of Expertise: The Campaign against Established Knowledge and Why it Matters*. New York, NY: Oxford University Press.

O'Neill, O. (1993) "Duties and Virtues." *Royal Institute of Philosophy Supplement* 35, 107–20.

O'Neill, O. (2007) "Experts, practitioners, and practical judgement." *Journal of Moral Philosophy* 4(2), 154–66.

O'Neill, O. (2012) "Global poverty and the limits of academic expertise." *Ethics & International Affairs* 26(2), 183–9.

Palacios-González, C., Pugh, J., Wilkinson, D., and Savulescu, J. (2022) "Ethical heuristics for pandemic allocation of ventilators across hospitals." *Developing World Bioethics* 22(1), 34–43.

Pellegrino, E. D. (1995) "Interests, obligations, and justice: Some notes toward an ethic of managed care." *The Journal of Clinical Ethics* 6(4), 312–17.

Pellegrino, E. D. and Thomasma, D. C. (1993) The Virtues in Medical Practice. NY: Oxford University Press; Davis, F. D. (1997) "Phronesis, clinical reasoning, and Pellegrino's philosophy of medicine." *Theoretical Medicine and Bioethics* 18, 173–95.

Pietrini, P., Lavazza, A., and Farina, M. (2022) "Covid-19 and biomedical experts: When epistemic authority is (probably) not enough." *Journal of Bioethical Inquiry* 19(1), 135–142.

Piscitello, G. M., Kapania, E. M., Miller, W. D., Rojas, J. C., et al. (2020) "Variation in ventilator allocation guidelines by US state during the coronavirus disease 2019 pandemic: A systematic review." *JAMA Network Open* 3(6), e2012606–e2012606.

Rawls, J. (1999) *A Theory of Justice*, (revised edn.) Cambridge, MA: Harvard University Press.

Rindermann, H. (2018) *Cognitive Capitalism: Human Capital and the Wellbeing of Nations*. Cambridge, UK: Cambridge University Press.

Roberts, R. C., Wood, W. J. (2007) *Intellectual Virtues: An Essay in Regulative Epistemology*. New York, NYC: Oxford University Press.

Samuelson, P. L. and Church, I. M. (2015) "When cognition turns vicious: Heuristics and biases in light of virtue epistemology." *Philosophical Psychology* 28(8), 1095–113.

Sandel, M. J. (2020) *The Tyranny of Merit: What's Become of the Common Good?*. New York, NYC: Farrar, Straus and Giroux.

Shook, J. and Giordano, J. (2014) "A principled, cosmopolitan neuroethics: Considerations for international relevance." *Philosophy, Ethics, and Humanities in Medicine* 9, 1–13.

Shook, J. and Giordano, J. (2017) "Neuroethical engagement on interdisciplinary and international scales," in *Debates About Neuroethics*, (eds.), E. Racine and J. Aspler, 225–45. Cham: Springer.

Shook, J. and Giordano, J. (2020) "Toward a new neuroethics in a multipolar and multicultural world." *Global-E*, 13(56), 2–10.

Singer, P. (1972) "Moral experts." *Analysis* 32(4), 115–17.

Stueber, K. (2019) "Empathy," in *The Stanford Encyclopedia of Philosophy*, (ed.) Edward N. Zalta, available at https://plato.stanford.edu/archives/fall2019/entries/empathy/.

Thagard, P. (2010) *The Brain and the Meaning of Life*. Princeton, NJ: Princeton University Press.

Tractenberg, R. E. (2013) "Ethical reasoning for quantitative scientists: A mastery rubric for developmental trajectories, professional identity, and portfolios that document both." *Proceeding of the Joint Statistical Meetings*. Montreal: QC.

Tractenberg, R. E., FitzGerald, K. T., and Giordano, J. (2015) "Engaging neuroethical issues generated by the use of neurotechnology in national security and defense: Toward process, methods, and paradigm," in *Neurotechnology in National Security and Defense: Practical Considerations, Neuroethical Concerns*, (ed.), Giordano, J., 259–77. Boca Raton: CRC Press.

Waring, D. (2000) "Why the practice of medicine is not a phronetic activity." *Theoretical Medicine and Bioethics*, 21, 139–51.

Williamson, T. (2002) *Knowledge and its Limits*. New York, NYC: Oxford University Press.

Zagzebski, L. T. (2012) *Epistemic Authority: A theory of Trust, Authority, and Autonomy in Belief*. New York: Oxford University Press.

PART 6
EXPERTISE ABOUT VALUE

12
Experts in Aesthetic Value Practices

Dominic McIver Lopes
University of British Columbia, Canada

Call them 'arbiters of taste', 'true judges', or 'influencers': there appear to be aesthetic experts. In particular, there appear to be experts in aesthetic value—in the measure of aesthetic goodness of at least some items. We routinely take advice on music, movies, and novels, wine and restaurants, travel destinations, home redecoration, and much else. From this it does not immediately follow that there are genuine experts in aesthetic value. We also take advice from stockbrokers and university admissions officers, who turn out not to be genuine experts (Shanteau 1992). This chapter defends a new theory stating what it is to be an aesthetic expert, such that there are aesthetic experts. On existing accounts, aesthetic experts are either guides to value or those whose responses constitute valuing practices. On the new theory, an aesthetic expert is someone whose performance determines which aesthetic practice is the one to which a group of people belongs.

1. Aesthetic Value

No case can be mounted for the existence of genuine experts in aesthetic value without a nod to some characteristics of aesthetic value. Philosophical work on aesthetic value enjoys a consensus on a number of points, five of which are relevant here.

First, aesthetic values are such features of items as their being elegant, glorious, and richly textured (Sibley 1959; Zangwill 1995; De Clercq 2002; Lopes 2018, ch. 2). Euler's identity is elegant, Chatterbox Falls is glorious during the spring melt, and Brìghde Chaimbeul's playing is richly textured. Nothing is aesthetically good (equivalently, beautiful) unless it is aesthetically good (that is, beautiful) in some specific way. Being elegant stands to being beautiful in the relation of determinate to determinable.

Second, an item always has a given aesthetic value because it has various, other, non-aesthetic features. The 'because' is non-causal; it stands for metaphysical explanation or grounding (Sibley 1965; Benovsky 2012; Lopes 2018, ch. 10). What makes Euler's identity elegant is it how represents a relationship between fundamental elements of mathematics. Chaimbeul's playing is richly textured in virtue of its sonic qualities: she can preserve its rich texture only by ensuring it continues to have those qualities, and she can change its aesthetic value only by changing its sound.

Third, aesthetic values are not confined to art; they are to be found in nature, ideas, and intellectual structures, and designed artefacts too. Just about any kind of item can bear aesthetic value, and there is no kind such that what it is to be a member of that kind fixes what it is to be aesthetically good. In particular, aesthetic value is not identical to artistic value, the value that items have qua works of art. Euler's identity and Chatterbox Falls are not works of art. Some values that items have qua works of art are not aesthetic (Lopes 2014).

Fourth, as the second point of consensus implies, almost all philosophers are aesthetic realists; they accept that there are aesthetic value facts (cf. Todd 2004). It is a fact that Euler's identity is elegant, and it is another fact that Chaimbeul's playing is richly textured. Cutting across the consensus, there are debates about whether or not aesthetic value facts are response-constituted (for references see Lopes 2018, 182 and 196). Facts about us are facts, nonetheless.

Fifth, aesthetic value facts are normative aesthetic reasons. Normative aesthetic reasons are not aesthetic evaluations. Evaluations are states of mind that attribute values, and aesthetic evaluations attribute aesthetic values. Not all attributions of aesthetic value are beliefs: I believe that Euler's identity is elegant, but I hear the rich texture of Chaimbeul's playing, and I thrill to the falls' glory. When all goes well, aesthetic evaluations accurately represent normative aesthetic reasons. Normative aesthetic reasons merit or lend weight to aesthetic evaluations. The fact that Euler's identity is elegant lends weight to the proposition that I should judge it elegant, the fact that Chaimbeul's playing is richly textured merits my listening to it, and the fact that the falls are glorious lends weight to the proposition that they should thrill me. Failing to judge, hear, or feel these things is failing to respond as I have normative aesthetic reason to respond.

Normative aesthetic reasons are also practical reasons, facts that lend weight to what an agent should do. The fact that the playing is richly textured is reason for me to queue it up on my playlist. That is, it lends weight to the proposition that I should queue it up. Failing to do so is failing to do what I have normative aesthetic reason to do. By the same token, the fact that the falls are glorious is reason for me to visit them in the spring, and failing to do that is failing to what I have a normative aesthetic reason to do.

In sum, aesthetic values are determinates of beauty that are grounded in other features of items (not only art works) and that figure in normative aesthetic reasons. These five points of consensus set the stage for theories of aesthetic value. A theory of aesthetic value might address one or both of two questions.

One is a demarcation question. Being elegant and glorious are aesthetic values. The cleverness of a distinction, the courage of a journalist on the front line, and an umbrella's sturdiness are not aesthetic values. The demarcation question asks, what makes it the case that some values are specifically aesthetic?

Another question concerns normativity. Suppose we know what demarcates aesthetic values. Knowing that leaves it mysterious why aesthetic value facts are reasons that lend weight to what we should think or do. Why do the facts that Euler's identity is elegant, and the falls are glorious lend weight to its being the case that anyone should respond to them in certain ways?

AMIRA: You should check out *The Reeling*.
MACKENZIE: Why?
A: Chaimbeul's playing is richly textured.
M: I'll take your word for it. But why does it follow that should I check it out?
A: Anyone always has reason, albeit not always decisive, to listen to beautiful performances.
M: Really? I don't see that.

Amira might now roll out an answer to the normative question.

We have now caught a glimpse of aesthetic expertise. MacKenzie is prepared to take Amira's advice on the album's aesthetic value, and perhaps she is warranted in so doing because Amira is an expert in aesthetic evaluation. That is, perhaps Amira is an expert either in what aesthetic reasons there are or in what reasons are aesthetic.

2. The Autonomy Challenge

Getting more than a glimpse of aesthetic expertise is going to mean getting into theories of aesthetic value that answer either the demarcation question or the normative question. To see why, consider a challenge to aesthetic value expertise that evokes an ideal of aesthetic autonomy.

Some have acquired a capacity to perform exceptionally well in making aesthetic evaluations (e.g. Kozbelt 2020). They need not be aesthetic experts, however. Suppose that aesthetic experts are those to whom non-experts should defer, just because they have acquired capacities to perform exceptionally well in aesthetic evaluation. Non-experts should defer to experts only if expert testimony is both available and may be used. The challenge is that expert aesthetic testimony is either unavailable or it may not be used. Consequently, in order to make an aesthetic evaluation, each must rely on themselves and not on anyone else. Each must evaluate autonomously. So, granting that there are people with acquired capacities to perform exceptionally well in aesthetic evaluation, they are not experts, because it is not the case that non-experts should defer to them. Amira might perform exceptionally well in assessing the value of the kind of music that

Chaimbeul plays, but she is not an expert because MacKenzie should not defer to her. MacKenzie should not defer to her because she should rely on her own competence. The background thought is that an expert plays a role or provides a service, but nobody can or should play the role or provide the service of aesthetic expert.

The challenge has roots in Immanuel Kant's aesthetics (2000[1790]). Kant is often quoted as holding that an aesthetic evaluation is 'not to be grounded on collecting votes and asking around among other people about the sort of sensations they have', and 'the subject [must] judge for himself, without having to grope about by means of experience among the judgments of others', for 'To make the judgments of others into the determining ground of one's own would be heteronomy' (2000[1790], 5, 281–2).

Its putative autonomy is one among several features of aesthetic evaluation that Kant proposes to explain. Very briefly, aesthetic evaluations are (or a grounded on) experiences of the harmonious free play between two cognitive faculties, imagination, and understanding. When they are in harmonious free play, the faculties operate unconstrained by anything except their constitutive principles. Moreover, their getting into harmonious free play is facilitated by the adoption of three policies of what Kant calls 'common sense', namely, to think for oneself, to think in the shoes of others, and to think consistently. Both the policies of common sense and the principles constitutive of the faculties are self-legislated, and autonomy is self-determination through self-legislation (Matherne 2019, Lopes 2021, Matherne 2021). Given a theory of aesthetic evaluation as the harmonious free play of imagination and cognition, and given the picture of autonomy as self-legislated principles and policies, the oft quoted passages can be read as a challenge to aesthetic expertise. In view of the nature of aesthetic evaluation, we should not to defer to anyone else, so there can be no aesthetic experts.

The spirit of the often-quoted passages continues to attract; not so much Kant's explanation of them. Robert Hopkins (2011) has reframed contemporary thinking by pointing out that the challenge can either be that expert aesthetic testimony is unavailable or that it may not be used.

Prior to Hopkins, most took the challenge to be that there can be no aesthetic knowledge by testimony—the knowledge is unavailable (e.g. Tormey 1973; Hopkins 2000; Meskin 2004; Laetz 2008; Lord 2016; McKinnon 2017; Ransom 2019; cf. Lopes 2014, ch. 8). The argument is not that there can be no aesthetic knowledge, tout court. To begin with, the first three points of consensus undercut any such premise. Moreover, the unavailability of aesthetic knowledge via testimony is supposed to support the autonomy of aesthetic evaluation, and the point of relying on oneself might well be to get aesthetic knowledge. Presumably, there is something special about aesthetic value or evaluation if aesthetic evaluations figure in a kind of knowledge that cannot be conveyed through testimony. What we need is what we got from Kant, namely an argument to the unavailability claim from a theory of aesthetic value or evaluation. (What we get are appeals to intuition that stalemate against contrary intuitions.)

Hopkins's alternative to the unavailability claim is that there are norms on aesthetic practices that prohibit the use of aesthetic testimony. The norms place the onus on aesthetic evaluators to rely only on themselves. Hopkins proposes two candidates (2011, 149–50). One states that licence to use an aesthetic evaluation requires one to grasp its grounds. The second states that licence to use an aesthetic evaluation requires one to have experienced for oneself the item evaluated. Either norm prohibits deference to others, though the prohibition can be overridden, specifically in circumstances where there are serious impediments to experiencing the item or grasping the grounds of the evaluation. Since an aesthetic expert is someone to whom we should defer, and the norm says that we should not defer to others in matters aesthetic, there are no aesthetic experts.

Stating the two norms does not vindicate the challenge, though. To do that, a case needs to be made that the norms exist and that they are well grounded. Hopkins concludes that we need answers to a question, which he puts four ways: 'Why is our aesthetic practice governed by that norm? What about that practice is responsible for its having the normative structure thus described? What is the source of that norm's authority over us?' (2011, 156). Put another way, what is it about aesthetic value that mandates norms of use such as these?

Thi Nguyen (2019) has floated an answer. To begin with, he proposes a norm that prohibits the use of aesthetic testimony (2019, 1130). Adopting the language of autonomy, the norm is that one should make aesthetic evaluations by using one's own faculties and abilities, without relying on testimony. Some have intuitions that such a norm exists, and Nguyen pumps the intuitions, but intuitions vary, so it is important that he also offers an explanation of the norm, one that appeals to a claim about the motivational structure of aesthetic engagement (see also King 2017; Van der Berg 2019). Let aesthetic engagement consist in those processes and acts, whatever they are, that yield aesthetic evaluations. Nguyen's claim is that we do not engage aesthetically in order to make correct aesthetic evaluations; instead, we try to make correct aesthetic evaluations in order to be able to engage aesthetically—that is, in order to be able to run the processes and perform the acts that go into aesthetic engagement. The priority placed on engagement as an end mandates adopting a means to achieve it that is inefficient: one must try to make correct aesthetic evaluations without hints or shortcuts as to which aesthetic evaluation is correct. 'The aesthetic appreciator who defers to testimony', Nguyen writes, 'is making the same mistake as the marathon runner who takes a taxi to the finish line' (2019, 1146). Value lies in using one's capabilities to try to make the correct evaluation, not in having made it. To get the value, we adopt the autonomy norm, prohibiting deference to experts, relying wholly on our own powers of engagement. Deference is not prohibited across the board in science: 'the reason we defer to expert testimony in the sciences, but not in aesthetic appreciation, is that getting correct judgments is the primary source of value in the sciences' (Nguyen 2019, 1138).

Again, we are left wondering whether the theory of aesthetic value is correct. Is the primary source of aesthetic value engagement? Does such a view answer the normative or demarcation questions? Why could there be no practice whose members engage aesthetically in order to make correct aesthetic evaluations? How can that not be good, too? Or why would the goodness not count as aesthetic?

The lesson is not that Hopkins's questions cannot be adequately answered, either along the lines sketched by Kant and Nguyen or along some other lines. The jury is out on the autonomy challenge. Rather, the lesson is that those who wish to press the challenge must argue, just as Hopkins sees, that the nature of aesthetic value makes it the case that our aesthetic value practices should prohibit deference to aesthetic experts, such that there are no aesthetic experts. A good place to start in mounting such an argument would be to confront views of aesthetic value that call for some evaluators to play the role or provide the service of aesthetic expert. Here are three.

3. Truffle Pigs, Our Best Barometers

That is how Jerrold Levinson describes the 'true judges' that headline in Hume's aesthetics (Levinson 2002, 234 on Hume 1993[1777]b). Levinson's conception of aesthetic expertise is a natural consequence of Hume's having a hedonic theory of aesthetic value that is meant to answer the normative question.

Aesthetic hedonism has been the go-to theory of aesthetic value at least since Plato, not only in European philosophy but in South Asian philosophy too (Pollock 2018; Van der Berg 2020). Until recently, the theory has been taken for granted without argument, and little effort has been devoted to crafting alternatives to it. Several variant formulations exist, but the following will do for present purposes—nothing in what follows hinges on the details. For aesthetic values, V,

aesthetic hedonism: x is V = (1) x has a capacity to please A in suitable circumstances, and (2) x is V because (1).

Two notes. To begin with, the 'because' signals metaphysical rather than causal explanation. Clause (2) says that what makes it the case that x is V is that x has some other features that ground a capacity to produce pleasure in suitable circumstances. In addition, 'pleasure' should be construed broadly. Pleasure is not always sensual or sensory; there is pleasure in solving a problem, receiving praise, achieving a goal, and acting effectively. There is also pleasure, broadly construed, in listening to sad songs and watching scary movies. To mark this, some prefer to talk of 'finally valuable experiences'.

Aesthetic hedonism suggests an obvious and compelling answer to the normative question. The question was, why does the fact that x is V lend weight to its being the case that anyone should aesthetically engage x? The answer is that for x to be V is for x to please, and everyone always has reason (albeit not always decisive) to do what yields pleasure. In response to MacKenzie's wondering why anyone always has reason to listen to beautiful performances, Amira can explain that beautiful performances please, and everyone always has reason to do what yields pleasure. If MacKenzie queries that, then Amira's best reply is a shrug.

With this, we only get a personally oriented normativity. That is, MacKenzie has reason to listen to Chaimbeul only if she would get pleasure were she to listen to Chaimbeul. Not everyone gets pleasure from the same things. For a scant handful of aesthetic hedonists, there is nothing more to say (e.g. Santayana 1896; Melchionne 2010; Melchionne 2015).

Bernard Williams once quipped, 'I simply don't like staying in good hotels' (1985, 125). If aesthetic normativity is personally oriented, then his quip is incoherent. If to be aesthetically good is just to have a capacity to please him, then he is saying that he fails to meet his own personal standard. But he is not saying that, for he stands by his disliking good hotels. He is saying the hotels are good by a standard that he fails to live up to and that is not his personal standard. He recognizes an aesthetic normativity that implicates a standard external to him, and he simply confesses that he does not meet that standard.

Most appeal to an interpersonal standard embodied in an ideal responder or, in Hume's phrase, a 'true judge'. The thought is that MacKenzie has reason to listen to Chaimbeul so long as Chaimbeul would please true judges. If MacKenzie, being the kind of responder she is, does not in fact get pleasure from Chaimbeul, then she has reason to make herself more like true judges, so that she will get the pleasure that they get. Aesthetic normativity is standardized.

This claim intensifies the normative question. As Levinson asks, 'why should one be moved by the fact that such and such things are approved or preferred by ideal critics, if one is not [an ideal critic] oneself?' (2002, 230). What reason does anyone have to switch one source of pleasure (what pleases them) for another (what pleases true judges)? Put another way, what makes the true judges a normative standard in the first place? What makes them experts in aesthetic goodness, such that non-experts have reason to defer to their evaluations?

Building on work by Mary Mothersill (1989), Levinson reconstructs Hume's reasoning. Some items (it is an empirical discovery what they are) appeal to people of very different times, cultures, and social groups. The best explanation of their appeal is that the items have high aesthetic value. So, the items have high aesthetic value. Let the true judges be those people, whoever they are, who in fact prefer those items to other items. If those familiar with two experiences prefer one to the other, then the one has greater hedonic value than the other. So the true judges prefer the better works. Anyone always has reason to maximize pleasure. So anyone has reason to do what it takes to have the experiences that the true judges have.

Aesthetic hedonists widely share in the spirit of Hume's answer to the normative question (although not all do, as we shall see in the next section). After all, if aesthetic value is hedonic value and we have reason to maximize pleasure, then it is natural to surmise that aesthetic experts, if any there be, are our best hedonic barometers, even truffle pigs. However, an advantage of Levinson's reconstruction of Hume's reasoning is that it maps out where the problems might be. At least some of those problems have been pointed out (esp. Shelley 2011; Riggle 2013; Lopes 2018, ch. 4). One problem with the generic Humean approach to the normative question will continue as a theme through the next two sections.

The problem concerns the approach's universalism, which implies that culturally located aesthetic value practices are contra-normative (Lopes 2024a, ch. 3). Suppose that aesthetic normativity is standardized. Each has most aesthetic reason to appreciate in alignment with true judges. My belonging to a social group where I learn to comply with a local hedonic standard very likely interferes with my acting on my true aesthetic reasons—the ones that align with the verdict of true judges. What I should do is break free of the yoke of my practice and cultivate my taste to coincide with that of true judges. Aesthetic hedonists who answer the normative question by appeal to a true judge standard must be prepared to accept that there would be nothing amiss were everyone to act on the aesthetic reasons they have, so that diverse aesthetic practices would dissolve, leaving behind an aesthetic monoculture.

Short of throwing in the towel on the true judges, three options are open. One is to make what might seem to be a simple fix: each aesthetic practice has its own normative standard, embodied in the responses of its proprietary true judges. However, this would require that a diverse range of aesthetic practices be on a hedonic par with one another. Only by a massive coincidence would a diverse range of aesthetic practices be on a hedonic par with one another. As long as migration from some practices to other practices would yield net hedonic benefits, we have aesthetic reason to migrate and thereby reduce the diversity of aesthetic practices. Another option is to reply that some non-aesthetic reasons are weighty enough to countermand aesthetic reasons that would lead to a monoculture, were we to act on them. That is a hard a bullet to bite as long as we are committed to thinking that aesthetic reasons sustain the diversity of the aesthetic field. Surely the aesthetic domain is inherently diverse; acting on our aesthetic reasons should promote diversity. Option three is to concede that the monoculture is to be embraced as the aesthetic ideal. That is a harder bullet to bite. As Alexander Nehamas comments, some have dreamt that 'we would all find beauty in the very same places' but 'that dream is a nightmare' (2007, 83).

The lesson is not that Levinson's reconstruction of the Humean tradition inadequately answers the normative question. Its problems might be solved. However, enough has been said both to model how to extract a conception of aesthetic expertise from a theory of aesthetic value but also to motivate a look at another model.

4. Pleasure, Society, and the Self

On one construal, Humean true judges are expert guides, equipped to nose out sources of aesthetic goodness. Paul Guyer (2005) offers an alternative. The alternative exploits the possibility that an aesthetic hedonist might give a non-hedonic answer to the normative question. That aesthetic values are something like capacities to please does not by itself entail that what makes it the case that x is V is reason to appreciate x is that we have reason to get pleasure. We might have non-hedonic reason to appreciate what yields pleasure. Historically, some aesthetic hedonists have offered non-hedonic theories of aesthetic normativity (see Zuckert 2010 and Lopes 2021 on Kant 2000[1790]; Lopes 2019 and Lawson and Lopes forthcoming on Bhattacharyya 2011[1930]; Lopes 2024b on Bolzano 2023[1843+1849]). Guyer places Hume in this tradition: true judges are experts to whom we should defer because they help to constitute communities of hedonic response to which we have non-hedonic reason to belong.

Guyer prepares the ground by cataloguing some forms of beauty. In some items, beauty is a capacity to please by appearance, hence by perception not augmented by sympathy or imagination. Three other forms of beauty are apprehended at least in part via sympathy or imagination. First, by means of sympathy, one can take pleasure in an object as one that is useful or pleasing to another person. As Hume explains in the *Treatise*,

> Here the object, which is denominated beautiful, pleases only by its tendency to produce a certain effect. That effect is the pleasure or advantage of some other person. Now the pleasure of a stranger, for whom we have no friendship, pleases us only by sympathy. To this principle, therefore, is owing the beauty, which we find in everything that is useful. (1739, 3.3.1.8)

Second, by means of imagination, one can take pleasure in something that does not actually benefit anyone, seeing it as apt to benefit people generally (Hume 1739, 3.3.1.20). Thus 'a fertile soil, and a happy climate, delight us by a reflection on the happiness which they wou'd afford the inhabitants, tho' , at present the country be desart and uninhabited' (Hume 1739, 3.3.1.20). Finally, imagination can be brought to bear on a work of art in order to take pleasure in how it is fitted to serve its internal or artistic purpose.

From this catalogue, we can infer that a full capacity to take aesthetic pleasure must be learned or cultivated. Moreover, capacities for sympathy and imagination both require and create a shared point of view (Guyer 2005, 67). Sympathetic pleasure in the usefulness of Euler's identity to the mathematician requires and creates a shared response. So does imagining the quality of life enjoyed by those who might come to live in the mountain meadows above Chatterbox Falls. Thus, Hume's true judges have a bundle of traits that equip them to have sympathetic and imaginative responses that others can share. Famously, they have 'strong sense, united to delicate sentiment, improved by practice, perfected by comparison, and cleared of all prejudice' (Hume 1993[1777]b, 147).

Those endowed with these traits are, as Hume immediately adds, the 'true standard of taste and beauty'. That is, they are experts to whom non-experts should defer. Yet, one might ask once again what reason non-experts have to defer to experts' evaluations. Granted, by deferring to experts, we converge on a shared response, but what reason do we have to converge on a shared response? Guyer writes that 'if we will issue similar judgements, then we must issue our judgements from similar positions or on the basis of similar conditions; but why must we wish to agree in judgements in the first place?' (2005, 69). What is wrong with '*à chacun son goût*'? Why prioritize traits conducive to sympathy and imagination?

Two answers to this question lie deep at the heart of Hume's concerns, and some of our concerns too. Only one appeals to hedonic normativity.

Creatures like us enhance our pleasures by sharing them. Hume observes that we have 'the most ardent desire of society' and 'a perfect solitude is, perhaps, the greatest punishment we can suffer', for 'every pleasure languishes when enjoy'd a-part from company, and every pain becomes more cruel and intolerable' (1739, 2.2.5.15). In Guyer's gloss, 'to every pleasure there may be added the pleasure of agreement in the enjoyment of that pleasure' (2005, 64). If Hume observes accurately, then the suggestion would be that we have a special hedonic reason to defer to true judges. After all, their traits equip them to exercise capacities of sympathy and imagination that require but also create shared responses. By deferring to them, we can share our hedonic responses. Shared pleasure in an item exceeds solo pleasure in the item. We always have reason to maximize pleasure. Ergo, we have hedonic reason to defer to them.

Guyer brings out the second answer, picking up on the italics in a passage from the *Treatise*, where Hume is discussing the sentiments and aesthetic pleasure:

> every particular man has a peculiar position with regard to others; and 'tis impossible we cou'd ever converse together on any reasonable terms, were each of us to consider characters and persons, only as they appear from his peculiar point of view. In order, therefore, to prevent those continual *contradictions*, and

arrive at a more *stable* judgment of things, we fix on some *steady* and *general* points of view; and always, in our thoughts, place ourselves in them, whatever may be our present situation. (1739, 3.3.1.15)

By implication, we have an interest in consistent and stable responses seated in steady and general points of view. Perhaps the worry is that pleasure in sharing can be obtained perfectly well by converging on unsettled and flighty responses—that is, by collectively following fads. The traits of the true judges seem to be calculated to secure less changeable responses.

What interest might be served by stable responses seated in the steady point of view that true judges afford? In part, we have an interest in self-control. Whereas the 'the good or ill accidents of life are very little at our disposal', we can choose opportunities for aesthetic engagement (Hume 1993[1777]a, 10). In part, we have an interest in durable responses. As Guyer reminds us, in a Humean world, 'the reality of objects but even more of self lies in nothing other than the continuity of mental content', so 'durable sentiments must make a fundamental contribution to the perception of one's own identity' (2005, 62). Simply endorsing our immediate responses to momentary appearances without referring them to more general points of view risks 'our very sense of self' (Guyer 2005, 69; see also Williams 2002, 191–8; Fricker 2007, 52–3). Recent empirical work indicates that aesthetic affiliation is as strong a factor in perceived self-identity as religious affiliation (Fingerhut et al. 2021). The proposal is that our interest in establishing ourselves as durable beings with a level of control over our experiences is non-hedonic reason to defer to true judges.

The proposal also explains why true judges would have a role in transmitting a canon of beauties from generation to generation (Guyer 2005, 74). The canon stabilizes pleasures and makes them durable. Some empirical studies of canon formation suggest the mechanism at play here. In the mere exposure effect, a positive attitude towards a stimulus varies in proportion to the frequency of mere exposure to it (Zajonc 1968). In an elegant series of experiments, James Cutting (2003) showed that preferences for the works in the Impressionist canon is largely a function of how frequently those works are reproduced. In fact, Cutting was able to manipulate his subjects' preferences by changing frequencies of appearance in their local environment. A Humean might jump on this: true judges provide for shared responses by establishing a canon of works that most often appear in the local aesthetic landscape.

Unlike hedonic answers to the normative question, this proposal does not represent the cultural differentiation of the aesthetic sphere as contra-normative. The problem was that, if reasons to defer to aesthetic experts are hedonic, then we should defer to those experts who will maximize our pleasures, and chances are that this will mean that we should abandon the aesthetic practices to which we have affiliated, so that we all converge on nearly the same aesthetic culture. The problem no longer arises if reasons to defer to aesthetic experts reduce to reasons to stabilize our personal pleasures in society with our neighbours. It suffices for stability that we coordinate with some others. Suppose, analogously, that a way to stabilize the doxastic self is by sharing a language: it would be enough if we each shared a language with some others. It is not required that we all coverage on a unique language. Likewise, we can act on our reasons to affiliate with others hedonically while keeping faith with local true judges.

Different answers to the normative question draw different portraits of aesthetic expertise. Our having aesthetic reasons can give us reason to defer to aesthetic experts. If aesthetic reasons are hedonic, and they are reasons to defer to aesthetic experts, then aesthetic experts are our best truffle pigs. If aesthetic reasons are reasons to stabilize the self, and they are reasons to defer to aesthetic experts, then aesthetic experts constitute communities of taste by exercising sympathy and imagination so as to create the conditions for stable and durable responses.

5. Performance in Practice

On existing accounts, aesthetic experts are either guides to value or those whose responses constitute valuing practices. On a new account, which develops out of an alternative to aesthetic hedonism, an aesthetic expert is someone whose performance determines which aesthetic practice, out of many possible ones, is the one to which a group of people belong. Begin with the alternative to aesthetic hedonism, the network theory (Lopes 2018).

Recall that aesthetic evaluations are states of mind that attribute aesthetic values, such as being elegant, glorious, and richly textured. Also recall that items have these features in virtue of their having other, non-aesthetic features. The rich texture of Chaimbeul's playing is grounded in its timbre, timing, and dynamics, for example. Let an aesthetic act be any act that is motivated by an aesthetic evaluation. Aesthetic acts include editing, curating, collecting, conserving, exhibiting, teaching, connecting audiences, making, and performing, as well as appreciating.

Competences in aesthetic acts specialize along two dimensions. On one hand, what it takes to appreciate Chaimbeul's playing is not the same as what it takes to record it, review it, or market it to its audience. Success in each of these act types requires a distinct competence. On the other hand, Chaimbeul's capacity for richly textured playing does not equip her to play richly textured Karnatic music on the *bansuri*. Likewise, competence in reviewing her music is not competence in reviewing restaurants, and competence in marketing her music is poor preparation for marketing Kehinde Wiley's paintings. So, the same aesthetic act-type can be performed in different aesthetic practices, and specialization within a practice means that competence need not transfer to other practices. To see the difference practices make, hold act-types constant. To see the difference act-types make, hold practices constant. These specificities give us something grippy to explain, aesthetic values figure in the doings of agents who are specialized by aesthetic practice and also by act-type.

For the division of labour into practices and act-types to work, all members of a practice must be on the same page aesthetically. Chaimbeul, her recording engineer, and her marketer must all share an understanding of what makes that specific kind of music richly textured. A sound engineer from the world of heavy metal is not going to make it easy for Chaimbeul or her marketing team to succeed in their respective endeavours. Likewise, someone adept at marketing Wiley's painting faces a steep learning curve before he can help market Chaimbeul's music.

A division of labour can be a product of agents acting on motives that accurately represent their reasons. Imagine a society that hits on the idea of dance. At first they simply perform dances, but as the dance culture begins to thrive, they also choreograph dances, compose suitable music, craft suitable masks and costumes, construct dance spaces, exchange critical advice, and document dances for posterity. Perhaps everyone is a generalist who can and does perform all these acts. Nonetheless, as long as each has reason to succeed, each has reason to develop some of their competences at the expense of other competences, provided that they can expect their neighbours to fill in the gaps. Critics have reason to leave choreography and performance to others, who have reason to leave criticism to those who have honed their critical skills. However, for the division of labour to work, all must be on the same page aesthetically. They must agree that this, not that, is a richly textured dance.

Being on the same page is a metaphor. Let an aesthetic value profile of a practice be the pattern of aesthetic values of items as grounded by the items' non-aesthetic features (Lopes 2018b, ch. 10). A dancer makes a movement. In the vocabulary of human kinetics, they move their foot in a direction with force and velocity. Executed in ballet by Misty Copeland, the move is vivid, but the very same gesture is lifeless when performed in Kathakali by Kanak Rele. Each dancer realizes an aesthetic value by making a movement, but the very same movement, as characterized in the vocabulary of human kinetics, grounds different aesthetic values. Equally, in different traditions, different movements might ground the very same aesthetic value. Vivid in Kathakali requires a gesture different from Copeland's. Ascending a level, the properties that ground a vivid dance move are different from the ones that ground a vivid proof, a vivid melody, or a vivid sunrise.

Now we can substitute the technicality for the metaphor. Agents in an aesthetic practice have reason to specialize provided that they can expect others act in ways that fill in the gaps in their competences. For each to fill in the gaps of others' competences, all must act in ways that often enough comply with the same aesthetic value profile. When they do so, all have reason to expect the others to comply with the same aesthetic value profile. With that expectation secure, they have reason to specialize, boosting each other's chances of success. As long as each has reason to expect enough other people to perform in compliance with the same aesthetic value profile, they belong to a social practice constituted both by the profile and by a norm to comply with the profile.

All elements of the network theory are now in place. The theory states what makes aesthetic values reason-giving:

> an aesthetic value, V, is reason-giving = the fact that x is V lends weight to the proposition that it would be an aesthetic achievement for some A to φ in C, where x is an item in an aesthetic practice, K, and A's competence to φ is aligned upon an aesthetic value profile that is constitutive of K.

Here achievement is nothing more than success out of competence (Sosa 2007). So, what makes an aesthetic value reason-giving is that its instantiation in the fact that x is V lends weight to the proposition that a member of an aesthetic practice would be more likely to succeed through their specialized competence by acting in accordance with the practice's aesthetic value profile.

So stated, the theory answers the normative question, what makes it the case that an item's having an aesthetic value gives anyone reason to do anything? An answer completes the schema:

> an aesthetic value, V, is reason-giving = the fact that x is V lends weight to the proposition that....

A hedonic answer to the normative question fills in the dots by appeal to the principle that anyone always has reason to maximize pleasure. Guyer's reading of Hume fills in the dots by appeal to our interest in making ourselves stable and durable. The network theory reduces aesthetic normativity to the normativity of achievement. To achieve is to perform an act successfully, as a result of competence. Anyone who acts at all thereby has reason to achieve. In other words, the network theory reduces aesthetic normativity to plain vanilla practical normativity: when you act (aesthetically), you have reason to act as someone acts who acts well (aesthetically).

Someone who acts well is not yet an expert—not even someone who acts reliably well. An expert is someone to whom non-experts should defer, because they act reliably well. The question is, what portrait of aesthetic experts is reflected in the network theory? Do they help us to find out how to achieve (as true judges might help us to see what brings the greatest pleasure)? Or, through their performances, do they constitute aesthetic value profiles (as true judges might constitute a standard of taste)? If neither, then what function is served by deference to them?

The answer is that an aesthetic expert is someone whose performance determines which aesthetic practice is the one to which a group of people belong.

Serving this function is not constituting the aesthetic value profile of the practice. An aesthetic value profile is an abstractum, a pattern that relates aesthetic values of items to the non-aesthetic features that ground the values. The pattern is not made what it is by anyone's responses or activities. Consider a language. Scots Gaelic is also an abstractum, a lexicon, and grammar. It is not made what it is by anyone's speaking it. Instead, the coordinated activities of members of a group explain why, out of all the possible languages, the language with that lexicon and grammar is the one that they speak (Lewis 1969). Likewise, the coordinated activity of members of a social group explain why, out of all the possible correlations between aesthetic values and their non-aesthetic grounds, this is the one that they use.

Serving this function does not help us to find out how to achieve. On the network theory, no aesthetic value profile is better, independent of the practice, than any other for coordinating agents with specialized aesthetic competences. Again, consider languages. Speaking a language in which '*feallsanachd*' refers to philosophy is neither better nor worse as a means of communication than a language in which 'philosophy' refers to philosophy. The most that we can say is that, once a group is up and running with a language, they acquire (non-decisive) practical reasons to stick with it. Likewise, aesthetic experts are not guides to help members of a group to discover which of all the possible aesthetic profiles is the one that will work best as a means of coordinating their activities. There is no such discovery to be made.

So the network theory proposes a model of aesthetic experts distinct from the two Humean models. Aesthetic experts determine which aesthetic practice is the one to which a group of people belong. The reason why they have this power is simply that they perform aesthetic acts reliably well. As a result, their aesthetic acts are often enough motivated by evaluations that accurately represent their aesthetic reasons. In accurately representing their aesthetic reasons, they are complying with the practice's aesthetic value profile, and compliance with the practice's aesthetic value profile is what enables any member of the practice to coordinate with other members. Members of the practice who are not experts are often enough motivated by aesthetic evaluations that do not represent their aesthetic reasons accurately. They should defer to aesthetic experts just because they should comply with their practice's aesthetic value profile.

To make vivid the idea, it might help to consider the point of faultless aesthetic disagreements, disagreements that are not due to any party's inattention, ignorance, insincerity, pretence, or inadvertence (Young 2017). For example,

AMIRA: Chaimbeul's playing on *The Reeling* is richly textured.
MÀIRI: No, it's not; it's too thin.

Assume that neither Amira nor Màiri is at fault in any way. Yet they might proceed deeper into their exchange. Why? What would be the point? Different answers to this question are suggested by different answers to the normative question. A hedonic answer to the normative question suggests that the point of their disagreement is to discover what is actually more pleasing. Guyer's reading of Hume suggests that the point of their disagreement is to constitute an enduring community of those who respond alike (Guyer 2005, 69–70; see also Egan 2010). The network theory suggests that faultless aesthetic disagreements are metalinguistic negotiations that determine what the aesthetic profile is (Sundell 2011; Lopes 2018, ch. 9). Take this non-aesthetic disagreement:

CAROLYN: This knife is plenty sharp.
NORIHIDE: No, it's not; it's too dull.

This might be a faultless disagreement where both speak truly, and neither is at fault. Yet it has a point, namely, to determine the context—is it French cooking or Japanese? Likewise, the point of the disagreement between Amira and Màiri is to determine what the aesthetic value profile is. Is it a practice whose profile is one where a performance with those non-aesthetic features is richly textured, or is it one where the performance is thin? The speaker, if any, who is an expert, has the upper hand. We have reason to defer to her because we have reason to comply with the aesthetic profile of the practice.

Return, in closing, to the autonomy challenge. Nguyen proposes that we do not engage aesthetically in order to make correct aesthetic evaluations; we try to make correct aesthetic evaluations in order to be able to engage aesthetically. As long as engagement amounts to trying to make aesthetic evaluations without hints or shortcuts as to which aesthetic evaluation is correct, deference to aesthetic experts is prohibited. However, it is compatible with this picture that aesthetic appreciators should defer to aesthetic experts about what the context is in which they engage. Deference on this matter is not like a marathon runner taking a taxi to the finish line; it is like a marathon runner taking direction from the officials about the race's route. The same goes for those who engage with Chaimbeul's music: in trying to make out what aesthetic qualities it has, they need to be listening to it as the kind of music that it is. In as much as Amira is trying to attune MacKenzie to the musical context by saying that Chambeul's playing is richly textured, MacKenzie should defer to her.

No argument has been given for the network theory or against aesthetic hedonism and its answers to the normative question. There remains plenty of work to do, but some progress has been made on three fronts. First, the task of deciding on autonomy is now clearer. Can the case be made that the testimony of aesthetic experts is either unavailable or not to be used, where experts play any of the three roles so far described (and there might be more)? Second, facts about the nature of expertise and our reliance on it in our aesthetic practices can be used to decide between different theories of aesthetic value and its normativity. Third, we now have three models of the kind of work that can be done by experts on value in general. A new and good question concerns what we can say, now, about experts on moral value, or epistemic value.

Acknowledgement

My thanks to an anonymous referee for very helpful comments.

References

Benovsky, Jiri. (2012) "Aesthetic supervenience vs aesthetic grounding." *Estetika: The European Journal of Aesthetics* 49(2), 166–78.

Bhattacharyya, K. C. (2011[1930]) "The concept of *rasa*," in *Indian Philosophy in English: From Renaissance to Independence*, (eds.), Nalini Bhushan and Jay L. Garfield, 195–206. Oxford: Oxford University Press.

Bolzano, Bernard. (2023[1843+1849]) *Essays on Beauty and the Arts*, (ed.) Dominic McIver Lopes, (trans.) Adam Bresnahan. Boston: Hackett.

Cutting, James E. (2003) "Gustave Caillebotte, French impressionism, and mere exposure." *Psychonomic Bulletin and Review* 10(2), 319–43.

De Clercq, Rafaël (2002) "The concept of an aesthetic property," *Journal of Aesthetics and Art Criticism* 60(2). 167–76.

Egan, Andy. (2010) "Disputing about taste," in *Disagreement*, (eds.), Richard Feldman and Ted A. Warfield, 247–92. Oxford: Oxford University Press.

Fingerhut Joerg, Javier Gomez-Lavin, Claudia Winklmayr, and Jesse J. Prinz. (2021) "The aesthetic self: The importance of aesthetic taste in music and art for our perceived identity." *Frontiers in Psychology* 11(577703), 1–18.

Fricker, Miranda. (2007) *Epistemic Injustice: Power and the Ethics of Knowing*. Oxford: Oxford University Press.

Guyer, Paul. (2005) "The standard of taste and the 'most ardent desire of society,'" in *Values of Beauty: Historical Essays in Aesthetics*, 37–74. Cambridge: Cambridge University Press.

Hopkins, Robert. (2000) "Beauty and testimony," in *Philosophy, the Good, the True, and the Beautiful*, (ed.), Anthony O'Hear, 209–36. Cambridge: Cambridge University Press.

Hopkins, Robert. (2011) "Hopkins, how to be a pessimist about aesthetic testimony." *Journal of Philosophy* 108(3), 138–57.

Hume, David. (1739) *A Treatise of Human Nature*. London: John Noon.

Hume, David. (1993[1777]a) "Of the delicacy of taste and passion," in *Selected Essays*, (eds.), Stephen Copley and Andrew Edgar, 10–13. Oxford University Press.

Hume, David. (1993[1777]b) "Of the standard of taste," in *Selected Essays*, (eds.), Stephen Copley and Andrew Edgar, 133–54. Oxford University Press.

Kant, Immanuel. (2000[1790]) *Critique of the Power of Judgment*, (ed.) Paul Guyer, (trans.) Paul Guyer and Eric Matthews. Cambridge University Press.

King, Alex (2017) "The virtue of subtlety and the vice of the heavy hand." *British Journal of Aesthetics* 57(2), 119–37.

Kozbelt, Aaron. (2020) "The influence of expertise on aesthetics," in *The Oxford Handbook of Empirical Aesthetics*, (eds.), Marcos Nadal and Oshin Vartanian, 787–819. Oxford University Press.

Laetz, Brian. (2008) "A modest defense of aesthetic testimony." *Journal of Aesthetics and Art Criticism* 66(4), 355–63.

Lawson, Emily and Dominic McIver Lopes (forthcoming). "Courageous love: K. C. Bhattacharyya on the paradox of sorrow." *Journal of the American Philosophical Association.*

Levinson, Jerrold. (2002) "Hume's standard of taste: The real problem." *Journal of Aesthetics and Art Criticism* 60(3), 227–38.

Lewis, David. (1969) *Convention*. Harvard University Press.

Lopes, Dominic McIver. (2014) *Beyond Art*. Oxford University Press.

Lopes, Dominic McIver. (2018) *Being for Beauty: Aesthetic Agency and Value*. Oxford: Oxford University Press.

Lopes, Dominic McIver. (2019) "Feeling for freedom: K. C. Bhattacharyya on Rasa." *British Journal of Aesthetics* 59(4), 465–77.

Lopes, Dominic McIver. (2021) "Beyond the pleasure principle: A Kantian aesthetics of autonomy." *Estetika: The European Journal of Aesthetics* 58(1), 1–18.

Lopes, Dominic McIver (2024a) *Aesthetic Injustice*. Oxford: Oxford University Press.

Lopes, Dominic McIver. (2024b) "Bolzano on aesthetic normativity." *British Journal of Aesthetics*, preprint pp. 1–14.

Lord, Errol. (2016) "On the rational power of aesthetic testimony." *British Journal of Aesthetics* 56(1), 1–13.

Matherne, Samantha. (2019) "Kant on aesthetic autonomy and common sense." *Philosophers' Imprint* 19, 1–22.

Matherne, Samantha. (2021) "Aesthetic autonomy and norms of exposure." *Pacific Philosophical Quarterly* 102(4), 686–711.

McKinnon, Rachel. (2017) "How to be an optimist about aesthetic testimony." *Episteme* 14(2), 177–96.

Melchionne, Kevin. (2010) "On the old saw 'I know nothing about art but I know what I like.'" *Journal of Aesthetics and Art Criticism* 68(2), 131–41.

Melchionne, Kevin. (2015) "Norms of cultivation." *Contemporary Aesthetics* 13.

Meskin, Aaron. (2004) "Aesthetic testimony: What can we learn from others about beauty and art?." *Philosophy and Phenomenological Research* 69(1), 65–91.

Mothersill, Mary. (1989) "Hume and the paradox of taste," in *Aesthetics: A Critical Anthology*, (eds.), George Dickie, Richard Sclafani, and Ronald Roblin, 269–86. New York: St Martin's.

Nehamas, Alexander. (2007) *Only a Promise of Happiness: The Place of Beauty in a World of Art*. Princeton: Princeton University Press.

Nguyen, Thi. (2019) "Autonomy and aesthetic engagement." *Mind* 129(516), 1127–56.

Pollock, Sheldon. (ed.) (2018) *The Rasa Reader: Classical Indian Aesthetics*. New York: Columbia University Press.

Ransom, Madeleine. (2019) "Frauds, posers, and sheep: A virtue theoretic solution to the acquaintance debate." *Philosophy and Phenomenological Research* 98(2), 417–34.

Riggle, Nicholas. (2013) "Levinson on the aesthetic ideal." *Journal of Aesthetics and Art Criticism* 71(3), 277–81.

Santayana, George. (1896) *The Sense of Beauty*. Scribner's.

Shanteau, James. (1992) "Competence in experts: The role of task characteristics." *Organizational Behavior and Human Decision Processes* 53(2), 252–62.

Shelley, James. (2011) "Hume and the value of the beautiful." *British Journal of Aesthetics* 51(2), 213–22.

Sibley, Frank. (1959) "Aesthetic concepts." *Philosophical Review* 68(4), 421–50.

Sibley, Frank. (1965) "Aesthetic and nonaesthetic." *Philosophical Review* 74(2), 135–59.

Sosa, Ernest. (2007) *A Virtue Epistemology: Apt Belief and Reflective Knowledge*. Oxford: Oxford University Press.

Sundell, Timothy. (2011) "Disagreements about taste." *Philosophical Studies* 155(2), 267–88.

Todd, Cain Samuel. (2004) "Quasi-Realism, acquaintance, and the normative claims of aesthetic judgement." *British Journal of Aesthetics* 44(3), 277–96.

Tormey, Alan. (1973) "Critical judgments." *Theoria* 39(1), 35–49.

Van der Berg, Servaas. (2019) "The motivational structure of appreciation." *Philosophical Quarterly* 69(276), 445–66.

Van der Berg, Servaas. (2020) "Aesthetic hedonism and its critics." *Philosophy Compass* 15(1), 1–15.

Williams, Bernard. (2002) *Truth and Truthfulness: An Essay in Genealogy*. Princeton: Princeton University Press.

Young, James O. (ed.) (2017) *Semantics of Aesthetic Judgement*. Oxford: Oxford University Press.

Zajonc, Robert B. (1968) "Attitudinal effects of mere exposure." *Journal of Personality and Social Psychology Monographs* 9(1), 1–27.

Zangwill, Nick. (1995) "The beautiful, the dainty, and the dumpy." *British Journal of Aesthetics* 35(4), 317–29.

Zuckert, Rachel. (2010) *Kant on Beauty and Biology: An Interpretation of the Critique of Judgment*. Cambridge: Cambridge University Press.

13
Moral Expertise and Socratic AI

Emma C. Gordon
University of Glasgow
emma.gordon@glasgow.ac.uk

1. Introduction

A central strand of research in contemporary bioethics concerns questions about whether—or to what extent—we should *enhance* ourselves. In this context, 'enhancement' refers to using medicine or technology to go beyond treating illness or injury, making ourselves *better than well*.[1] When exploring ethical issues that arise around enhancement, bioethicists are often interested specifically in questions about some particular sub-type of enhancement. For example, those who focus on cognitive enhancement might consider whether achievements accomplished with the aid of, 'smart drugs' or cognitive performance-enhancing technology[2] are in some sense less valuable or might worry about the possibility that we are ill-equipped to handle new responsibilities that certain enhancements[3] could bring.[4]

In this paper, we'll be focusing specifically on *moral* enhancement, and its connection with questions of moral expertise. Much of the existing literature on moral enhancement comes from Ingmar Persson and Julian Savulescu's seminal series of works (e.g., Persson and Savulescu 2008, 2012) in which they argue that our species has a moral imperative to pursue biotechnological moral enhancement to avoid what they call *ultimate harm*. They think this for a couple of reasons.

---

[1] The question of how to distinguish 'enhancement' from related concepts in bioethics is itself contentious; for a recent critical overview, see Gordon (2022b, ch. 1).

[2] For an overview of various kinds of cognitive enhancement, see Bostrom and Sandberg (2009).

[3] For work on the connection between cognitive enhancement and responsibility, see e.g., Sandel (2007) and Maslen et al. (2015). For various bioconservative takes on the connection between enhancement and the value of achievement, see e.g., Kass (2002) and Harris (2011), and for more optimistic perspectives on the impact of cognitive enhancement on achievement see e.g., Carter and Pritchard (2019); Wang (2021); and Gordon and Willis (2023).

[4] In a related debate that focuses on the enhancement of emotions and interpersonal relationships, advocates of enhancement hold that child's right to be loved leaves at least some parents with a responsibility to use 'love drugs' while critics insist that love lies outside of what can be obtained through the assistance of biotechnological interventions. For work in defence of (qualified) use of love drugs, e.g., Liao (2011), regarding parental love, and e.g., Earp et al. (2012) in the case of romantic love. For some prominent critiques of 'love drugs' e.g., Gupta (2012) and Nyholm (2015).

Emma C. Gordon, *Moral Expertise and Socratic AI* In: *Expertise: Philosophical Perspectives.* Edited by: Mirko Farina, Andrea Lavazza, and Duncan Pritchard, Oxford University Press. © Emma C. Gordon 2024.
DOI: 10.1093/oso/9780198877301.003.0013

Firstly, moral enhancement offers a way of protecting ourselves from those who would do us harm (a subset of the population that becomes increasingly aware of hurting others the more cognitively enhanced our species becomes). And secondly, Persson and Savulescu think our common-sense moral psychology is unfit for the modern context in which we can all cause harm (at least in the long run) by not caring enough about relevant environmental threats such as climate change. Consequently, we need to change our nature.[5] As will see, most of the literature on moral enhancement focuses on drug-based interventions (e.g., oxytocin,[6] psilocybin,[7] etc.), but more recent work (Lara and Deckers 2020) asks whether the most promising route to moral enhancement might be via *artificial intelligence* rather than medication. That said, we'll see that the use of artificial intelligence for moral enhancement poses hitherto unappreciated issues regarding *moral expertise*. In particular, (i) it's not obvious that such moral enhancement is compatible with manifesting genuine moral expertise, and (ii) the capacity of artificial intelligence (AI) to improve moral reasoning might influence our criteria for identifying moral experts.

Here is the plan for what follows. Section 2 will get the idea of enhancement through artificial intelligence in view, contrasting several different approaches and situating these proposals within the broader context of the moral enhancement debate in bioethics. Following this, Section 3 will ask whether AI-assisted moral enhancement is compatible with genuine moral expertise, and several desiderata for a positive answer will be canvassed. Next, in Section 4, we'll consider how the capacity of Socratic AI to improve moral reasoning might influence, given its satisfaction of (many of) these desiderata, our criteria for identifying moral experts.

2. Moral Enhancement Through Artificial Intelligence

We already rely on technology to pursue many of our goals. Apps like Duolingo and Babbel gamify language learning, Google Maps and satellite navigation systems (SATNAV)s can lead us straight to the conference venue where we'll be presenting our next paper, and our watches and phones can monitor everything from our sleep cycles to our blood oxygen levels to help us pursue better health.[8] These are just a few example cases in which our use of technology will often make us able to reach our objectives at a faster pace.[9] For example, contrast the ease with

---

[5] For some key responses to Persson and Savulescu's moral enhancement proposal, see e.g., Harris (2011); Hardcastle (2018); Jotterand and Levin (2019); and de Melo-Martin (2018).
[6] See Hurlemann et al. (2010); Mikolajczak et al. (2010); though cf., Earp (2018).
[7] See, e.g., Pokorny et al. (2017) and Earp (2018).   [8] For discussion, see Wilson (2020).
[9] For some related results detailing the cognitive efficiency of cognitive offloading, see, e.g., Berry et al. 2019; Gilbert et al. (2020) and Grinschgl et al. (2020).

which one can do customised vocabulary exercises on a phone (at any time and in any place) with one having to travel across town to attend a weekly in-person class, and consider the simplicity of saying "Hey Google, where is the Hotel Lovec?" as compared to wrestling with a paper map that needs to be unfolded and rotated as you try to figure out exactly where in Slovenia you need to be.

As noted in Section 1, one recent debate in bioethics concerns how to best pursue the goal of *moral* improvement. Traditional moral enhancement—as Persson and Savulescu (2008, 168) call it—involves 'the transmission of moral instruction and knowledge from earlier to subsequent generations'. However, as technology and medicine continue to advance, so too does the potential for us to use drugs, apps, brain-computer interfaces and so on for specifically moral development. In the early years of the moral enhancement debate, there was intense debate on the 'target' of moral enhancement—in other words, what should we want a moral enhancement drug to increase (or decrease) in us? Persson and Savulescu proposed what they hoped would be an uncontroversial proposal for moral enhancement to make us more altruistic and to improve our sense of justice. However, many different lines of objection followed. Some worried that an empathetic person might do *more* harm in response to, for example, being deeply moved by persecution of their marginalised group. Meanwhile, others argued that even if Persson and Savulescu are focusing on the right traits to enhance, it's hard to imagine a version of real form of moral bioenhancement that is precise and fine-grained enough to make us (for example) only as altruistic as we *should* be, reducing responses things like problematic rage while also allowing us to feel appropriate anger in certain contexts.[10]

While here is not the place to adjudicate the matter of whether some particular proposal for pharmacological moral enhancement is superior to others, what the above suggests is that simply trying to induce the right thoughts and feelings in people faces a number of philosophical and practical hurdles. As one emerging line of thought has it, advocates of moral enhancement might do better to focus on bringing about moral improvement that is more akin (analogously, in the case of cognitive enhancement) to the use of an app to learn a language or to reach a conference venue; that is, perhaps the most effective form of moral enhancement will not take the form of a brain-chemistry altering pill but rather artificial intelligence that engages us in communication to the end of bettering our moral character and reasoning.

There are different ways of fleshing out the core idea of reliance on AI in order to improve our moral decision-making. Alberto Giubilini and Julian Savulescu (2018) envision one notable kind of approach according to which, when an AI

---

[10] See e.g., de Melo-Martín (2018) for a version of the empathetic terrorist objection, and see Harris (2010) for a compelling discussion of why moral enhancement of the sort Persson and Savulescu originally envision is likely to be inadequately precise.

user is struggling to make a moral choice, the user might be prompted to rank a set of values that the AI would then use to generate a moral choice that is compatible with this ranking.[11] However, one might wonder if this—i.e., reliance on *moral coherence-prompting AI*—constitutes genuine moral *enhancement*. There is, after all, nothing in Giubilini and Savulescu's scenario that would normatively constrain our initial ranking of values. To see why the kind of internal consistency achieved by such a proposal could falls short of moral improvement, imagine someone inputs into the AI the values that reflect a particular religious view the user antecedently accepts. The AI will not encourage the agent to critically evaluate those values or that worldview, but rather simply point at various decision points what would be coherently recommended by (or at minimum be logically consistent with) those values. Plausibly, part of becoming morally enhanced or morally improved will involve (or at least be open to involving) some consistent, honest reflection *on one's values* (and not just engagement with what follows from those values), and that aspect of moral improvement is conspicuously absent from such an AI model.

A similar kind of criticism looks applicable to a closely related form of AI moral enhancement discussed in recent work by Alberto Tassella et al. (2023). Notice that one limitation to the Giubilini-Savulescu coherence-maximising moral AI is that—regardless of what values users identify to the AI as their own through an initial ranking—the user might be mistaken about what her own values in fact are. Empirical results on expressive reporting indicate that our self-reported values are often sensitive to non-epistemic reasons (e.g., social signalling) which can distort accuracy.[12] More generally, we might not have reliable access to what our values are (Nisbett and Wilson 1977; cf., Schwitzgebel 2008), even apart from concerns about non-epistemic factors. A form of AI-moral enhancement suggested by Tassella et al. (2023) offers some initial promise for controlling for this kind of unreliability, which could go as far as to automatically derive user preferences from their 'data ... up to constantly observing the users' everyday life' (2023, 5).[13] This kind of proposal, which replaces the 'reported preferences' with 'revealed preferences', aims to maximise accuracy in making recommendations in line with actual values; but—and here is the concern—just as with the Giubilini-Savulescu approach, it is debatable whether what is described here is genuine moral enhancement given that what is 'extracted' from revealed preferences and behaviour data by the AI might be a ranking of values that we take to be problematic.

---

[11] See Lara and Deckers (2020) for a detailed version of this sort of proposal, and see Constantinescu et al. (2022) as well as Lara (2021) for discussion of potential objections to this approach moral enhancement.

[12] See Hannon (2021) for discussion.

[13] This is one of several forms of potential AI moral enhancement proposed by Tassella et al. (2023). They note that the form that relies on revealed preferences extracted from data is not yet viable without ensuring the AI has some reliable understanding of how to interpret behaviour.

Recommendations from the AI would then be recommendations of actions consistent with undesirable revealed preferences.

Notice that one shared limitation of proposed AI moral enhancement programmes suggested by Giubilini, Savulescu, and Tassella, et al. is that they are, in short, compatible with the promotion of 'bad values', and in such a way as to make it not obvious that relying on them would 'enhance' us morally. This limitation, however, is—by the lights of some AI ethicists—a kind of 'feature' rather than a bug. Consider, for a moment, what might seem like the most obvious alternative to such approaches—viz., an AI that didn't *ask* your values (and then merely make recommendations consistent with them), but which started out with something like a 'good list' of moral values, and then made recommendations in accordance of those, and regardless of whether the recommendations line up with the user's values. *If* the values 'programmed in' to the AI are in fact good ones, then one might at least initially think that being always and everywhere guided by the AI morally might in some way constitute moral *enhancement* in a way that being guided by merely 'conditional prescriptions' logically consistent with your existing values (whatever they may be) might not be.

Pursuing this route—call it *strong prescriptive moral AI*—raises its own problems. First, it faces a version of what is called in AI ethics the *alignment problem*[14]—viz., the practical problem of aligning the behaviour of the AI with the human objectives. The difficulty of the alignment problem is already apparent outside of moral enhancement AI specifically, as is evident in the difficulty of 'programming in' values that govern how self-driving cars react at certain morally significant decision points. Given the lack of agreement about what values should be given most weight when assessing how a *human* should respond in, e.g., variations of trolley problems, it is at best challenging to know what values should be programmed into the AI. But even setting this point aside, there are other limitations to strong morally prescriptive AI. Suppose that the alignment problem in the case of constructing a moral enhancement AI could be 'solved';[15] even on that assumption, it is not obvious that *users* of the AI would be 'morally enhanced' by following the advice of such a prescriptive AI.

On this point consider David Archard's remarks about a related kind of situation: taking moral advice from *moral philosophers*. Should ordinary folks simply *defer* to the moral advice of moral philosophers, even if (albeit controversially) moral philosophers are 'experts' in the area? Archard maintains that even we grant a starting premise that moral philosophers have some degree of moral expertise, 'moral philosophers should not wish non-philosophers to defer to their

---

[14] See, e.g., Christian (2020) and Yudkowsky (2016).
[15] Perhaps, through some combination of cognitively enhanced moral philosophers working in collaboration with advanced machine learning algorithms, we discover some optimal set of moral values to 'programme in' to a moral enhancement AI, such that the prescriptive advice that the AI would give would be optimal (never mind how), thus in line with our objectives.

putative expertise' (Archard 2011). Why not? The worry is that if there is some valuable quality to autonomous thinking *behind* our moral judgments (and not just that the content of them is correct), then that won't be attained by blind deference to even the most accurate prescriptive moral AI. Consider, for example, current version of a prescriptive AI, Allen Institute for AI's research prototype 'Ask Delphi' (Jiang et al. 2022), which is described as 'an experimental framework based on deep neural networks trained directly to reason about descriptive ethical judgments'.[16] There is a text box on the webpage, where the user can ask any moral question, and it offers an answer (without explanation), within seconds. In experimenting with 'Ask Delphi' in July 2023 (using version 1.0.4),[17] I asked it some test questions: "Is it OK to rob a bank?" Its response was "It's wrong." I then asked it "Is it OK to help a friend?". Its response (as expected): "It's good." I then asked it, "It is OK to help a friend by robbing a bank?" This is, of course, more complicated. Its answer: "It's bad." Why is it bad? 'Ask Delphi' doesn't explain itself. In this respect, the 'autonomy' worry Archard raises for the idea of deferring to moral philosophers might not only carry over to strong prescriptive AI such as 'Ask Delphi', but carry over more substantially.[18]

Against the backdrop of the above example types of prospective AI moral enhancement, what emerges is a kind of dilemma. On the one hand, prospective AI moral enhancement that consists in recommending to the user what is consistent with her pre-existing reported (or revealed) values is not clearly genuine *enhancement*. On the other hand, attempting to overcome this limitation by making prospective AI more robustly prescriptive (as in the case of 'Ask Delphi') then runs in to a challenging version of the alignment problem and also faces (a version of) Archard's autonomy problem.

With the above concerns in mind, consider a proposal that has the basic kind of structure that could in principle at least navigate the above dilemma, and which will be our working focus (in connection with the possibility of AI-enhanced moral expertise) in Sections 3–4. This is Francisco Lara and Jan Deckers (2019)'s proposal for moral enhancement via *Socratic AI*, sometimes just called SocrAI (Lara 2021).

Unlike AI that recommends choices that fit with an inputted hierarchy of values (or, like strong prescriptive AI, which simply prescribes recommendations in

---

[16] See also Bang et al. (2023).   [17] https://delphi.allenai.org
[18] An interesting and broadly related earlier approach is developed by Klincewicz (2016), which uses an artificial moral reasoning engine to present moral arguments that are based in first-order normative theories (e.g., such as utilitarianism and Kantianism). In so far as the engine would be showing what consistent reasoning looks like from premises of such theory that one already accepts; the view might seem akin to the Giubilini-Savulescu coherence maximising approach. However, Klincewicz's moral reasoning engine is designed to play a stronger "normative" role, by potentially persuading users to accept its arguments conclusions. For instance, as Klincewicz says, "[...] it can, if prompted to do so, give answers to first-order normative questions, such as "should I report this to the authorities?" with a definite "yes" or "no" and then also provide reasons in support of that answer." In some respects, this latter characteristic of Klinewicz's proposal shares prescriptive commonalities with, e.g., "Ask Delphi."

accordance with predefined values), the Lara and Deckers model focuses on how AI can teach improve our *moral reasoning*. With Socratic AI, the AI (e.g., imagine running an Large Language Model [LLM] such as a variant of ChatGPT) asks the agent questions that aim e.g., to help to clarify existing beliefs and values and encourage the agent to uncover option space they may have missed thus far. This type of activity, Lara and Deckers suspect, will not only help the agent exercise and hone their moral reasoning skills but also *motivate* agents to make choices that align with what they think is right, after reasoned reflection.[19] Notice that Socratic AI avoids the second horn of the dilemma in that (without requiring built in values) it avoids the kind of alignment problem that faces 'Ask Delphi'; likewise, by not *recommending* particular courses of action, it sidesteps (also on the second horn) Archard's autonomy problem; Socratic AI is not encouraging you to *accept propositional moral content as true*. At the same time, there is some scope for dodging the first horn: to see why, compare Socratic AI versus, e.g., Giubilini/Savulescu's coherence maximising moral AI, in the case of someone who *begins with* (to simplify things here) bad values. Where will they end up, after interacting with these AIs, respectively? Whereas coherence-maximising moral AI simply recommends behaviour consistent with the bad values, Socratic AI prompts critical reflection that might potentially result in the *modification* of those values, including through improved moral reasoning skills that the Socratic AI aims to facilitate. In this way, we can see how the imagined individual might be genuinely morally better off via interaction with Socratic AI.

3. Is AI-Assisted Moral Enhancement Compatible with Genuine Moral Expertise?

Let's briefly take stock. We've seen that Socratic AI offers us at least one *prima facie* promising approach to AI-based moral enhancement, in that it offers a way we can rely on information (even if via sheer deference to prescriptive advice) from an AI assistant in a way that both (i) looks capable of leading to genuine moral improvement in a user; and (ii) does so without inviting the kinds of objections applicable to, e.g., strong prescriptive moral AIs.

Here is not the place to take any kind of definitive stance on whether Socratic AI *should* be pursued, or whether it is even successful (all things considered) as a moral enhancement strategy. Rather, the fact that we've seen (Section 2) that Socratic AI holds some particular promise compared to other would-be AI-based moral enhancement strategies makes Socratic AI a particularly fruitful form of AI-based moral enhancement.

---

[19] For a recent and related kind of proposal, see Volkman and Gabriels (2023).

The question we turn to now is whether *dependence on Socratic AI is compatible with possessing moral expertise?* The question of who the moral experts are, and in connection with wider discussions of the value of moral expertise, make the possibility that technology such as Socratic AI could help generate moral expertise in users especially salient. It also is suggestive of the idea of a kind of 'democratisation' of moral expertise: if Socratic AI were widely available then, ceteris paribus, so will the capacity to attain moral expertise.

This section will attempt to bring this guiding question of whether dependence on Socratic AI is compatible with attaining moral expertise under intellectual control, by identifying some key desiderata that any kind of AI-based moral enhancement would do well to satisfy if that enhancement should be thought to give rise to moral expertise Section 4 then, with reference to these desiderata, looks at how Socratic AI holds up.

Before getting into these desiderata, though, I want to first make a few simplifying assumptions that will need made to get the rest of the discussion off the ground. Is there really moral expertise? Some philosophers have thought not—pointing to reasons ranging from the fact that, as Ryle noted, it's not obvious that the difference between right and wrong is something capable of being 'forgotten', to the fact that (as C. D. Broad (1952) put it), moral philosophers aren't in the business of telling people what to do[20] (even if experts in general are in such a position, in their relevant domains). Others, such as Driver (2013) and J.S. Gordon (2023), think that arguments that have dismissed the possibility of moral expertise have been too quick.

Let's assume there is such a thing as moral expertise—viz., expertise in the domain of morality.[21] What would be the *nature* of such expertise be? What would it plausibly involve, on the assumption it is a real phenomenon humans can aspire to?

While it might be tempting to venture few plausible necessary conditions on moral expertise, I want to try to avoid doing so; this is because conditions sufficient for moral expertise might not also be necessary. Regardless of whether we think (as per Weinstein 1993) that epistemic expertise (roughly: expertise consisting in 'providing strong justifications for a range of proposition in a domain') and performative expertise (roughly: expertise consisting in 'the capacity to perform a skill well according to the rules and virtues of a practice') mark out genuinely different *types* of expertise (or whether they are beset understood as

---

[20] As Broad (1952) puts it, 'It is no part of the professional business of moral philosophers to tell people what they ought or ought not to do.... Moral philosophers, as such, have no special information not available to the general public, about what is right and what is wrong; nor have they any call to undertake those hortatory functions which are so adequately performed by clergymen, politicians, leader-writers'.

[21] While some discussions of expertise relative expertise to a skill as well as a domain (where we might think of a skill as a trait manifested within a wider domain—see Stichter (2015)), for simplicity I'll talk of expertise in connection with domains.

different realisations of the same type), we have empirical evidence that suggests that experts who perform well might not always be knowledgeable in the way that would seem sufficient in some contexts of being an expert. As Matt Stichter (2015, 113) puts it: 'Even when experts are able to articulate an explanation, the explanations are often inconsistent with the observed behaviour of the experts'.

Rather than to assume then that if moral expertise exists it would involve meeting any necessary conditions, let's—for our purposes here—simply take note of a cluster of *dimensions* that are widely taken to track expertise. On the assumption that expertise (like cognate notions of skill and know-how) is a matter of degree, we might then think that an individual is a better candidate for moral expertise the more of these dimensions they meet (without taking the failure of any particular dimension to be disqualifying).

These simplifying assumptions made, what are some of the dimensions of moral expertise? Literature on expertise generally (and moral expertise specifically) suggests a varying range. Here I want to note *six key dimensions* which different philosophers have identified, and in Section 4 we'll use these as provisional criteria to return in a more organised way to our question about Socratic AI and moral expertise.

One straightforward metric on which expertise is attributed is *knowledgeableness* (e.g., (Goldman 2001, 91)); we expect an expert ornithologist to know a lot about birds. In a similar vein, a moral expert should be knowledgeable about the domain of morality.[22]

A separate metric however tracks *performance*: we expect, e.g., an expert in the law to not merely engage knowledgeably in reflection on legal facts, but to *manifest* this knowledge in their performance (Weinstein 1993)—this might involve following laws, pointing out when a law is broken and what counts as the breaking of particular laws, and so on. In the moral case, it could envision—analogously—the attribution of moral expertise tracking something like *manifesting* moral knowledge in action.

A third expertise metric acknowledged in the literature is *automaticity*. As Stichter (2015) puts it, 'Experts do not need to devote much conscious attention to what they are doing, and this lack of conscious attention does not lead to any reduction in their performance' (2015, 64). Translated to specifically the moral case: we might expect a moral expert might, as Aristotle would expect of the morally virtuous, to do the right thing without prior deliberation on, e.g., moral principles. This point lines up likewise with thinking about expertise due to Hubert Dreyfus, who denies that expertise involves the conscious following of rules (moral or otherwise).

---

[22] This knowledge might take different forms (know-how, propositional knowledge, occurrent, tacit, etc.).

A fourth dimension that's been associated with expertise is *rational autonomy*. As Finnur Dellsén (2020, 358) argues in a recent paper: ""Experts should make up their own minds about issues that fall within their domain of expertise, as opposed to following the opinions of their fellow experts;"; "moral expertise, by this metric, would be associated with forming moral views in a way that is not merely deferential, including, by deference to other experts.

A fifth metric associated with expertise is *principle unification*—as defended in work on expertise and virtue by Julia Annas (1995). As Annas sees it, domains that admit of skill and expertise are governed by various unifying principles, and the expert must have a grasp of not only part of the field. A chess expert grasps principles governing good openings playing black and playing white, openings and end games, etc. and not just some principles lining up with a 'part' of chess strategy. By parity of reasoning, we might associate moral expertise with having more than a mere partial grasp of moral principles, viz., more than just a grasp of *some* principles.

A sixth and final metric of expertise, discussed in various ways by Alvin Goldman (2018) and Christian Quast (2018), though cf., Michel Croce (2019), is *helpfulness*. As Goldman puts it, in attributing expertise, we consider not just the expert's own traits in isolation from their community, but also 'what experts can *do* for laypersons by means of their special knowledge or skill'. Part of genuine expertise, for Goldman, is having 'the capacity to help others (especially laypersons) solve a variety of problems' in the domain in which one is an expert. A medical expert can help a patient treat an illness; and by parity of reasoning, a moral expert by this metric will be positioned to be morally *helpful*—viz., to (perhaps) help others think through moral problems well.

4. Socratic AI and Moral Expertise, Revisited

What we've gained now from the previous section is that (i) on the simplifying assumption that there is such a thing as moral expertise, we can expect that (ii) the extent to which one attains moral expertise could be reasonably expected to track the six different recognised expertise metrics detailed in Section 3, and we saw roughly what those metrics, in the case of moral expertise specifically, might look like if satisfied.

Against that background, let's return now to Socratic AI. By way of reminder, the question under investigation here isn't whether Socratic AI *itself* might aspire to expertise. That question takes us well beyond what I'm aiming to cover here, as it raises questions (including questions of increasing interest in bioethics) about whether artificial agents might be literal bearers of agency properties,[23] like

---

[23] For recent discussion on this point, see, e.g., Cervantes et al. (2020) and Bryson, Kime, and Zürich (2011).

knowledge and expertise. (See, however, Rodríguez-López and Rueda 2023 for a recent case defending at least some kinds of AI as bona fide moral experts.)

Rather, let's consider the extent to which dependence on Socratic AI-based moral enhancement might track the possession of genuine moral expertise, taking each of the six metrics in turn.

First, consider *knowledgeableness*. There is well-known scepticism in the literature on moral deference about whether one can gain moral *knowledge* via testimony,[24] and this is a serious strike against the thought that one could gain moral expertise (at least by this metric) simply through consistent reliance on even the best kind of strong prescriptive AI (e.g., such as 'Ask Delphi'). The concern is, roughly, that there is something distinctive about, e.g., moral and aesthetic beliefs, which is that we are in the market for knowledge of such beliefs only by in some sense appreciating for ourselves why they are true when they are. In so far as sheer deference allows one to believe a proposition in the absence of such appreciation for its grounds, sheer deference isn't a route to moral (and aesthetic) knowledge, or so the argument goes.

Interestingly for our purposes, notice how this line of argument does not carry over from a reliance on strong prescriptive moral AI to a reliance on Socratic AI. Here an analogy between knowledge gained through psychotherapy and Socratic AI is instructive. Let's take as a starting point the premise that cognitive behavioural therapy (CBT) is a potential avenue for gaining self-knowledge—in so far as it facilitates change in one's beliefs about oneself in par through targeting and changing irrational aspects of an agent's thinking about herself. Consider now that interacting with Socratic AI appears to be akin to interacting with a very narrowly focused cognitive behavioural therapist, insofar as both CBT and Socratic AI work to find inconsistent and irrational aspects of an agent's thinking (e.g., Overholser 2010). However, there's at least one way in which a practitioner of Socratic AI is likely to be more effective than a CBT practitioner. While the cognitive behavioural therapist has to work to set aside their own values, the Socratic AI simply doesn't come with particular values or choices of values—Lara and Deckers (281–2) propose the system have 'no previous lists or systems of values from which to improve the morality of the agent', with algorithms designed to avoid the machine being biased towards any particular values or normative ethical theories. Instead 'through the constant interaction between the agent and the system, the possibility that the agent's values would be changed through their dialogue with the machine is increased' (Lara and Deckers 2019, 281). If (non-prescriptive) CBT is a route to self-knowledge (even if not by prescribing an agent accept beliefs on sheer deference), then we have reason to think Socratic AI would likewise be an analogous kind of route to knowledge in the moral domain.

---

[24] See, e.g., McGrath (2009) and Hills (2009).

Suppose then that a user of AI gains moral knowledge through her dependence on Socratic AI. Once this much is granted, is there any barrier to the user of Socratic AI manifesting moral knowledge acquired in action? A critic here might suggest that Socratic AI, even if capable of inculcating moral knowledge (in a way broadly analogous to the way CBT might do so) the knowledge is not going to be as situation specific or actionable as, e.g., the kind of moral *information* that we could expect form a strong prescriptive moral AI. Take for example an imagined case where you are in doubt about whether to give a particular friend who has fallen short in the past a second chance with a high-responsibility task. Whereas a strong prescriptive moral AI might simply communicate action-guiding information here, Socratic AI will not do so. Is this much a genuine barrier to a user of Socratic AI attaining moral expertise along as captured by a performance metric on expertise? There's a good case here for pressing back with some optimism. Imagine here a 'good case' where one's moral values improve over time through the use of Socratic AI, through reflective and thoughtful interactions with the AI, which prompt moral belief revision in a way where rationally supported moral knowledge (e.g., perhaps of certain moral principles) is attained. *Manifesting* moral knowledge in one's actions might very well manifest known *principles* rather than merely known (situation specific) moral information one might have in a particular situation. For example, your repaying a loan might manifest your general knowledge that loans should be repaid, rather than any specific knowledge about whether you should repay a particular loan on occasion. Once this point is appreciated, though, initial reservations for doubting that Socratic AI interactions might support expert moral action seems overstated.

What about *automaticity*? This might look, initially, like the most serious disanalogy between what we'd expect of a domain expert and what we'd envision in the case of a user (in the moral domain) of Socratic AI. Here is perhaps the strongest form of the challenge: following Stichter, 'Experts do not need to devote much conscious attention to what they are doing' (2015, 64). A user of Socratic AI, however, is consciously engaging with the AI, depending on its responses, in a way we might conceptualise as a kind of intermediate and conscious step, a 'thinking and an acting' that's incongruous with the automaticity of an expert. This is a fair point of criticism. However, there is perhaps an equally compelling line of reply— one that is concessionary in that it simply grants that the *consultation* of Socratic AI is a conscious activity, one involving conscious attention, which is as such an activity that is at odds with automaticity. However, as this line of thought goes, we should distinguish between the learning phase (when one is interacting with Socratic AI), and whatever moral expertise would be the *result* of such learning. If we think of Socratic AI as a kind of 'scaffolding', whereby one (through interaction with the AI) learns morally over time, then we can grant the critic that there is no automaticity during the learning stage. *But*, as the thought goes, this is just what we should expect of learning in other domains, where conscious attention to one's action during learning *precedes* expertise. In the case of moral expertise attained through Socratic AI, then, the thought would be that (in connection with automaticity) the conscious attention to *learning* is not disanalogous with what we find in other domains of expertise; and further, following a period of moral leaning through Socratic AI, we can expect one's moral knowledge-manifesting action would be similarly automatic.

What about *rational autonomy?* In so far as (as per Dellsén 2020) experts as such should 'make up their own minds', moral expertise not only would preclude, e.g., either deference *or* agnosticism, but it would also be such that it could be improved through gaining skills to improve at making up one's mind in a moral matter in a rational way. With this in mind, not only would moral enhancement via Socratic AI avoid the promotion of moral deference (as already noted) but it functions so as to improve one morally specifically by empowering the user's rational autonomy via improved moral reasoning skills. On this point, Lara and Deckers give several example illustrations of how Socratic AI users might become better moral reasoners:

- By being assisted in anticipating and explaining the likely consequences of particular choices.
- By learning about how aspects of our environment impact decision-making.
- By being made aware of how human biology impacts decision-making.
- By having ambiguous language use pointed out.
- By having clarity highlighted and encouraged.
- By learning about empirical support for particular beliefs (and learning about the *lack* of empirical support for particular beliefs). (Lara and Deckers 2020)

The only viable line of rejoinder I see in the case of rational autonomy, on behalf of the critic of the prospects of gaining moral expertise through Socratic AI, would hold that one's rational autonomy might be in some way diminished by her *epistemic dependence* on the Socratic AI itself. This objection overgeneralises, however. If relying on dialogue to facilitate moral reasoning is rational autonomy-undermining, then presumably it will likewise be autonomy undermining in more traditional forms of education that avoid indoctrination by simply aiming to facilitate critical thinking skills. Put another way, the bare dependence on Socratic AI isn't plausible rational autonomy undermining unless we accept, implausibly, trivialise the undermining of rational autonomy so as to grant that it occurs in more standard cases of acquiring critical thinking skills in learning.

What about Annas's 'principle unification' metric of expertise? On the assumption that domains that admit of expertise are governed by constitutive principles, should we think non-prescriptive character of Socratic AI would somehow prevent one from coming to learn such principles? On this point, it's worth considering how this argument might be turned completely around. If we take seriously Annas's suggestion that the expert has a 'grasp' of the relevant principles in the domain, then there is scope to think that prescriptive AI not only isn't needed to gain knowledge of moral principles, but that it would not be a candidate for disseminating such knowledge. Consider here, as is suggested by the moral deference literature, that coming to believe prescribed principles is not a route to grasping (of the sort we might identify with *understanding*). If that is right, then rather than to view Socratic dialogue of the sort facilitated by Socratic AI as not sufficient for generating knowledge of moral principles, we should think that if one can gain knowledge of such principles at all, the way in which Socratic dialogue might facilitate grasping—a point I've argued for elsewhere[25]—suggests Socratic AI might be particularly well placed to help a user gain such knowledge.

Let's consider now the expertise metric of 'helpfulness', defended variously by Goldman and Quast. One might think—imagining a critical stance here—that a user is prevented from giving useful moral guidance if she is depending herself on external scaffolding. As the thought might go, the credit for the helpfulness goes to the Socratic AI, not to the user. There are two lines of response here. First, the general principle underlying the reasoning overgeneralises. We don't say that expert air traffic controllers, for instance, lack the kind of 'helpfulness' apposite to expertise given that their helpfulness is predicated upon their dependence on the computers they require to track airplane patterns in real time. Second, and perhaps more importantly, even if we granted that dependence on something external to one is in some way at odds with the kind of helpfulness to laypersons befitting of expertise, the objection here would be at most applicable to initial learning stages with Socratic AI. Recall the respond to the anticipated objection from automaticity: after prolonged use of Socratic AI, one might (as with other learning processes) be well positioned to manifest her moral knowledge through the offering of guidance. If the premise that one can gain moral knowledge through one's interactions with Socratic AI over time is granted, then so should be the associated idea that the downstream manifestation of that knowledge could be put to the service of assisting others in a way we'd expect an expert in any domain to be in a position to do.

Summing up, then, it looks like we have every reason to be optimistic that Socratic AI is a genuine route to moral expertise.[26] And even more than that, on the assumption that expertise is a matter of degree, and that the six metrics identified are plausible cluster criteria for identifying experts (even if not individually necessary conditions), we've got cause to be optimistic that the kind of

---

[25] See on this point Gordon (2016).
[26] It's worth noting that I take the optimism here to simply apply to Socratic AI as one, among potentially many, routes to moral expertise. For instance, it might also be that, entirely independently of the benefits one can attain through Socratic AI, engaging with exemplars or role models can have important moral benefit. Thanks to a referee for raising this point.

moral enhancement that would be facilitated through Socratic AI is capable of leading to a significant level of moral expertise—that is, moral expertise that lines up with an array of the most typical markers of expertise we should expect of in any domain.

This result has an interesting bearing on the connection between moral enhancement and moral expertise, which is that—to put it simply—the very idea of enhanced expertise in the moral domain is about as viable as the idea of moral enhancement through AI more generally. That is, in so far as reliance on Socratic AI (even if not necessarily on other versions of AI-based moral enhancement) is a genuine form of moral enhancement (a conclusion that looked promising in Section 2), we have good cause to think that there is no barrier between the enhancement one would attain and the inculcation of genuine moral expertise.

This is a welcome result at least from the perspective where ceteris paribus the democratisation of expertise is valued. Whereas in some domains, expertise might be available only to certain privileged or elite, we've got good reason to think that in so far as Socratic AI could be widely available—as we already expect it might be given the recent emergence of LLMs when can be accessible (e.g., much like OpenAI's ChatGPT from a mobile phone)—a more egalitarian distribution of moral expertise is not implausible, at least for those who have the desire (as with therapy) to put in the reflective work.

5. Concluding Remarks

Artificial intelligence offers one of the most promising new routes to moral enhancement. Separately, the question of *moral expertise* has remained an important one at the forefront of ethics and its intersection with social epistemology. The aim here has been to bring these debates together, to show that a viable form of moral enhancement via artificial intelligence is at the same time a viable route to moral expertise. Section 1 introduced some of the guiding themes of these debates, Section 2 distinguished between several varieties of AI-based moral enhancement, and showed how Socratic AI has some potential advantages over other proposals in so far as it offers a genuine route to moral enhancement. Sections 3–4 then evaluated whether Socratic AI might be in the market not only for supporting moral enhancement but full-blown moral expertise. Section 3 outlined six metrics associated with expertise in a given domain generally, with discussion of what this would look like in the moral domain specifically; Section 4 then considered how users depending on Socratic AI measure up to these metrics, concluding with some optimism that in so far as Socratic AI is a route to moral enhancement, it is likewise a route to moral expertise. What follows more generally is welcome result about the potential democratisation of moral expertise.

References

Annas, Julia. (1995) "Virtue as a skill." *International Journal of Philosophical Studies* 3(2), 227–43.

Archard, David. (2011) "Why moral philosophers are not and should not be moral experts." *Bioethics* 25(3), 119–27.

Bang, Yejin, Nayeon Lee, Tiezheng Yu, Leila Khalatbari, et al. (2023) "Towards Answering Open-Ended Ethical Quandary Questions." arXiv, available at https://doi.org/10.48550/arXiv.2205.05989.

Berry, Ed D. J., Richard J. Allen, Mark Mon-Williams, and Amanda H. Waterman. (2019) "Cognitive offloading: Structuring the environment to improve children's working memory task performance." *Cognitive Science* 43(8), e12770, available at https://doi.org/10.1111/cogs.12770.

Bostrom, Nick and Anders Sandberg. (2009) "Cognitive enhancement: Methods, ethics, regulatory challenges." *Science and Engineering Ethics* 15(3), 311–41.

Broad, C. D. (1952) *"Ethics and the History of Philosophy: Selected Essays."* London, UK: Routledge.

Bryson, Joanna J., Philip P. Kime, and C. Zürich. (2011) "Just an artifact: Why machines are perceived as moral agents." *IJCAI Proceedings-International Joint Conference on Artificial Intelligence* 22, 1641. DOI:10.5591/978-1-57735-516-8/IJCAI11-276.

Carter, J. Adam and Duncan Pritchard. (2019) "The epistemology of cognitive enhancement." *The Journal of Medicine and Philosophy: A Forum for Bioethics and Philosophy of Medicine*, 44, 220–42. US:.Oxford University Press.

Cervantes, José-Antonio, Sonia López, Luis-Felipe Rodríguez, Salvador Cervantes, et al. (2020) "Artificial moral agents: A survey of the current status." *Science and Engineering Ethics* 26, 501–32.

Christian, Brian. (2020) *The Alignment Problem: Machine Learning and Human Values.* New York, NYC: WW Norton & Company.

Constantinescu, Mihaela, Constantin Vic\ua, Radu Uszkai, and Cristina Voinea. (2022) "Blame it on the AI? On the moral responsibility of artificial moral advisors." *Philosophy and Technology* 35(2), 1–26, available at https://doi.org/10.1007/s13347-022-00529-z.

Croce, Michel. (2019) "Objective expertise and functionalist constraints." *Social Epistemology Review and Reply Collective* 8(5), 25–35.

Dellsén, Finnur. (2020) "The epistemic value of expert autonomy." *Philosophy and Phenomenological Research* 100(2), 344–61.

Driver, Julia. (2013) "Moral expertise: Judgment, practice, and analysis." *Social Philosophy and Policy* 30(1–2), 280–96, available at https://doi.org/10.1017/S0265052513000137.

Earp, Brian D. (2018) "Psychedelic moral enhancement." *Royal Institute of Philosophy Supplements* 83(October), 415–39, available at https://doi.org/10.1017/S1358246118000474.

Earp, Brian D., Anders Sandberg, and Julian Savulescu. (2012) "Natural selection, childrearing, and the ethics of marriage (and divorce): Building a case for the neuroenhancement of human relationships." *Philosophy & Technology* 25(4), 561–87 available at https://doi.org/10.1007/s13347-012-0081-8.

Gilbert, Sam J., Arabella Bird, Jason M. Carpenter, Stephen M. Fleming, et al. (2020) "Optimal use of reminders: Metacognition, effort, and cognitive offloading." *Journal of Experimental Psychology: General* 149(3), 501.

Giubilini, Alberto and Julian Savulescu. (2018) "The artificial moral advisor. the ideal observer meets artificial intelligence." *Philosophy and Technology* 31(2), 169–88, available at https://doi.org/10.1007/s13347-017-0285-z.

Goldman, Alvin I. (2001) "Experts: Which ones should you trust?" *Philosophy and Phenomenological Research* 63(1), 85–110, available at https://doi.org/10.1111/j.1933-1592.2001.tb00093.x.

Goldman, Alvin I. (2018) " Expertise." *Topoi* 37(1), 3–10.

Gordon, Emma. (2016) " Social Epistemology and the Acquisition of Understanding," in *Explaining Understanding: New Perspectives from Epistemology and Philosophy of Science*, 293–317. Routledge.

Gordon, Emma C. (2022b) *Human Enhancement and Well-Being: A Case for Optimism*. London, UK: Routledge.

Gordon, Emma C. and Rebecca J. Willis. (2023) "Pharmacological cognitive enhancement and the value of achievements: An intervention." *Bioethics* 37(2), 130–34.

Gordon, John-Stewart. (2023) "Moral expertise revisited." *Bioethics* 37(6), 533–42, available at https://doi.org/10.1111/bioe.13172.

Grinschgl, Sandra, Hauke S. Meyerhoff, and Frank Papenmeier. (2020) "Interface and interaction design: How mobile touch devices foster cognitive offloading." *Computers in Human Behavior* 108(July), 106317, available at https://doi.org/10.1016/j.chb.2020.106317.

Gupta, Kristina. (2012) "Protecting sexual diversity: Rethinking the use of neurotechnological interventions to alter sexuality." *AJOB Neuroscience* 3 (3), 24–8.

Hannon, Michael. (2021) "Disagreement or badmouthing? The role of expressive discourse in politics," in *Political Epistemology*, (eds.), Elizabeth Edenberg and Michael Hannon. Oxford: Oxford University Press.

Hardcastle, Valerie Gray. (2018) "Lone wolf terrorists and the impotence of moral enhancement." *Royal Institute of Philosophy Supplements* 83, 271–91.

Harris, John. (2011) "Moral enhancement and freedom." *Bioethics* 25(2), 102–11.

Hills, Alison. (2009) "Moral testimony and moral epistemology." *Ethics* 120(1), 94–127.

Hurlemann, René, Alexandra Patin, Oezguer A. Onur, Michael X. Cohen, et al. (2010) "Oxytocin enhances amygdala-dependent, socially reinforced learning and emotional empathy in humans." *Journal of Neuroscience* 30(14), 4999–5007.

Jiang, Liwei, Jena D. Hwang, Chandra Bhagavatula, Ronan Le Bras, et al. (2022) "Can machines learn morality? The Delphi experiment." arXiv, available at https://doi.org/10.48550/arXiv.2110.07574.

Jotterand, Fabrice and Susan B. Levin. (2019) "Moral deficits, moral motivation and the feasibility of moral bioenhancement." *Topoi* 38, 63–71.

Kass, Leon. (2002) *Life, Liberty and the Defense of Dignity: The Challenge for Bioethics.* New York: Encounter Books.

Klincewicz, Michał. (2016) "Artificial intelligence as a means to moral enhancement." *Studies in Logic, Grammar and Rhetoric* 48(1), 171–87.

Lara, Francisco. (2021) "Why a virtual assistant for moral enhancement when we could have a Socrates?." *Science and Engineering Ethics* 27(4), 42, available at https://doi.org/10.1007/s11948-021-00318-5.

Lara, Francisco and Jan Deckers. (2020) "Artificial intelligence as a Socratic assistant for moral enhancement." *Neuroethics* 13(3), 275–87, available at https://doi.org/10.1007/s12152-019-09401-y.

Liao, S. Matthew. (2011) "Parental love drugs: Some ethical considerations." *Bioethics* 25(9), 489–94, available at https://doi.org/10.1111/j.1467-8519.2009.01796.x.

Maslen, Hannah, Filippo Santoni de Sio, and Nadira Faber. (2015) "With cognitive enhancement comes great responsibility?." *Responsible Innovation 2: Concepts, Approaches, and Applications*, 121–38. Springer, Cham. https://doi.org/10.1007/978-3-319-17308-5_7.

McGrath, Sarah. (2009) "The puzzle of pure moral deference." *Philosophical Perspectives* 23, 321–44.

Melo-Martín, Inmaculada de. (2018) "The trouble with moral enhancement." *Royal Institute of Philosophy Supplement* 83, 19–33, available at https://doi.org/10.1017/s1358246118000279.

Mikolajczak, Moïra, Nicolas Pinon, Anthony Lane, Philippe de Timary, and Olivier Luminet. (2010) "Oxytocin not only increases trust when money is at stake, but also when confidential information is in the balance." *Biological Psychology* 85(1), 182–4, available at https://doi.org/10.1016/j.biopsycho.2010.05.010.

Nisbett, Richard E. and Timothy D. Wilson. (1977) "Telling more than we can know: Verbal reports on mental processes." *Psychological Review* 84(3), 231.

Nyholm, Sven. (2015) "Love troubles: Human attachment and biomedical enhancements." *Journal of Applied Philosophy* 32(2), 190–202.

Overholser, James C. (2010) "Psychotherapy according to the Socratic method: Integrating ancient philosophy with contemporary cognitive therapy." *Journal of Cognitive Psychotherapy* 24(4), 354–63.

Persson, Ingmar and Julian Savulescu. (2008) "The perils of cognitive enhancement and the urgent imperative to enhance the moral character of humanity." *Journal of Applied Philosophy* 25(3), 162–77, available at https://doi.org/10.1111/j.1468-5930.2008.00410.x.

Persson, Ingmar and Julian Savulescu. (2012) *Unfit for the Future: The Need for Moral Enhancement.* Oxford: Oxford University Press.

Pokorny, Thomas, Katrin H Preller, Michael Kometer, Isabel Dziobek, et al. (2017) "Effect of psilocybin on empathy and moral decision-making." *International*

Journal of Neuropsychopharmacology 20(9), 747–57, available at https://doi.org/10.1093/ijnp/pyx047.

Quast, Christian. (2018) "Expertise: A practical explication." *Topoi* 37(1), 11–27, available at https://doi.org/10.1007/s11245-016-9411-2.

Rodríguez-López, Blanca and Jon Rueda. (2023) "Artificial moral experts: Asking for ethical advice to artificial intelligent assistants." *AI and Ethics* 1–9.

Sandel, Michael J. (2007) *The Case against Perfection: Ethics in the Age of Genetic Engineering.* Cambridge, MA: Harvard University Press.

Schwitzgebel, Eric. (2008) "The unreliability of naive introspection." *Philosophical Review* 117(2), 245–73.

Stichter, Matt. (2015) "Philosophical and psychological accounts of expertise and experts." *HUMANA.MENTE Journal of Philosophical Studies* 8(28), 105–28, available at https://www.humanamente.eu/index.php/HM/article/view/83.

Tassella, Marco, Rémy Chaput, and Mathieu Guillermin. (2023) "Artificial moral advisors: Enhancing human ethical decision-making," in, 1. *IEEE*, available at https://doi.org/10.1109/ETHICS57328.2023.10155026.

Volkman, Richard and Katleen Gabriels. (2023) "AI moral enhancement: Upgrading the socio-technical system of moral engagement." *Science and Engineering Ethics* 29(2), 11, available at https://doi.org/10.1007/s11948-023-00428-2.

Wang, Ju. (2021) "Cognitive enhancement and the value of cognitive achievement." *Journal of Applied Philosophy* 38(1), 121–35.

Weinstein, Bruce D. (1993) "What is an expert?' *Theoretical Medicine* 14(1), 57–73, available at https://doi.org/10.1007/BF00993988.

Wilson, Clare. (2020) *New Apple Watch Monitors Blood Oxygen–Is That Useful?* Amsterdam, The Netherlands: Elsevier.

Yudkowsky, Eliezer. (2016) "The AI alignment problem: Why it is hard, and where to start." *Symbolic Systems Distinguished Speaker* 4. https://intelligence.org/2016/12/28/ai-alignment-why-its-hard-and-where-to-start/.

PART 7
NEW DIRECTIONS

14
Decolonising Experts

Veli Mitova
African Centre for Epistemology and Philosophy of Science
University of Johannesburg

1. Introduction

What would happen to our notion of expertise if we took seriously decolonial theorists' insight that former colonial subjects continue to be epistemically marginalised and illegitimately stripped of epistemic authority? We should revise our accounts of expertise. Or so I argue in this paper.

If the argument works, it will advance both scholarship on epistemic decolonisation and the epistemology of expertise. Concerning the former, epistemic decolonisation is, in the first instance, a project of recentring the knowledge enterprise onto one's current geo-cultural location (Ndlovu-Gatsheni 2018). This crucially involves restoring the rightful epistemic authority to the previously colonised (Mitova 2020). An epistemology of expertise that ignores the acknowledged experts of the colonised is thus both theoretically incomplete—since it has no room for certain kinds of experts—and morally suspect—since it enforces the epistemic marginalisation of the oppressed.[1]

In this paper, I develop the beginnings of an account of expertise that redresses these theoretical and moral shortcomings. The argument develops in four steps. First, I show that the experts of the marginalised—for instance, the traditional healers of Southern Africa—are indeed experts Section 2). Next, I argue that existing accounts of expertise cannot accommodate this claim (Section 3). Third, I sketch a view of expertise—I call it 'communitarian functionalism'—that allows for the experts of the marginalised to count as experts (Section 4). Very roughly: a person counts as an expert in domain D by virtue of their role in their epistemic community and whether they live up to this role in an epistemically responsible manner. The final step of the argument is to refine communitarian functionalism by defending it against objections (Section 5).

---

[1] See, for instance, Okeja (2013) for how this plays out with African experts in literature and philosophy.

2. Marginalised Experts Are Experts

Epistemically marginalised communities are communities that don't have an authoritative voice in the global knowledge economy. The epistemic injustice literature distinguishes many ways in which such marginalisation can play itself out. Members of a certain community may, for instance, not have their testimony believed because of prejudice against their race or other aspects of their social identity. (Miranda Fricker 2007 calls this 'testimonial injustice'.) This typically goes hand in hand with what Fricker (2007) calls 'hermeneutical marginalisation'— the sidelining of the epistemic resources of this community. Consequently, this community tends to be denied the right to shape the knowledge economy in ways that benefit it epistemically—where its experiences are understood, its interests are fostered, and so on. (Kristie Dotson 2012 dubs some versions of this phenomenon 'contributory injustice'.)

2.1 Sangomas

Throughout this paper, the phrase 'marginalised experts' will be shorthand for the experts of the marginalised—those individuals in an epistemically marginalised community who are considered by this community to be its experts and are marginalised in virtue of being the epistemic authorities of this community. The marginalisation is a feature of the community in the first instance, and the expert's marginalisation is a direct product of that. In this paper, I will use the traditional healers of Southern Africa as examples of marginalised experts. These are known as Sangomas in South Africa, a label that I will use throughout.[2] In this sub-section I give a very brief sketch of these experts, and in the next I argue that they are indeed experts.

Traditionally labelled as witch doctors by the white colonisers and derided as part of the general 'savagery' that the 'Civilising Mission' was meant to eradicate, Sangomas are to this day on the epistemic margins. Take South Africa today, thirty years after the end of Apartheid, and fifteen years after an Act of Parliament was passed to supposedly restore to them their rightful epistemic and legal authority.[3] Despite the fact that many people consult them, Sangomas' sick notes are not accepted by employers. Despite the arduousness of the training they need to undergo, employers don't give them study leave for it. And although an official Traditional Healers' Council has been under negotiation for some

---

[2] Strictly speaking, Sangomas are diviners, while Inyangas are herbalists. I will follow common usage to refer to both as Sangomas, to keep things simple. For an argument that this isn't misleading, see Cumes (2013, 58).
[3] South African Government (2007).

time, it has not yet been constituted with the legal or epistemic authority of the Medical Council.

The basic function of a Sangoma is to take care of the physical, social, and spiritual wellbeing of a patient. Notably, this includes the patient's relationships with their community, ancestors, and other supernatural beings (Sogolo 1998). I say 'supernatural' for the benefit of the Western reader. African cosmologies don't draw a distinction between the natural and non-natural in the way the West does. These two spheres are seamlessly one, with causal relations running in both directions. Supernatural causal explanations complement, rather than compete with, natural ones (e.g. Sogolo 1998, 234).

Although the *basic* function of a Sangoma is simple enough, their skills and more specific functions are extremely complex and often hard to understand without a grasp of the full African cosmological picture. For instance, a Sangoma 'uses altered states of consciousness, spirit possession, and sometimes out-of-body spirit flight to gain knowledge about any problem at hand' (Cumes 2013, 58). Another nice dramatisation of this complexity is the Sangoma's relation to the substances they use to heal, so called *muthi*:

> *Muthi* refers to substances fabricated by an expert hand, substances designed by persons possessing secret knowledge to achieve either positive ends of healing, involving cleansing, strengthening, and protecting persons from evil forces, or negative ends of witchcraft, bringing illness, misfortune, and death to others or illicit wealth and power to the witch...*Muthi* also plays a part in communications between humans and spirits. Spirits both activate the powers inherent in *muthi* and empower these substances with new force.
> (Ashforth 2005, 212–3, italics in original)

The *Sangoma* is, thus, someone who both produces the *muthi*, channelling the right spirits into it, and the one who mediates between patient and these spirits in the realisation of its powers. According to David Cumes (2013, 60), there are four kinds of spirits that such mediating might involve—ancestral, cosmic, terrestrial, and water.

Suppose that you have a cosmology which needs such inter-spirit mediation. Clearly, the mediating skills would be priceless and certainly an excellent candidate for conferring the title of expert on their owner. Or so I now argue.

2.2 Sangomas Are Experts

There are at least three rationales for thinking of Sangomas as experts—one moral and two epistemic. The moral one is that crediting marginalised experts with expertise is an imperative of epistemic decolonisation. As already suggested, such decolonisation is precisely about restoring the rightful epistemic authority to the marginalised. Not crediting with expertise what they consider their most authoritative voices would be to deny—rather than restore—epistemic authority to the marginalised. This moral rationale is tentative. First, just because something would be morally good it doesn't mean that it is true. Second, it may be that restoring epistemic authority results in greater moral harm. I don't think it would, but in the absence of weighing all relevant moral considerations, the current rationale is inconclusive.

The second rationale is that the skills involved in *muthi* alone seem really rare and complex (again on the assumption that one accepts the relevant cosmology). This was already implicit in the language of the earlier cited passage: *muthi* are 'substances produced by an *expert* hand, substances designed by persons possessing *secret knowledge*'.[4] Here is an even more explicit statement of the expertise involved:

> Ordinary people do not possess the means of distinguishing among the agency inherent in the *muthi* (whether it be harmful or healthful), the agency of the human principal, and the agency of the victim or patient. *For this they must rely upon experts*, healers.   (Ashforth 2005, 216, my italics)

Finally, the most persuasive rationale for why we should consider Sangomas experts is that they meet standard conditions for responsibly placed trust in scientific experts. According to Grasswick (2018), for example, in order for our trust in scientific experts to be responsibly placed, the expert needs to meet what she calls 'the Competence condition' and 'the Sincerity/Care condition'.

The competence condition is the requirement that we should only trust scientists and scientific communities who are competent in the relevant domain. And that means that they need to be able to provide us with *significant* knowledge, to 'filter information for us, determining what the best understandings of the day are and omitting poorer quality, less important, or outdated research' (Grasswick 2018, 78). Competence, then, amounts to a cluster of competencies that enable experts to provide such information (ibid.).

I take it as read that Sangomas meet this condition in straightforward ways: given their extensive training and unique specialisation in multiple and holistic aspects of healing, we don't in fact have any more competent candidates for the expert job.[5] The closest anyone comes is a Western medical practitioner, but their

---

[4] Talk of 'knowledge' here can't be taken in the technical, factive sense of the term by people—like me—who don't believe in supernatural entities. But then can we really talk about expertise on non-existent entities? A natural way to respond here is to take expert beliefs involving such entities as a special case of false beliefs, which I later (Section 3.1) argue are no obstacle to expertise.

[5] Not all of '*Bungoma* education' is made transparent, largely due to the secret nature of much of the knowledge Sangomas are meant to possess. But the two basic stages (at least in some training) are to do with 'the *emadloti*, the '*Nguni*' spirits, who are the immediate and local ancestors of the practitioners' and with 'the *lindzawe*—the foreign spirits associated with water in rivers and streams, who offer

training is not nearly as extensive: it only covers the physical-biological aspects of healing and doesn't touch on psychological or social wellbeing (except for a small bunch of specialists), let alone the health of one's relationships with one's ancestors or other spirits.

What of the Sincerity/Care condition? Let me split this condition in two, to avoid unnecessary confusion. First, and obviously, we are only responsible in trusting an expert if she is honest with us (ibid., 76). If my doctor keeps lying to me, clearly I would be irresponsible in trusting her. Of course, there could be special circumstances in which such a lie would help me recover, but a doctor is still the *type* of expert who is honest about diagnosis, treatment, and so on. Similarly, we are only responsible in trusting a community of experts if it doesn't have a history of deceiving us.

Second, the Care condition requires that the expert has our interests at heart (ibid.). This is a straightforward consequence of our situatedness as knowers. Our social situation determines what is considered significant knowledge—the thing that experts are supposed to provide (as per the Competence condition)—and whose interests are understood and how well (ibid.). Similarly, given that part of the scientific expert's job is to filter *relevant* information for us (again as per the Competence condition), this filtering requires taking our actual situation and interests into account (ibid., 82). For example (not Grasswick's), in a colonial environment, the kind of knowledge relevant to Black citizens' health and wellbeing is not considered significant; nor is understanding of their interests a priority for the scientific community. Or consider other historical violations of these conditions: Nazis experimenting on hypothermia with Jewish prisoners, or how little was known about African diseases (ibid., 88) or female sexuality until recently (ibid., 85). It is obviously irresponsible to trust experts with such a history.

Do Sangomas meet the Sincerity/Care condition? Plainly, yes. Recall the first *muthi* quote: the Sangomas who are healers meet them by definition, since this is what distinguishes them from witches: healers use their skills for the benefit of the patient, whereas witches use theirs to harm people. Of course, there could be exceptions on a case-by-case basis, but the general point stands: Sangomas are the *type* of candidate-expert who are trustworthy in the rich senses outlined by Grasswick: they are custodians of significant knowledge,[6] which they provide in line with the interests of their patients, and they don't have a history of actively ignoring the interests of certain groups or of unethically prioritising their scientific interests to the detriment of the layperson.

---

special kinds of healing and power' (Thornton 2017, 193). This is done through an 'intense apprenticeship', an example of which is offered in Thornton (2017) (see 196 for the physical routine, 257–65 for the spiritual training, and 201–56 for the graduation ritual).

[6] Again, 'knowledge' here can't be understood in the technical sense. See fn 5.

It may be objected here that this argument is much too quick. After all, these are empirical points that cannot just be asserted without suitable historical evidence. Moreover, there is prima facie reason to be suspicious of these claims: given some of the patriarchal practices prevalent in Africa, it is unlikely that traditional healing has not at least ignored the interests of women. I think this objection is fair, but as I lack both the space and expertise to go into historical detail, let me slightly modify my claim. There is no reason to think that Sangomas, as a scientific body, have violated the Sincerity/Care condition *any more than* Western medical practitioners. So, if it is responsible to trust Western medical practitioners so it is to trust Sangomas. But since the conditions of trust concerned trusting *expertise* and Western medical practitioners are uncontroversially experts, it is plausible to think of Sangomas as experts.

This isn't conclusive support for the claim that Sangomas are experts, of course, since Grasswick's account doesn't tell us when someone is an expert but just *indicates* expertise by telling us when it is appropriate to trust someone as expert. But hopefully, in conjunction with the other two rationales I have made a good *prima facie* case. *Sangomas*, then, have as strong a claim to being experts as Western medical practitioners.

3. Standard Accounts of Expertise Can't Count Marginalised Experts as Experts

In this section, I argue that mainstream accounts of expertise can't agree with this claim. Two provisos before I start. First, I will focus here exclusively on accounts of what is known as cognitive expertise, since if Sangomas are experts, they are this kind of expert. This is to be contrasted with the expertise of, say, a football player or an abstract painter. So 'expert' in what follows is shorthand for 'cognitive expert'. Second, I don't by any means aim to provide either an exhaustive overview of existing accounts of cognitive expertise nor conclusive arguments against them, but just to give a taster for available views and the prima facie problems with them.

3.1 Truth-Centred Accounts

One of the most influential accounts of expertise is Alvin Goldman's veritism (2001, 2018). On this view, someone is an expert just in case they have a larger number of true beliefs than a certain comparison class (typically lay people), fewer false beliefs than that class, and a significant number of true beliefs in the domain, regardless of comparison classes (2108, 4). Some veritists reject some elements. For instance, David Coady (2012, 29) thinks neither the second nor the third necessary for expertise. But these details won't matter here, since the account is in trouble when it comes to marginalised experts in virtue of the requirement on which all veritists agree—that the expert has more true beliefs in the relevant domain than the non-expert.

The problem is that many Sangomas' beliefs clash with our best scientific theories. Here is a nice example of such a cluster of beliefs:

Muthi can also *work as a material force through the medium of a dream.* A person can dream he is eating something without being aware that a witch has poisoned the food in his dream. When he awakes, *the muthi will cause afflictions just as real as if the food had been consumed while awake.*

(Ashforth 2005, 213, my italics)

The highlighted beliefs are in clear violation of one of our most fundamental scientific principles—causal closure. But if Sangomas' beliefs clash with such fundamental principles, they are plausibly false. And since the Sangoma is going to have a lot more beliefs in the domain of their expertise than the lay person— after all that's the point of training—the Sangoma would end up having more false beliefs than the lay person. Thus, veritism can't accommodate the claim that a Sangoma is an expert.

OK, you might think, so Sangomas don't satisfy truth-conditions for expertise. Given how circumstantial the argument from the previous section was, perhaps this is just a sign that they really aren't experts. It should be obvious, however, that Sangomas aren't the only candidate-experts who breach this condition. For instance, the account has the implication that we have not had scientific experts until very recently: much of pre-Copernican cosmology informed experts' thinking; and, again, experts would have more false beliefs than lay people since lay people wouldn't have any beliefs on such abstruse topics.[7] Since scientific expertise is the paradigm for expertise, the truth account can't be right.

A similar fate awaits accounts of expertise spelt out in terms of truth-related concepts such as evidence. Goldman (2018, 5), for instance, considers (and dismisses) the following criterion of expertise: the expert is someone who 'possesses substantially more and/or better evidence concerning propositions in [the domain of expertise] than most people in the relevant comparison class' (ibid., 5). Once again, the sorts of things that a Sangoma considers evidence is in stark contradiction of what counts as evidence according to our best scientific practices. Take, for instance, divination, in which many Sangomas are experts. It involves drawing conclusions about a person's health on the basis of where on their body certain sticks 'leap' when thrown, or about a person's past, ancestry, and so on, based on the kind of arrangement certain bones produce when thrown (Peek

---

[7] This point is inspired by Rybko (MS, 7).

1991, 23). Everything I said of the truth account applies here: if the quality or amount of one's evidence and justification are criterial of expertise, Sangomas aren't experts and we have not had scientific experts until very recently.[8]

It could be objected that the culprit here isn't veritism but my fixing truth- and evidence-standards by reference to our best scientific theories. I must confess that I can't see any more plausible standards to adopt, but suppose we fixed the standard elsewhere, say with the best African scientific theories. Then, since they still clash with Western scientific theories (given their non-naturalistic assumptions), present-day Western scientific experts will turn out to be non-experts. This conclusion is, surely, as implausible as those that we did not have scientific experts until recently or that Sangomas are not experts. The problem, then, is with truth-related accounts themselves, not the framework we adopt as a measure of truth or evidence.

3.2 Skill Accounts

Perhaps a more skill-centred account will do. Using again Goldman's formulation as shorthand:

> the skill capacities of an expert—e.g., in golf or tennis—would enable an expert to teach or show run-of-the-mill players how to execute certain athletic maneuvers that they are initially unable to execute. (2018, 4)

I am not sure that anyone holds this view in quite such stark terms (certainly Goldman doesn't), but some participants in the debate posit something like it, suitably modified, as a necessary condition on *cognitive* expertise. For instance, Christian Quast (2018a) thinks that an expert isn't just someone in possession of certain skills but also someone with the ability at the very least to explain to the non-expert the kind of service this skill is providing to the lay community.

Neither the teaching nor the explaining versions of the skill condition works for classifying the Sangoma as an expert. Concerning teaching, this can't be a requirement on Sangomas: their job is to heal and divine, not to teach. And concerning the ability to explain, many of the things that a Sangoma does have an irreducibly revelational aspect (Shaw 1991, 142), which makes them in principle incommunicable to the uninitiated. This is what gives Sangomas their occult status (to a Western eye anyway), and in some ways makes their training considerably more

---

[8] A more subjectivist view of evidence would avoid this consequence. Such subjectivism about evidence is incompatible with the factivity of evidence for which I have argued at length elsewhere (Mitova 2017, ch. 7). However, I do allow for a more subjective notion of *justification* to play a role in fixing expertise when I argue for the responsibility condition. (See Sections 4.1 and 5.)

complex than Western medical training which only concerns the physical. Recall, too, that a Sangoma's knowledge is in fact meant to be *not* communicated. Thus *muthi*, in the opening quote was defined as 'substances designed by persons possessing *secret* knowledge'. (See also Cumes 2013, 58.)

Notice that the failure to satisfy these conditions, once again, isn't unique to Sangomas. Those familiar with the knowledge-how literature, will recall the most cited expert there—the chicken sexer. Here is an uncontroversial expert who can neither explain nor teach the skill she has. Indeed, as Nguyen (2022) has recently argued, much expertise involves skills and knowledge that are of necessity inaccessible to the lay person. Moreover, insisting that we make them accessible, Nguyen argues, undermines the full exercise of expert understanding and practice. One doesn't need to embrace this more controversial second part of the argument to acknowledge that at least sometimes it's OK for expertise to involve incommunicable elements.

But what about an account of expertise that posits the skill as necessary without requiring the ability to teach or explain it? Let me start by conceding that skill is probably necessary for expertise if we take 'skill' to mean broadly a competence of some kind or other, including very abstract theoretical ones. But it is clearly not sufficient: suppose that I have a rare gift of blowing bubble-gum bubbles—they get huge, never burst over my glasses, and are just the most amazing bubbles in the— world. I think that we can agree that this is a skill, but no one would be tempted to call me an expert. So, skill is not a sufficient condition for expertise. Of course, this isn't cognitive expertise, but the idea is the same: if I had a rare skill in counting, and recording the numbers of, grains of sand, presumably our verdict would be the same. The question then is what we need to add to skill to get expertise.

3.3 Service Accounts

A natural recent answer is that we need to add that the skill is used in the service of the expert's community. The most worked out version of this is Quast's (2018a, 2018b). He argues, very roughly, that an expert is someone who reliably provides a service (he calls this primary expertise), is able to explain this service to the service-recipient (he calls this secondary expertise), and has the right dispositions in so doing—i.e. is willing to exercise both kinds of expertise.[9] The account is intricate, and I cannot possibly do it justice here. Let me just point to two aspects which prevent it from accommodating Sangomas as experts. I have already gestured at the first: if second-order expertise is a requirement on expertise, then Sangomas aren't experts.

---

[9] This clear way of distilling the three conditions from what is otherwise quite a laborious argument is due to Croce (2019b).

The second is that willingness to provide a service can't be a requirement on expertise, regardless of the issue of marginalisation.[10] I take myself to be a philosophy expert. Until today I have been willing to exercise both my primary and secondary expertise. Suppose tonight I stopped being willing to do so forever. Will I suddenly wake up a non-expert tomorrow morning?

So, a service account should give up both the secondary expertise and the willingness requirements. Quast offers such an account elsewhere.

> Someone is an expert for a range of products (or tasks) r if and only if she is an authority concerning r and competent enough to reliably and creditably fulfill difficult service-activities within r accurately for which she is particularly responsible. (Quast 2018b, 24)

This account overcomes the previous objections and seems to be able to accommodate the claim that Sangomas are experts: they perform such tasks accurately, reliably, and creditably. However, the account is implausible for other reasons. For lack of space, let me just mention the crucial one at which Quast himself gestures (ibid., 25). Consider, for example (not his), a genius IT geek or a highly talented philosopher tinkering away their whole lives in their mum's garage. Or notch it up and imagine them on desert islands. These are hardly service-activities and they will never benefit anyone. Yet intuitively, the IT geek and philosopher have as good a claim to being experts in what they do as uncontroversial experts. This consideration also speaks against a neighbouring account motivated by a critique of Quast—Michel Croce's (2019a). He argues that it is a necessary condition on expertise that it contribute to the progress of one's expertise-domain. Neither the geek nor the philosopher meets this condition, since by hypothesis neither will share with their peers.

4. Communitarian Functionalism

If the arguments so far worked, an account of expertise should meet at least two desiderata:

No Marginalisation: It should count as experts the experts of the marginalised even though they may breach the truth condition, and

Standard experts: it should count as experts uncontroversially acknowledged experts.

---

[10] See also Croce (2018b, 31) for a more elaborate example.

The first desideratum is a straightforward consequence of the argument that marginalised experts are indeed experts. The second is motivated by the independent problems with the accounts I discussed. But it is also common sense: if an account of expertise has the consequence that medical doctors or nuclear physicists aren't experts, then it has unacceptably parted ways with our concept of expertise.[11] In this section, I sketch an account of expertise that meets these desiderata—what I will call communitarian functionalism.

4.1 The Core Conditions

The core conditions of communitarian functionalism are inspired by the threefold motivation for *No Marginalisation*: counting them as experts is an imperative of epistemic decolonisation; they possess valuable epistemic skills; and they meet core conditions for responsibly placed trust in experts. Building on these ideas, we should expect at least some kinds of experts to be people who have unique skills or knowledge that are valuable to their community, and which they use responsibly.

Slightly more formally:

Communitarian functionalism
S is a cognitive expert in domain D at time T if and only if S has epistemic competencies in D such that:

(i) D is of value to S's epistemic community C at T (*Valuable*); and
(ii) The competencies are largely unique to S rather than had by many members of C at T (*Unique*); and
(iii) They are in line with the best scientific theories and practices available in C at T (*Scientific*); and
(iv) They are formed and exercised by S in an epistemically responsible way in C at T (*Responsible*).

I have called this view communitarian functionalism to foreground the central role one's position in an epistemic community plays in determining whether one is an expert (hence, communitarianism), and the fact that this position is a matter of one's role in this community (hence, functionalism). In both these ways

---

[11] A reasonable third desideratum is that an account of expertise shouldn't have the consequence that charlatans count as experts. I don't discuss it in the main text since none of the accounts considered here breach it. My account meets it by virtue of conditions (i) and (iii) below: the charlatan's skills are neither in a valuable domain nor consonant with our best scientific theories.

the account is aligned with the service accounts, though I will show how it avoids their problems.[12]

4.1.1 Epistemic competencies

The first thing in need of explanation is what kind of epistemic competencies the schema features. I think we should be as liberal as possible here. Such competencies would include propositional beliefs; diagnostic skills, such as the ability to diagnose a problem in domain D; problem-solving skills in D; and testimonial skills, such as the ability to give expert testimony in D. The precise nature of the propositions and skills involved would be fixed by the kind of domain we have in mind. And the exact number of true propositions and the minimum threshold of competencies needed for someone to qualify as an expert will be of necessity a matter of vagueness which it isn't the job of a schema to settle.

The important thing to note at this stage is that the schema posits these competencies as necessary and jointly sufficient conditions for *cognitive* expertise. This allows room for many other kinds of expert. For instance, a virtuoso pianist or a good highly abstract painter are clearly experts in their field by all accounts, but they equally clearly don't meet the above conditions.

4.1.2 Epistemic community

All the conditions are relative to an epistemic community. What is that? The term is used quite loosely in the social epistemology literature, to denote the community with which one shares epistemic resources, and on which one depends for these resources. I will retain this looseness, in the hope that we all have enough of an intuitive grip on the notion to make it serviceable in this context. The most crucial idea about it is that of situatedness and interdependence (Pohlhaus 2012). An epistemic community is something that is partly a function of one's social identity and the position in social and power structures that this affords one. For instance, as a marginally situated knower, one finds that the knowledge economy isn't geared towards making sense of one's experiences, but rather to making sense of the experiences of the powerful. A nice example is Miranda Fricker's description of how the concept of sexual harassment needed to be literally invented—against the pushback of the powerful—in order to make sense of the traumatic experiences of many women. (Whether Fricker got right *who* invented it is a separate issue. For a convincing argument that she got it wrong—in epistemically unjust ways—see Berenstain 2020.)

The point here is that shared experiences and the need for shared resources to understand them create the boundaries of one's epistemic community. The point can be extended to one's worldview more generally. For instance, the epistemic

---

[12] I should also mention that I encountered Quast and Croce after I already had the basics of my view.

community of—crudely—the West is different from the epistemic community of those with a predominantly African metaphysic. The latter does not distinguish the natural and supernatural in the way Western worldviews do, but conceives of both as equally and in the same way causally involved in our lives. Such views fix the boundaries of *epistemic* communities—rather than just cultural ones—because worldviews need epistemic resources in order to be articulated and expressed.

Notice that given the link to social identities, epistemic communities are fairly big, structured, and largely involuntary in the sense that our social identities go with certain worldviews and power structures that aren't up to us. Thus, a group of conspiracy or flat-Earth theorists doesn't form an epistemic community in the relevant sense, and nor does a bunch of homeopathic practitioners in the West. This will be important for disallowing scientific standards to float arbitrarily.

4.1.3 Valuable
The first condition of communitarian functionalism is that the relevant competencies are in an area that matters to one's epistemic community. Suppose again that I had an amazing gift of counting grains of sand and motes of dust. And suppose I exercised this gift regularly, responsibly, and accurately. Presumably, no one would be tempted to call me an expert. We appeal to experts when we need a question answered or a problem solved. But the fact that we need an answer or solution presupposes that the topic of inquiry matters to us. This means that the domain in which S is an expert must be valuable to S's community. Most often the way the value of the domain manifests itself is in the fact that we value the competencies of the expert. But this is not always so. There are cases in which we would want to call someone an expert even though their competencies are disvaluable. Take for example an accomplished economist whose work aims to exploit a certain group, or a fine tactician in an unjust war. If we insisted that the competencies themselves must be valuable, we would have trouble calling these experts. Hence the first condition concerns the value of the domain.

What kind of value exactly is involved here? I think it can be of many kinds, and as long as the domain in question is valuable in one way, *Valuable* is met. First, the value can be either instrumental or final. Thus, I can appeal to an economist to help me with my financial embarrassments. Presumably the value of the financial domain is instrumental. But I can also appeal to an expert for something that I am simply curious about for its own sake, such as when I ask a mathematician to update me on whether anyone has yet proved Fermat's last theorem.

Second, the value in question can be of any normative domain—the moral, aesthetic, practical, or epistemic. But for the cognitive kinds of expert I have isolated here it will be most likely practical or epistemic. Although a scientific domain can be of great moral value or disvalue—if it saves or harms lives, for instance—this is not why we value such skills in the first instance. The value comes as a bonus, so to speak.

4.1.4 Unique

At the risk of labouring the obvious, experts are people who serve a particular role in an epistemic community: they provide knowledge or skills which are not possessed by the rest of the community. Hence, we phone our doctor or architect for an *expert* opinion, and courts appoint forensic experts to help settle the jury's mind. If many of us had these competencies, this would be unnecessary: we could just call a friend; and the jury could just ask the scribe.

4.1.5 Scientific

The third condition is motivated by my earlier objection to veritistic accounts of expertise that they imply that the world has not had experts until recently. The way to credit pre-Copernican experts with expertise is by making expertise relative to the best theories of one's epistemic community at the time. Given the way I proposed to fix epistemic communities, the relevant community—and hence scientific standard—is the pre-Copernican one. Similarly, and still in light of the way social identity partly fixes the boundaries of the community—the relevant community for homeopaths is the same as the one for medical doctors—crudely, the West. Since only the latter's competencies are in line with this community's scientific practices, only the latter are experts. These claims are, of course, fuzzy around the edges, but hopefully they are sharp enough to give us the basics of an intuitively plausible picture.

4.1.6 Responsible

The final condition is inspired by the argument in Section 2.3 that we should think of Sangomas as experts because they meet the conditions for responsibly placed trust in experts. The thought now is this. Suppose that I am a sloppy believer who forms her beliefs at will and by guesswork. Would it be epistemically responsible of you to accept my testimony or use my skills to fix your problems? Plainly, not. But if so, for my trust in an expert to be responsibly placed, the expert themselves must *form* their beliefs and other competencies in epistemically responsible ways.

A similar thought holds with how the expert *exercises* their competencies: I would be epistemically irresponsible to believe them if I knew they were irresponsible with giving testimony (say they gave it when their belief is barely justified) or with exercising their skills (say they tried things out on me that have been barely tested).

I should note that this requirement of responsibility isn't one on expert-tokens but on expert *types*. That is, an expert is the *type* of professional who is epistemically responsible for the acquisition and exercise of their competencies. This is what distinguishes them from a charlatan. That is not to say that an individual expert might not slip in their responsibilities; it is just to say that responsibility is a requirement on expertise as such.

4.2 Communitarian Functionalism Meets the Desiderata on Expertise Accounts

4.2.1 No marginalisation

The first desideratum is that an account of expertise should count as experts the experts of the marginalised. Since communitarian functionalism was motivated by the thought that other accounts breach this condition, it would be surprising if the view itself breached it. But let me make sure just in case.

First, the domain of the healers' competencies—physical and spiritual wellbeing—is clearly valuable to their epistemic community. The relevant community is those who share the worldview which forms the backdrop of traditional healers' work—at least some of the core African metaphysic. Unlike Western medical doctors, who only treat bodily ailments, healers provide far more comprehensive care of one's whole being, including one's relationships with one's community, ancestors, and gods (Sogolo 1998).

Second, healers have epistemic competencies which are clearly unique to them rather than had by their epistemic community. Such competencies are acquired in one of three ways: 'inheritance of abilities from ancestors; training by skilled masters; and direct communication with higher powers such as the ancestors, spirits...or the Holy Spirit' (Ashforth 2005, 215). Clearly, such competencies aren't exactly one a penny, hence the veneration felt towards healers.

Third, healers' epistemic competencies tend to be in line with the best scientific practices of the relevant epistemic community. Most healers undergo difficult and rigorous training into these age-old practices. Notice that the fact that such practices clash with best *Western* scientific practice (as argued in Section 3.1) is irrelevant for healers' status as experts, since Western scientific communities are not part of the healer's epistemic community.

Finally, the kinds of healers that I have been discussing here are, by definition, the well-intentioned and responsible ones. This, indeed, is what distinguishes them from witches, as emphasised in the discussion of the Sincerity/Care condition (Section 2.2). The responsibility that distinguishes the healer from the witch is, of course, in the first instance moral. But one cannot be *morally* responsible in the acquisition and exercise of one's abilities if one is *epistemically* careless. Epistemic carelessness means leaving it up to chance whether my testimony or skill heals you rather than kills you. But such leaving it up to chance is clearly morally irresponsible. Hence one can't be morally responsible in the exercise or acquisition of one's competencies without being epistemically responsible.[13]

---

[13] Admittedly, this claim might be more controversial than I am letting on. (Thanks to an anonymous reviewer for this point.) In particular, it will be more easily accepted by philosophers who conceive of the epistemic normative domain as a subspecies of the moral (e.g. Zagzebski 1996) than those who reject this conception (e.g. Baehr 2011). So called moral encroachers (e.g. Basu and Schroeder 2019) will also be sympathetic to my claim. However, the claim itself does not depend on

4.2.2 Standard experts

It should be obvious that the proposal also meets the second desideratum on a theory of expertise: it counts as experts uncontroversially acknowledged experts. Think of the forensic expert called into court to adjudicate in a difficult case. First, these competencies are in a domain that is valued by her epistemic community: that's why we are asking her to give testimony and why we have spent taxes on her forensics degree. Second, her competencies are unique to her: that's why she has undergone prolonged training, and we are asking her rather than the scribe or the judge's teenage daughter. Third, her competencies are in line with our best scientific practices: a forensic expert's testimony typically appeals to the latest and greatest in forensic science. Finally, given the grave consequences of the verdict, it is assumed on all sides that the forensics expert has done due diligence in acquiring these skills and testimony.

4.3 Theoretical Advantages

Obviously, a fully-fledged account of the position would have a lot more to say on all these points. But due to space constraints, all I can do here is to show that this will be a worthwhile exercise by arguing that the account enjoys the advantages of its competitors while not incurring their liabilities (this subsection); and addressing the most obvious potential objections (Section 5).

4.3.1 Truth-centred accounts

The problem with the truth-condition on expertise was that it disqualified marginalised experts, since many of their beliefs are false. This meant that this view can't deliver the first desideratum on a theory of expertise—*No Marginalisation*. But it also meant that the experts of yore were not in fact experts, hence breaching *Standard Experts*.

Communitarian functionalism overcomes both problems. I have already shown how it meets *No Marginalisation*. It should also be obvious that it avoids the problem of no experts: since expertise is relative to a community and time, and the competencies of experts of yore were in a valuable domain, unique, in tune with the best theories of the day, and responsibly exercised, they turn out to be experts according to my proposal.

At the same time, my view accommodates what is attractive about other truth-related views such as those that think of the expert as someone who has better or more evidence for her beliefs. By making it a requirement on expertise that the beliefs and competencies of the expert be acquired in an *epistemically* responsible

---

taking sides in these debates. It only requires the idea that *when it comes to exercising one's skills* if one has not caused any moral harm or wrong, one will inevitably have also been epistemically responsible in exercising these skills.

way, we make sure that the beliefs are justified at least on an internalist view of justification. On one such view, at least, it is sufficient for justification that the believer has done her epistemic bit when forming her beliefs (e.g. Foley 2005).[14] But the view doesn't suffer from the problem which bedevilled the evidence view—that the kind of evidence on which some healers base their belief is in fact not good evidence since it clashes with our best scientific theories.

4.3.2 Skills accounts

The skills account was plausible when 'skill' was understood broadly to include abstract competencies. Talk of competencies in my account is inspired by this consideration. At the same time, the proposal overcomes the objection that skills are insufficient for expertise. Communitarian functionalism specifies that these skills need to be in a domain valuable to the community as well as responsibly acquired and exercised. This allows it to overcome counterexamples like the bubble-gum and sand-counting 'experts'.

4.3.3 Service account

Finally, communitarian functionalism allows us to accommodate the good things about the service views while avoiding their pitfalls. What the service views had right is that an expert is providing something to their community. What it got wrong were three of the conditions it posited on expertise: the ability to *explain* what one is doing, the *willingness* to exercise one's expertise, and the provision of a service or something *useful* to the community. Communitarian functionalism doesn't posit any of these elements. Consider the last: the counterexamples were the isolated IT geek and lone philosopher. IT and philosophy are, of course, valuable to our community, even if they don't provide a service or have practical benefits in particular instances. Communitarian functionalism allows that if a community ascribes final—rather than instrumental—value to certain domains, then competencies in this domain are sufficient for expertise (provided they meet the other conditions). This is the kind of value we ascribe to philosophical and lone IT activities when performed for their own sake.

5. Objections and Refinements

If these thoughts are on the right track, I have shown communitarian functionalism to at least be a serious alternative to existing views: it meets core desiderata

---

[14] Objection: Sangomas are not epistemically responsible given their easy access to our best scientific theories. Reply: the objector forgets that the relevant scientific theories are the African ones, not the Western. Otherwise, Western medical practitioners and scientists are being epistemically irresponsible for not taking into consideration African (and other non-Western) scientific and cosmological views.

on a theory of expertise; and enjoys the advantages of contenders without suffering from their shortcomings. But it may be thought to suffer from problems of its own. In this section, I consider the two most obvious ones, by way of strengthening the proposal.

5.1 Relativism?

The first worry is that communitarian functionalism is unacceptably relativistic. After all, do we really want to say that witches and all sorts of people supposedly wielding magic are experts? There seems to be something terribly, well, unscientific about that. And worse, if we sever the connection between expertise and truth, what is expertise worth to us?

Let me start by saying that I am no friend to relativism in any form. Indeed, I have spent some time trying to work out how to accommodate similar worries in an account of epistemic decolonisation that takes seriously the epistemic authority of victims of colonialism (Mitova 2021). That said, my reply here is concessive: to the extent that any of the views considered here are plausible, they must dispense with truth or factivity as a requirement on expertise. As discussed, such a requirement automatically renders inexpert most of the experts throughout history. Once we want to accommodate such expertise, we must make our notion of expertise relative to a community in one way or another. And once there, the spectre of relativism seems to loom large.[15]

But perhaps this anxiety comes from thinking of expertise too much in propositional terms, of the expert as someone who spits out answers to our questions and we then see if these are true or false. Making truth relative to a community—as we are implicitly doing through making success a matter of fit with best scientific theories in this community—has a disturbing ring to it. But once we make it a matter of what the community values, things cheer up. For it is pretty clear that despite certain beliefs that a Western audience might find difficult to swallow, healers do help their patients get better. So, they have competencies which they exercise successfully. And what can be objectionable about that?

5.2 Science

The second objection I will consider is that the proposed account seems to lionise science as the be-all and end-all of expertise. But this is highly problematic in the

---

[15] Croce (2019b, 28) argues that the functionalism in such accounts is also a source of relativity: if we fix expertise to the kind of service the expert provides to a relevant community, the service can be so quirky and needed by such a small community, that we'd have an implausibly vast number of experts. I think that this charge doesn't ultimately go beyond the relativity that is introduced by making experts relative to communities.

context of an argument which purports to advance epistemic decolonisation. After all, science as we know it is not only responsible for the racialisation of our species and the colonisation of some of us, but also has the kind of unearned hegemonic epistemic authority that it is precisely the aim of epistemic decolonisation to dismantle (e.g. Mitova 2020). For this reason, some proponents of epistemic decolonisation have gone so far as to argue that a properly decolonised knowledge economy altogether dispenses with science. This, for instance, was the war-cry of the *Science Must Fall* movement in South Africa in the late 2010s. Thus, measuring the expertise of marginalised experts by scientific standards is self-subversive, at best, and morally and theoretically pernicious, at worst.

A first pass here is to remind the objector that the conditions for scientific expertise are in fact satisfied by marginalised experts, and hence they are at least *dialectically* unimpeachable. But this response doesn't take the objection as seriously as we should. The right response, I think, is to imagine that we didn't posit the condition *Science*. That is, that we do not treat the expertise of the marginalised, *when it clearly comes to medical matters*, as scientific. Where does that leave us? To me that sounds like precisely what colonisation did: to deny scientific authority to the scientific voices of the marginalised. Such a denial is clearly far more at odds with the project of epistemic decolonisation than using a scientific standard for (cognitive) expertise and discovering that the experts of the marginalised meet it. What needs to happen is not that science must be rejected but that we must develop the right, non-marginalising notion of science. This is what communitarian functionalism allows us to do.

Acknowledgements

Huge thanks to Mirko Farina, Niels de Haan, Karl Landström, Dimpho Maponya, Caitlin Rybko, Sebastian Schmidt, Abraham Tobi, and an anonymous OUP reviewer for really helpful feedback on earlier drafts. Thanks, too, to my audiences at the African Epistemologies Advanced Seminar Series (University of Cape Town, 2022), the Annual PSSA conference (University of Johannesburg 2023), the *Canada Chair for Epistemic Injustice Speaker Series* (UQAM, 2023), and the *Constructing Social Hierarchies* workshop (University of Melbourne, 2023).

References

Ashforth (2005) "Muthi, Medicine and Witchcraft: Regulating 'African Science' in Post Apartheid South Africa?" *Social Dynamics* 31 (2), 211–42.

Baehr, J. (2011) *The Inquiring Mind: On Intellectual Virtues and Virtue Epistemology.* New York: Oxford University Press.

Basu, R. and Schroeder, M. (2019) "Doxastic Wrongings," in *Pragmatic Encroachment in Epistemology*, (eds.), Kim, B. and McGrath, M., 181–205. New York: Routledge.

Berenstain, N. (2020) "White feminist gaslighting." *Hypatia* 35(4), 733–58.

Coady, D. (2012) "Experts and the Laity," in D. Coady, *What to Believe Now: Applying Epistemology to Contemporary Issues*, 27–58). Oxford: Blackwell.

Croce, M. (2019a) "On what it takes to be an expert." *The Philosophical Quarterly* 69(274), 1–21.

Croce, M. (2019b) "Objective expertise and functionalist constraints." *Social Epistemology Review and Reply Collective* 8(5), 25–35.

Cumes, D. (2013) "South African Indigenous medicine: How it works." *Explore* 9(1), 58–65.

Dotson, K. (2012) "A cautionary tale: On limiting epistemic oppression." *Frontiers—A Journal of Women's Studies* 33(1), 24–47.

Foley, R. (2005) "Justified Belief as Responsible Belief," in *Contemporary Debates in Epistemology*, (eds.), Sosa, E. and Steup, M., 313–26. London: Blackwell.

Fricker, M. (2007) *Epistemic injustice: Power and the ethics of knowing*. New York: Oxford University Press.

Goldman, A. (2018) "Expertise." *Topoi* 37(1), 1–8.

Goldman, A. (2001) "Experts: Which ones should you trust?." *Philosophy and Phenomenological Research* 63(1), 85–113.

Grasswick, H. (2018) "Understanding epistemic trust injustices and their harms." *Royal Institute of Philosophy Supplement* 84, 69–91.

Mitova, V. (2021) "How to decolonise knowledge without too much relativism," in *Decolonisation as Democratisation*, (ed.), Khumalo, S., 24–47. Cape Town: HSRC Press.

Mitova, V. (2020) "Decolonising knowledge here and now." *Philosophical Papers* 49(2), 191–212.

Mitova, V. (2017) *Believable Evidence*. Cambridge: Cambridge University Press.

Ndlovu-Gatsheni, S. J. (2018) *Epistemic Freedom in Africa: Deprovincialization and Decolonisation*. London and New York: Routledge.

Nguyen, C. T. (2022) "Transparency is surveillance." *Philosophy and Phenomenological Research* 105 (2), 259–503.

Okeja, U. (2013) "Postcolonial discourses and the equivocation of expertise." *Philosophia Africana* 15(2), 107–16.

Peek, P. M. (1991) "Becoming a diviner," in *African Divination Systems: Ways of Knowing*, (ed.), Peek, P. M., 23–6. Indiana University Press.

Pohlhaus, G. (2012) "Relational knowing and epistemic injustice: Toward a theory of willful hermeneutical ignorance." *Hypatia* 27(4), 715–35.

Quast, C. (2018a) "Towards a balanced account of expertise." *Social Epistemology* 32(6), 397–419.

Quast, C. (2018b) "Expertise: A practical explication." *Topoi* 37, 11–27.

Rybko, C. (MS) "Expertise. Towards an epistemology of Google's knowledge Graphs." PhD dissertation, University of Johannesburg.

Shaw, R. (1991) "Splitting truths from darkness: Epistemological aspects of temne divination," in *African Divination Systems: Ways of Knowing*, (ed.), Peek, P. M., 137–52. Indiana University Press.

Sogolo, G. (1998) "The concept of cause in African thought," in *The African Philosophy Reader*, (eds.), Coetzee, P. H. and Roux A. P. J., 228–37. London: Routledge.

South African Government. (2007) "Traditional Health Practitioners Act of 2007," available at https://www.gov.za/documents/traditional-health-practitioners-act.

Thornton, R. (2017) *Healing the Exposed Being: The Ngoma healing tradition in South Africa*. Wits University Press.

Zagzebski, L. T. (1996) *Virtues of the Mind: An Inquiry into the Nature of Virtue and the Ethical Foundations of Knowledge*. Cambridge: Cambridge University Press.

15
Public Expertise and Ignorance

Duncan Pritchard
University of California, Irvine
dhpritch@uci.edu

1. Manufacturing Ignorance

Call *public experts* those experts who offer their expertise to benefit the general public. Clearly not all experts are public experts. Some expertise is only relevant to a very narrow audience (think, for example, of the expertise of train-spotters), while some expertise is not meant to benefit the general public, even if would be of benefit to them (think, for example, of how certain kinds of economic expertise is geared only towards the profits of those who fund it). Our interest is specifically in public experts, so understood, and their relationship to ignorance.

It is natural to cast public experts as combating ignorance in the populace at large. Think, for example, about experts who raise awareness of medical issues (such as the dangers of undiagnosed type-2 diabetes) or the experts who brought the climate crisis to the fore of our attention. On this model, public experts are bringing the light of knowledge to bear on the dark corners of public life where it has not shone yet, thereby extinguishing ignorance.

I want to suggest that this way of thinking about the relationship between public expertise and ignorance is not the complete story. In particular, it misses out a crucial element of the role that public experts often play in this regard, which is not just to combat but also to *manufacture* ignorance. This might sound like a very puzzling claim to make: why would experts be concerned to manufacture ignorance? In order to understand this claim we need to think a bit more deeply about the nature of ignorance, since it is this notion that is not being fully appreciated here.

As a crude first pass, one might think that ignorance is just the absence of knowledge (or, failing that, the absence of some other epistemic standing, like true belief), but I will be suggesting that this can't be the right way to think about this notion. If ignorance were just the lack of knowledge, then it really would be the case that the relationship between expertise and ignorance is simply one of the former combatting the other. But there is a further dimension to ignorance, as it is also concerned with a normative standing. As I will be arguing, while ascribing a lack of knowledge can be normatively neutral (one can be blamelessly lacking in

Duncan Pritchard, *Public Expertise and Ignorance* In: *Expertise: Philosophical Perspectives.* Edited by: Mirko Farina, Andrea Lavazza, and Duncan Pritchard, Oxford University Press. © Duncan Pritchard 2024.
DOI: 10.1093/oso/9780198877301.003.0015

knowledge, for example), ascriptions of ignorance always have a negative normative valence. In short, to be ignorant is to lack knowledge that one ought to have. If that's right, then a crucial part of the role of public experts, before we even get to the stage of combatting ignorance, is often to first elevate the public's lack of knowledge of a certain subject matter, such as climate change or the dwindling numbers of black rhinos, so that it constitutes ignorance in the first place. In this way, a certain kind of public expertise is concerned to manufacture ignorance before it combats it. We will call public expertise that plays this social role *informative public expertise*.

Once we understand this aspect of the role of public experts, however, it also enables us to bring into sharper relief a second way in which public expertise can function. This is not to manufacture and then combat ignorance as informative public expertise does but rather to reveal hitherto hidden ignorance. Call this *critical public expertise*. This is a role that public experts play that usually has an explicitly political purpose. The idea is to reveal that our lack of knowledge is not blameless at all, as might have otherwise supposed, but rather manifests our intellectual culpability. In this way, public expertise can also perform an important socially critical function of highlighting concealed ignorance. As we will see, it is only with the contrast with informative public expertise in place that we can understand the nature of critical public expertise.

2. The Negative Normative Valence of Ignorance

Most philosophical treatments of ignorance regard it as simply the absence of knowledge.[1] Where philosophers depart from this traditional account of ignorance, the claim is usually only that a different epistemic standing is lacking, such as true belief.[2] What most views of ignorance thus share is the idea that it is simply the absence of an epistemic standing. For convenience, in what follows we will stick with convention and treat the target epistemic standing that is lacking when one is ignorant as knowledge.[3] Nonetheless, I think it is clear from further reflection on the matter that ignorance cannot simply be a lack of knowledge.

---

[1] For some key defences of the so-called traditional account of ignorance in terms of lack of knowledge, see Zimmerman (2008); Le Morvan (2011a, 2011b, 2012, 2013); and DeNicola (2018).

[2] For some of the main defences of the so-called 'new view' of ignorance as the lack of true belief, see Goldman and Olsson (2009); van Woudenberg (2009); and Peels (2010, 2023; cf. Peels 2011, 2012). This account, and the traditional account of ignorance in terms of the lack of knowledge (see endnote 1), are usefully surveyed in Le Morvan and Peels (2016). Note that we are here focusing on propositional forms of ignorance—see Nottelmann (2015) and El Kassar (2018) for two recent discussions of non-propositional forms of ignorance and how they relate to the propositional variety.

[3] I discuss this issue in some detail in Pritchard (2021b) and argue that the proper object of assessment when it comes to ignorance is a kind of lack of awareness, an epistemic standing that can fall short of knowledge. It will be harmless to set this complication to one side for our purposes here, however, and just focus on the absence of knowledge.

Consider, for example, facts that one could easily know right now but which one has no possible reason to find out, such as how many spoons one has in one's kitchen cupboard. This is something that one does not know, but it would be odd to describe this lack of knowledge as ignorance: I am not ignorant of the number of spoons in my kitchen cupboard, even though this is clearly unknown. The crux of the matter is that there is no reasonable expectation that one should know this truth. Indeed, in the usual run of events, there is every rational expectation that one should *not* know it, as what possible reason would there be to go to the trouble of finding this out? Hence, although this truth is clearly unknown, it is not an instance of ignorance.

Once one starts to reflect on the matter one realizes that the cases multiply. What about truths that one could not possibly know, for example because they are well beyond anyone's capacity to know? I don't know what coloured underpants Caesar was wearing on the day he crossed the Rubicon (I assume there is no historical record of this fact available). I guess no-one will ever know this now: this truth is simply epistemically inaccessible to us. It would be odd to describe this lack of knowledge as ignorance though. How could I be ignorant of a claim that I could never know?

Or consider cases where one actively decides, for good reasons, to not know a particular truth. For example, consider a lawyer who is aware that being cognisant of certain facts about her client might create problems for that client. In exercising her professional responsibilities to her client as a lawyer, she might therefore ensure that there are specific truths that she does not know, even though they could have easily been known, had she been so inclined. But would we really say that our lawyer is *ignorant* of these truths that she elects not to know? After all, she chooses not to know them, and does so for very good reasons; ignorance does not seem applicable here.

There are many other cases of this kind, as I've described elsewhere.[4] What they reveal is that ignorance has a normative status associated with it. This gives an ascription of ignorance a normative valence that a mere ascription of a lack of knowledge lacks. In particular, while one can specify that someone lacks knowledge without implying any kind of intellectual fault on their part, in ascribing ignorance to an agent one is implying that it is a lack of knowledge that is due to one's intellectual fault. In short, this is because to be ignorant of something is to fail to know something that one ought to have known.[5]

We can see this point at work in the examples we have looked at. In failing to know the number of spoons in my cupboard drawer I am not thereby ignorant because this is not something that I am reasonably expected to know. It is not a

---

[4] See, for example, Pritchard (2021a, 2021b). See also Pritchard (2021c, 2022a, 2022b).
[5] Strictly speaking—see endnote 3—my favoured view of ignorance is that one is ignorant when one is unaware of a fact that one ought to be aware of. But lack of knowledge will suit our purposes just fine.

truth that I ought to have known (indeed, if anything, I ought not to know it), which is why it is not in the market for ignorance. The same goes for truths that I cannot possibly know, such as what colour underpants Caesar wore on the day he crossed the Rubicon. Finally, even propositions that I could easily know, and which might be worth knowing (unlike the truth about the spoons), might nonetheless be unknown for entirely appropriate reasons, just like in our lawyer case, If so, then this is not something that one ought to know either, or hence it wouldn't count as ignorance.

The idea that ignorance has this specifically normative dimension should not be surprising. After all, to call someone ignorant is clearly to disparage them, in a way that merely noting that they lack knowledge need not do. Moreover, although we are here talking of *propositional ignorance*—i.e. ignorance as it relates to specific propositions—one would expect propositional ignorance to be related to the more general phenomena of *character ignorance*. That is, we don't just ascribe ignorance of particular propositions to people, but we also ascribe ignorance to their characters—one can be an ignorant *person*. Calling someone an ignorant person is clearly a highly disparaging thing to do. One can, of course, be ignorant of particular propositions without thereby being an ignorant person. Significantly, however, the converse does not hold: if one is an ignorant person, then one is certainly going to be disposed to be propositionally ignorant. In general, ignorant people manifest a range of intellectual vices, from arrogance to intellectual carelessness, and this leads them to fail to know propositions that they ought to know. It is no wonder then that propositional ignorance inherits the negative normative valence that is already clearly present in character ignorance.[6]

I've termed the idea that ignorance is not merely the absence of an epistemic standing but also involves this negative normative valence the *normative account of ignorance*. I've defended this view at length elsewhere, and don't propose to defend it further here.[7] My goal is rather to show how bringing this proposal to bear on our understanding of the role of public expertise is important to properly capturing two fundamental ways in which public expertise functions.

3. Informative Public Expertise

A notable consequence of the normative account of ignorance is that whether one counts as ignorant of a particular proposition can depend on a range of social factors that might have no bearing at all on whether one has knowledge of this

---

[6] For more on this connection between character and propositional ignorance, see Pritchard (2021b, 2022b).

[7] See especially Pritchard (2021a, 2021b). For some related discussions of the normative account of ignorance, see also Pritchard (2021c, 2022a, 2022b).

proposition. For example, we don't regard small children as being ignorant for failing to know key facts about contemporary economics or politics, but we might well regard their parents for being ignorant for lacking this same knowledge. This is because there is no reasonable expectation that children should know these truths, unlike their parents. Similarly, what one can be reasonably expected to know can vary in terms of one's social roles. If one is a consultant neurologist, for example, then one is expected to be aware of the latest research relevant to one's field and would be judged ignorant for failing to know this information. In contrast, there is no corresponding expectation that a member of the public should know these arcane facts, which is why an ascription of ignorance here would not be appropriate.[8]

This social feature of the notion of ignorance is salient to the role of public experts. To begin with, I want to focus on a specific kind of public expert whose role it is to make the public aware of important information that they might otherwise be unaware. This is informative public expertise. A great variety of experts perform this public role, from geologists informing us about earthquake risks to economists informing us about the likelihood that mortgage rates will dramatically rise in the near future.[9] It is natural to conceive of what such public experts are doing as helping us to be less ignorant by spreading the knowledge that results of their expertise. While this is generally true, it is often not the full story.

Consider one important role that public experts can play, which is to make the public cognisant of an important phenomenon that they would have been otherwise unaware. Think, for example, of the public information campaigns regarding acquired immune deficiency syndrome (AIDS) in the 1980s or the way that scientists have raised public consciousness of the climate change crisis. In these cases we as a society are relying on the experts to provide us with knowledge about a particular subject matter (AIDS, climate change) that we would likely have otherwise lacked. As we might put it, expertise is shining the light of knowledge and thereby removing the darkness of ignorance.

The problem with this way of characterizing the situation, however, is that it misses out a crucial element. Take a public health campaign like that associated with AIDS in the 1980s. It is certainly true that prior to these campaigns most of the general public lacked knowledge of what AIDS was. But were people at this point *ignorant* of this illness? It is hard to see why their lack of knowledge would be given this normatively loaded status. The issue is that unless one were a member of the specific communities that were being devastated by this illness,

---

[8] See Goldberg (e.g. 2011, 2017, 2018, passim) for some interesting work unpacking the kind of normative epistemic expectations that would be relevant here, such as what one should have known.

[9] Note that in what follows I will be taking it as given that the experts we are concerned with are genuine. I will also be setting to one side what is involved in being an expert. For further discussion of the notion of an expert, including the epistemic stock we should invest in them, see Goldman (2001, 2018); Coady (2012, ch., 2); Croce (2019); Watson (2020); and Pritchard (forthcoming).

or one was part of the medical establishment dealing with it, then there would be no realistic way of being aware of it. That was precisely why national ad campaigns were required to raise awareness of the illness.

I am suggesting that what such an ad campaign is thus doing is effectively making use of expert advice to *manufacture* ignorance. Prior to the ad campaign, the general public were simply unaware of AIDS. Crucially, however, this lack of knowledge doesn't amount to ignorance (for most people anyway), given that there is no reasonable expectation that an ordinary member of the public should know about this illness. By making people aware of AIDS, however, the ad campaign thereby creates the reasonable expectation that one should know about this illness, and hence ensures that anyone who doesn't know about it is therefore ignorant of it. Indeed, the tagline for the UK advertising campaign was 'AIDS: Don't Die of Ignorance'. The message is clear: we are making you aware of something that you ought to know, so make sure you know it, since if you don't (and hence are ignorant), then you are in danger.[10,11]

Indeed, cases of informative public expertise are often explicitly characterized in terms of making the general public aware of something of which they were—quite reasonably—hitherto unaware. Such public information campaigns are typically presented in terms of 'raising awareness' in just this fashion. Consider one of the most successful public information campaigns of recent times: the efforts to combat dangerous levels of ozone depletion in the atmosphere. The problem posed by ozone depletion wasn't discovered by atmospheric scientists until the mid-1980s, when expeditions to Antarctica identified a growing hole in the ozone layer. It was quickly determined that this problem was being caused by chemicals called chlorofluorocarbons (CFCs) that are used, for example, in many personal hygiene products. Within a few years, many advanced countries had begun public information campaigns to make the general public aware of the

---

[10] It shouldn't be thought problematic that on this conception of public expertise the public knowing more can entail that they are in a (narrow) sense more ignorant. This is, after all, a familiar aspect of how epistemic defeaters can work, whereby being aware of a defeater can undermine the epistemic standing of a belief that was previously in good order. This general phenomenon does mean that sometimes knowing less can put one in an epistemically superior position (at least with regard to one's beliefs in specific propositions anyway). For a classic discussion of this general phenomenon, see Elgin (1988). For more on defeaters in contemporary epistemology, see Grundmann (2011) and Moretti and Piazza (2018).

[11] As Dan DeNicola has pointed out to me, on the traditional conception of ignorance as lack of knowledge there is also a sense in which experts can 'manufacture' ignorance. This is because one role of experts can be to create new forms of specialized knowledge. Inevitably, the non-experts will lack this new knowledge and hence, on this view, just in virtue of creating this specialized knowledge, the experts are also creating ignorance in the non-experts. This scenario is clearly very different to the notion of manufacturing ignorance in play here, not least because this a case where ignorance is the result of the expertise itself rather than being the result of the public dissemination of expert *advice*, which is what we are envisaging. Moreover, I think it is also clear that this alternative notion of manufacturing ignorance is not plausible. Indeed, it reveals the deficiencies in the account of ignorance that underlies it. Why should the mere fact that I lack specialized expert knowledge that there is no reasonable expectation that I should be aware of entail that I am ignorant?

problem and to convince them to make changes to their use of products that would limit CFC emissions. A secondary goal of such campaigns was also to create a political climate whereby government treaties could be established to formally limit CFC emissions. On both fronts these campaigns were incredibly successful, as within a relatively short space of time the general public became aware both of the problem and its seriousness. This in turn led to changes in personal consumption and to public support for the governmental treaties that followed. The campaign to prevent ozone depletion is these days highlighted as a model example of successful global climate coordination.[12]

Such a campaign clearly satisfies the rubric of informative public expertise that I have set out. Consider first the claim about ignorance being manufactured by the public expertise. Given the obscure and technical nature of the facts in play, one could hardly reasonably expect members of the public to be aware of this climate problem. Indeed, initially it was only scientists with quite specialized expertise who would be aware of the nature of this difficulty. The general public's widespread lack of knowledge of the depletion of the ozone layer was thus not a plausible candidate at this point to be treated as ignorance. The success of this information campaign in raising public awareness of the hole in the ozone layer changed all that, however. Now this has become something that one ought to know, and hence if one did not know it one would count as ignorant of it. Ignorance is thus being manufactured by the campaign, as it is now no longer possible for members of the public to be unaware of this problem without that being due to their intellectual fault.

Recognizing this aspect of the role of informative public expertise is I think crucial to properly understanding how it functions. Public information campaigns of this kind must first raise awareness, and thereby manufacture ignorance, in order to then combat the ignorance that has been manufactured.

4. Critical Public Expertise

I've argued that a core form of public expertise is informative public expertise, where this involves both manufacturing and combating ignorance. I now want to contrast this form of public expertise with a second variety—*critical public expertise*—that is devoted to highlighting as opposed to manufacturing ignorance. As we will see, we can bring this second form of public expertise into sharper relief by setting it alongside the more familiar informative public expertise.

---

[12] See, for example, this 2019 article in *The Guardian* newspaper, available at https://www.theguardian.com/environment/2019/jan/20/how-to-stop-the-climate-crisis-six-lessons-from-the-campaign-that-saved-the-ozone.

Critical public expertise is often directed towards explicitly political ends. In terms of the framework provided by the normative account of ignorance, the aim of the expertise is to demonstrate that the public's lack of knowledge of a certain subject matter already amounts to ignorance, even despite the fact that this lack of knowledge is not generally considered to be ignorance. More precisely, the claim is that this lack of knowledge manifests intellectual fault, and hence should be treated as ignorance. Accordingly, rather than manufacturing ignorance as informative public expertise does, critical public expertise instead maintains that what was hitherto not regarded as ignorance (but merely lack of knowledge) is in fact ignorance, and so should be treated as such.

Consider, for example, cases where climate change experts charge the public with complacently ignoring the seriousness of the climate change emergency, or Black Lives Matters activists complaining that police brutality against young Black men doesn't get properly dealt with.[13] These cases are different from the AIDS or ozone depletion campaigns as it is built into the presentation of the expertise that the target audience should be aware of the facts they are highlighting. The charge is that there is intellectual fault on the part of the public in not knowing the relevant facts.

The climate change movement is instructive in this regard. Whereas it began as a form of informative public expertise of the kind that we considered previously, whereby part of the job of the public expert is to turn widespread lack of knowledge into ignorance, over time it has morphed into our second kind of critical public expertise aimed at highlighting ignorance. The point is that now the information about the climate change emergency has been widely disseminated this has become the kind of domain where everyone ought to have knowledge, and hence a lack of knowledge amounts to ignorance.

When it comes to an issue like climate change, where the experts have been telling us about the problem for decades, the variety of ignorance in play among those living in the present day is inevitably a motivated, or *wilful*, ignorance. The information is fully available to the general public to know about climate change, which entails that the refusal to believe it requires one to engage in intellectually culpable behaviours. This might mean simply ignoring uncomfortable truths. In the case of Black Lives Matters, for example, it seems very plausible that some people simply refuse to listen to the unpleasant truths about, for instance, the incarceration rates of young Black males in US jails or their high rate of police homicide. Ignoring information in this way is clearly an intellectual vice; it certainly manifests cognitive bias.

---

[13] Interestingly, the contemporary discussion of the decolonization of philosophy runs along similar lines, in that the charge is that there are important facts that are salient to our discipline that practitioners should be aware of but often are not. For some helpful recent discussions of the issues in play here, see Allais (2016); Gordon (2019); and Mitova (2020). I set this particular debate to one side here, given that our concerns are with public expertise (i.e. expertise of interest to the general public).

A more serious kind of wilful ignorance is, of course, conspiracy theories. Simply ignoring widely publicized truths is difficult, and can lead to cognitive dissonance as one is regularly exposed to these truths that one refuses to believe. In order to avoid such dissonance, it is thus preferable (if still epistemically suboptimal) to indulge in alternative explanations of the information being presented that suit one's interests. In this way, one can be led to conspiracy theories, such as the many conspiracy theories that motivate climate change denial. (Similar conspiracy theories are, of course, available that oppose the kind of critical public expertise that promotes awareness of the police and judicial ill-treatment of young Black men.)[14] Either way, we have a lack of knowledge that is intellectually culpable and hence amounts to ignorance. The previous rounds of informative public expertise to raise the profile of the target facts have thus done their work, as they have ensured that lack of knowledge in this regard now amounts to ignorance.

Notice that this way of thinking of the political dimension to ignorance is very different to how it is usually understood in the contemporary literature. Ignorance is usually understood in this literature along conventional lines as either lack of knowledge (e.g. Fricker 2016, 144) or lack of true belief (e.g. Mills 2007, 16), with a further pernicious sub-class of ignorance delineated that involves a kind of intellectual culpability (e.g. Medina 2017). But this ignores the fact that the very notion of ignorance already includes an element of intellectual culpability. Ignorance is not simply the intellectually benign absence of knowledge. On the alternative way of thinking about political ignorance advocated here, we are able to capture the sense in which appeals to ignorance of this kind are meant to be exposing how a lack of knowledge of the target facts on the part of the folk amounts to ignorance due to the intellectual culpability involved. The role of critical public expertise is to accuse the public of a failure to knowledge what they should know.

5. Concluding Remarks

Once we recognize the normative dimension to ignorance we are able to gain a better understanding of two of the fundamental ways that public expertise functions. In its most basic form, public expertise performs an informative role. As we have seen, this does not merely involve combating ignorance, but also has a crucial component of raising public awareness of the target facts so that their lack of knowledge now amounts to ignorance. This is the sense in which I have argued that public expertise can have a role to play in manufacturing ignorance. With

---

[14] There is now a wealth of epistemological work on conspiracy theories. For two recent monographs on this topic, see Dentith (2014) and Cassam (2019).

informative public expertise delineated, we are then in a position to capture a very different role that public expertise plays, where the goal is less to inform than to highlight the public culpability in failing to know the salient facts. This is critical public expertise, a form of expertise which, as we have seen, typically performs a political function. The distinction between these two forms of public expertise is, inevitably, not sharp. We have noted, for example, that what begins as an expression of informative public expertise can over time transform into an expression of critical public expertise, as the experts rail against the fact that elements of the public are resistant to the information presented to them, such as by formulating conspiracy theories to explain it or away or simply doing all they can to ignore it. Nonetheless, the distinction is important to understanding public expertise and its relationship to ignorance.

Acknowledgements

I am grateful to Dan DeNicola for detailed comments on an earlier version of this piece. This essay was written while a Senior Research Associate of the *African Centre for Epistemology and Philosophy of Science* at the University of Johannesburg.

References

Allais, L. (2016) "Problematising Western philosophy as one part of Africanising the curriculum." *South African Journal of Philosophy* 35, 537–45.

Cassam, Q. (2019) *Conspiracy Theories*. London: Polity.

Coady, D. (2012) *What to Believe Now: Applying Epistemology to Contemporary Issues*. Oxford: Wiley-Blackwell.

Croce, M. (2019) "On what it takes to be an expert." *Philosophical Quarterly* 69, 1–21.

DeNicola, D. R. (2018) *Understanding Ignorance: The Surprising Impact of What We Don't Know*. Cambridge, MA: MIT Press.

Dentith, M. R. X. (2014) *The Philosophy of Conspiracy Theories*. London: Palgrave Macmillan.

Elgin, C. (1988) "The epistemic efficacy of stupidity." *Synthese* 74, 297–311.

El Kassar, N. (2018) "What ignorance really is: Examining the foundations of epistemology of ignorance." *Social Epistemology* 32, 300–10.

Fricker, M. (2016) "Epistemic injustice and the preservation of ignorance," in *The Epistemic Dimensions of Ignorance*, (eds.), R. Peels and M. Blaauw, 144–59. Cambridge: Cambridge University Press.

Goldberg, S. (2018) *To the Best of Our Knowledge: Social Expectations and Epistemic Normativity*, Oxford: Oxford University Press.

Goldberg, S. (2017) 'Should Have Known', *Synthese* 194, 2863–94.

Goldberg, S. (2011) "If that were true I would have heard about it by now," in *Social Epistemology: Essential Readings*, (eds.), A. Goldman and D. Whitcomb, 92–108. Oxford: Oxford University Press.

Goldman, A. (2001) "Experts: Which ones should you trust?." *Philosophy and Phenomenological Research* 63, 85–110.

Goldman, A. (2018) "Expertise." *Topoi* 37, 3–10.

Goldman, A. and Olsson, E. (2009) "Reliabilism and the value of knowledge," in *Epistemic Value*, (eds.), A. Haddock, A. Millar, and D. H. Pritchard, 19–41, Oxford: Oxford University Press.

Gordon, L. (2019) "Decolonizing philosophy." *Southern Journal of Philosophy* 57, 16–36.

Grundmann, T. (2011) "Defeasibility theories," in *Routledge Companion to Epistemology*, (eds.), S. Bernecker and D. H. Pritchard, 156–256. London: Routledge.

Le Morvan, P. (2011a) "Knowledge, ignorance and true belief." *Theoria* 77, 32–41.

Le Morvan, P. (2011b) "On ignorance: A reply to Peels." *Philosophia* 39, 335–44.

Le Morvan, P. (2012) "On ignorance: A vindication of the standard view." *Philosophia* 40, 379–93.

Le Morvan, P. (2013) "Why the standard conception of ignorance prevails." *Philosophia* 41, 239–56.

Le Morvan, P. and Peels, R. (2016) "The nature of ignorance: Two views," in *The Epistemic Dimensions of Ignorance*, (eds.), R. Peels and M. Blaauw, 12–32. Cambridge: Cambridge University Press.

Medina, J. (2017) "Epistemic injustice and epistemologies of ignorance," in *Routledge Companion to the Philosophy of Race*, (eds.), P. C. Taylor, L. M. Alcoff, and L. Anderson, 247–60. London: Routledge.

Mills, C. (2007) "White ignorance," in *Race and Epistemologies of Ignorance*, (eds.), S. Sullivan and N. Tuana, 13–38. Albany, NY: SUNY Press.

Mitova, V. (2020) "Decolonizing knowledge here and now." *Philosophical Papers* 49, 191–212.

Moretti, L. and Piazza, T. (2018) "Defeaters in current epistemology." *Synthese* 195, 2845–54.

Nottelmann, N. (2015) "Ignorance," in *Cambridge Dictionary of Philosophy*, 3rd edn, (ed.), R. Audi, 497–98. Cambridge: Cambridge University Press.

Peels, R. (2010) "What is ignorance?." *Philosophia* 38, 57–67.

Peels, R. (2011) "Ignorance is lack of true belief: A rejoinder to le Morvan." *Philosophia* 39, 344–55.

Peels, R. (2012) "The new view on ignorance undefeated." *Philosophia* 40, 741–50.

Peels, R. (2023) *Ignorance: A Philosophical Study*. Oxford: Oxford University Press.

Pritchard, D. H. (2021a) "Ignorance and inquiry." *American Philosophical Quarterly* 58, 111–23.

Pritchard, D. H. (2021b) "Ignorance and normativity." *Philosophical Topics* 49, 225–43.

Pritchard, D. H. (2021c) "Omniscience and ignorance." *Veritas* 66, 1–11.

Pritchard, D. H. (2022a) "Extended ignorance," in *Embodied, Extended, Ignorant Minds: New Studies on the Nature of Not-Knowing*, (eds.), S. Arfini and L. Magnani, ch. 4. Dordrecht, Holland: Springer.

Pritchard, D. H. (2022b) "Intellectual Virtue and Its Role in Epistemology." *Asian Journal of Philosophy* 1. Available at DOI: 10.1007/s44204-022-00024-4.

Pritchard, D. H. (forthcoming) "Hypocritical expertise," in *Overcoming the Myth of Neutrality: Expertise for a New World*, (eds.), M. Farina and A. Lavazza. London: Routledge.

van Woudenberg, R. (2009) "Ignorance and force: Two excusing conditions for false beliefs." *American Philosophical Quarterly* 46, 373–86.

Watson, J. C. (2020) *Expertise: A Philosophical Introduction*, London: Bloomsbury.

Zimmerman, M. J. (2008) *Living with Uncertainty: The Moral Significance of Ignorance*. Cambridge: Cambridge University Press.

Index

For the benefit of digital users, indexed terms that span two pages (e.g., 52–53) may, on occasion, appear on only one of those pages.

Aagaard, Jesper 87
acquired immune deficiency syndrome (AIDS) 278–9
aesthetic
  autonomy 215
  beliefs 242
  culture 223–4
  hedonists 218–21, 224, 228–9
  knowledge 216–17, 242
  normativity 219–21, 226
  reasons 214–15, 220, 224, 227
  value practices, experts in 213–29
  values 7, 213–29
Affordable Care Act 190–2
Africa, Southern 7, 253–4
agential stance, developing 77–8
alchemy 21, 26
Alfano, Mark 201–2
Allen Institute 236–7
American Psychological Association 68
Annas, Julia 241
Aquinas, Thomas 200
Archard, David 236–8
Aristotle 180–1, 200, 203, 240
arrogance 38, 41–2, 60, 173–4, 181, 185, 277
artificial intelligence (AI)
  "Ask Delphi" 236–8, 242
  moral enhancement 233, 235–41
  prescriptive 236–8, 242, 244–5
  Socratic AI 7, 232–46
astrology 25–7
authority
  epistemic 33–4, 37–8, 44–5, 47–9, 52, 54–5, 59–60, 162, 171–2, 189–204, 253–6, 270–1
  experts, of 36
  intellectual 171–2
  legitimacy, and the expert-layman problem 31–43
  moral 197–8
  political 31–4
  pseudo-authority 36–7
  trust, as indicator of 36
  aesthetic 215

challenge 215–18
problem, Archard's 236–8
putative 216
rational 241, 244
thought, of 37–8

Baehr, Jason 181, 185–6
Barrington Declaration 177
Battaly, Heather 181
Bayesians 113–15, 117, 120–1
Behe, Michael 22
beliefs
  aesthetic 242
  bad 42–3
  categorical 121–2
  false 19–20, 108–9, 139, 194, 198, 258–9
  fringe 4–5, 13–28
  scientific 108
  true 19–20, 171–2, 194, 198, 258–9, 274–5, 282
Bell Curve, The 179–80
benevolence 6–7, 200
Bhattacharya, Dr Jay 177
bias 4–5, 20, 22–3, 35, 38, 60, 72–5, 79, 173, 183, 195, 242, 281
bioethics 7, 232–4, 241–2
Birch, Jonathan 88–9
Black Lives Matters 281
Bledsoe, T.A. 192–3
Boyer, Dominic 137–8
Brandolini's Law 26
Brennan, Johnny 24
Brexit 53, 55–6
bullshit, problem of 15, 24–8

Cabrera, Frank 28
Chalmers, David 99
Chang, Hasok 135
Charmides, The; *see also* Plato.
Christen, Markus 201–2
Clark, Andy 99
climate change 3, 27, 71, 113, 147–8, 160–4, 232–3, 274–5, 278, 281–2
  crisis 278

climate change (*cont.*)
  denial 36, 282
  experts 281
  Intergovernmental Panel on Climate Change (IPCC) 36, 147, 161-2
  movement 281
  science 71, 147, 161-2
Coady, David 258-9
cognition
  collaborative 87-8, 90-3
  distributed 87-8
cognitive
  artifacts 99-100
  behavioural therapy (CBT) 242-3
  bias 60, 281
  dissonance 109, 282
  labour, division of 3, 33-4
  psychology 91-2
  skill, ecologies of 93-9
collaborative
  cognition 87-8, 90-3
  inhibition 90
  performance 86, 89-90, 93
  processes 92-3
  recall 91-2
  skills, embodied 5, 85-100
Collins, Francis 177-8
Collins, Harry 138-40
common sense 38-43, 52-3, 216, 232-3, 263
communication, microprocesses of 92-3
communitarian functionalism 7, 253, 262-71
communities
  cultural 131-2
  epistemic 7, 19, 45-51, 57, 60, 130-5, 139-40, 144, 197-8, 253, 263-8
  geopolitical 131-2
  linguistic 131-2
  marginalized 254
  open-loop 46
  religious 131-2
"competence condition" 256
complementarity, mechanisms of 99-100
compliance 51-2, 57, 59, 225-7
consent 31-4
conspiracy theories 54, 143, 282-3
Conway, Erik 161-2, 172-4
COVID-19 pandemic 3, 32-3, 41-2, 120-1, 147-8, 160-3, 171-3, 175, 177-80, 189-92, 196
  lockdowns 177, 189-90
creationists 18-19, 21-3, 25-6
credences 107-24
critical inquiry 127, 139
Croce, Michel 241, 262

CUDOS norms (Communism, Universality, Disinterestedness, and Organized Scepticism) 153
cultural rights 126-7, 134, 136, 144
Cumes, David 255
Cutting, James 223

Daley, Barbara 17
Darwin Day in America (book) 23
De Cruz, Helen 24
decision-making
  group 66
  moral 234-5
  political 6, 40
  science-based 35
  technical 138, 153
  technological 154-5
Deckers, Jan 237-8
"Defense of Common Sense" (article) 39-40; *see also* Moore, G.E.
Dellsén, Finnur 241
demarcation problem 4-5, 139-40
Dembski, Bill 22
democratic
  checks and balances 158-60
  epistemology, egalitarianism of 52-3
  equalitarianism and expertise 5
  equality, grounded in 33
  knowledge, access to 37-8
  populism 44-5
  representation 193
  science, and technocracy 40-2
  science, role of 147-64
  societies 31-4, 127, 147, 158, 163-4, 178-9, 184-5
Diagnostic and Statistical Manual of Mental Disorders 72
dialogue
  constructive 80-1
  integrating perspectives in 79-81
  Open Dialogue model 68-9
  Platonic 14
  Socratic 7, 14, 244-5
Discovery Institute 22-3
disinterestedness 40, 153-4, 159
disrespect 174, 178, 184-5
distrust 35, 40, 44-5, 53-5, 58, 172-4, 192-3
Douglas, Heather 112-13
Dreyfus, Hubert 240
Driver, Julia 239

education
  higher 48-9, 52-3, 199-200
  professional 50-1, 56, 58-60
  public 33-4, 59-60

elite
  performers 94
  team sports 95
Elliott, Kevin 123
epistemic
  authority 33–4, 37–8, 44–5, 47–9, 52, 54–5, 59–60, 162, 171–2, 189–204, 253–6, 270–1
  carelessness 267
  communities 7, 19, 45–51, 57, 60, 130–5, 139–40, 144, 197–8, 253, 263–8
  competencies 263–4, 267
  crisis 45
  decolonization 253, 255–6, 263, 270–1
  environment, polluted 56
  expertise, asymmetry of 58–60
  gap 58–9
  humility 79–80, 201–2
  injustice 49, 60, 142–3, 149–50, 254, 271
  jurisdictions 138, 140
  malpractice 25–6
  marginalization 253
  populism 37–43
  relationships, asymmetric 60
  responsibility 58, 60, 184, 196, 198–9
  trespassing 55, 60
  trust 33–4, 171–2, 178
  value 36, 228–9
  vices 25, 60
  virtues 45, 60, 194
epistemology
  bifurcated 56
  classical 19
  democratic, egalitarianism of 52–3
  discipline of 45
  expertise, of 4–5, 253
  intersectional 75
equality
  democracy, and expertise 33–4
  problem of 31
Ericcson, Anders 16–18
Esprit de Corps 96; *see also* Protevi, John.
ethics
  bioethics 7, 232–4, 241–2
  medical 71
ethical
  expertise 4, 191–2, 197
  issues 191, 198, 232
  meta-ethical observations 107–8
Evans, Robert 138–40
evolutionary theory 14, 21
experience
  evidence, as 74–6
  experience-based expertise 65–7, 137, 159, 163
  lived 65–7, 71–8
  narratives, production of 75
expertise
  abandonment of 56–7
  anti-expertise sentiments 5
  certified and uncertified 137–8, 140
  clarifying notion of 66–7
  classification of 150–1
  collective achievement, as 6
  concept and definition of 67–70, 263
  confidence in 53–4
  consensus 35
  contemporary debate, in 3–4
  contributory 138–40, 150–2, 159, 161–2, 193
  critical public 280–2
  democracy and, tension between 40, 44–5
  direct 77
  epistemic asymmetry of 58–60
  epistemology of 4–5, 253
  equalitarianism and 5
  ethical 4, 191–2, 197
  experience-based 65–7, 71, 76–81, 137, 159, 163
  expertise-in-action 189–204
  five-point guide to evaluate 19–24
  group 5, 89, 100
  institutionalized 44–60
  interactional 150
  inter-perspectival 139–43
  joint 5, 85–100
  judgement without values 107–24
  lay 65, 137–40
  literature on 137–8, 240
  lived experience, as 76
  meta 152
  moral 7, 197, 232–46
  performative 45–6, 85–7, 95, 239–40
  Periodic Table of 149, 151, 160
  perspectival 65–81
  philosophical debate about 4
  populist suspicion regarding 5
  "principle unification" metric of 244–5
  public 3, 7, 274–83
  question of 65–7
  rejection of 52–6
  Right to Expertise (RtE) 143–4
  scientific research on 15
  "scientization of" 140
  specialist 149–52, 156–7, 164
  status, attribution of 150
  Studies of Expertise and Experience (SEE) 137–8, 149, 152–60, 162–4
  systems 16

expertise (*cont.*)
  types and typology of 45–6, 149–50
  ubiquitous 151–2, 156–7, 159–60, 164
experts
  aesthetic 7, 213–29
  authorities and 171–2
  becoming 15–19
  climate-change 281
  decolonizing 253–71
  distrust in 35, 58, 172–4
  epistemic authority of 189–204
  expert-layman problem 31–43
  experts-in-action 192–203
  humility for 171–86
  lay 65
  marginalized 6–7, 254–63, 270–1
  responsibilities of 174–80
  trust and distrust in 172–4

fact-checking 27–8
fake news 28, 123
fallacies, logical 13, 26
falsehoods 46, 143–4, 185
falsifiability 127, 139
Fauci, Dr Anthony 176–9
Feyerabend, Paul 31–2, 72
fidelity to trust 6–7, 201
"flipflopping" 175
fortitude 6–7, 202–3
Frankfurt, Harry 24–7
Fricker, Miranda 254, 264
Fritts, Megan 28
functionalism, communitarian 7, 253, 262–71

Gaertig, Celia 175
Galileo 22
genetic engineering 33–4
genetic fallacy 23
Giubilini, Alberto 234–8
global warming; *see* climate change.
Goldman, Alvin 15, 19–24, 27, 194, 197, 241, 245, 258–60
Goodwin, Doris Kearns 185–6
Gordon, J.S. 239
Gove, Michael 55–6
Grasswick, Heidi 256–8
Gupta, Dr Sunetra 177
Guyer, Paul 221–3

Habermas, Jürgen 31–2, 40
Habgood-Coote, Joshua 89
Harding, Sandra 140
"harmony", dogma of 87, 98–9
Harris, Celia 93

Herrnstein, Richard 179–80
heuristics 18, 36, 42–3, 152, 201–2
historians 55–6, 185–6
homeopathy 25–6
Hopkins, Robert 216–18
Howard-Snyder, Daniel 181
human rights 6, 126, 131
Hume, David 25, 218–23, 226–8
humility, intellectual 6–7, 171–86, 201–3

ignorance
  expertise and 274–5
  manufacturing 274–5, 279–83
  negative normative valence of 275–7
  normative account of 7, 277–8
  philosophical treatments of 275
  public expertise, and 274–83
  trust, rejection, and 44–60
  wilful 54–6, 60, 281–2
Imitation Game 150
individualism 34–5, 154
influencers 54, 213
information
  deluge 37
  moral 243
  public 278–80
  social 5, 36–7, 42–3
injustice, epistemic 49, 60, 142–3, 149–50, 254, 271
institutionalized expertise 44–60
institutions, decline of public confidence in 53–4
intellectual
  apathy 60
  humility 6–7, 171–86, 201–3
  vices 277
  virtue 182–5, 189–204
intelligence, collective 91–2
intelligent design 21, 23
Intergovernmental Panel on Climate Change (IPCC) 36, 147, 161–2
International Covenant on Economic, Social and Cultural Rights (ICESCR) 126–31
Iskander, Natasha 85

Jasanoff, Sheila 31
Jeffrey, Richard 113–14, 117–20
joint action 5, 85–9, 97–9
joint expertise 5, 85–100
joint intelligence 85–6, 100
joint know-how 86–92, 100
Journal of the American Medical Association (JAMA) 22–3
judgements, higher-order 117–19

justice
  commutative and distributive 192
  contributory injustice 254
justification, direct-argumentative 20; *see also* Goldman, Alvin.

Kant, Immanuel 34–5, 216–18
Kitcher, Philip 31–2, 112, 199
knowledge
  aesthetic 216–17, 242
  experiential 39–40, 72, 75–6
  explicit 150–1
  inequality of 33–4
  local 73, 138–9, 141–4
  moral 240, 242–5
  narratives as form of 76
  objective 71, 199–200
  practical 86–7, 89
  reliable 136, 139, 143–4
  scientific 6, 33–5, 40, 50–1, 73, 127, 130–9, 141, 143–4, 147, 149
  secret 255–6, 260–1
  self-knowledge 242
  situated 131–2, 140–2
  specialized 44–5, 47, 49–50, 171, 198
  systems of 136, 138–9
  tacit 149–52
  technical 70–1, 199
  traditional 127
  wisdom, practical 6–7, 59, 184–5, 203
Kuhn, Thomas 72, 132
Kulldorff, Dr Martin 177

Landemore, Helene 193
Lara, Francisco 237–8
Large Language Model 237–8
legitimacy
  authority of scientific expertise, and 35
  epistemic authority, and 31–3
  expertise by experience, of 66
  problem of 31
  scientific method, of 32–3
Lehigh University 22
Leinhardt, Gaea 17
Levinson, Jerrold 218–19
lived experience 65–7, 71–8
lockdowns; *see* COVID-19 pandemic.
logical fallacies 13, 26

MacIntyre, Alasdair 200, 203
majoritarian principle 193
malpractice 25–6, 48–9, 55–6
Marcuse, Herbert 31–2

marginalization 253–4, 262
  communities 254
  experts 6, 254–63, 270–1
  "hermeneutical" 254
  No Marginalization 262–3, 267–8
markets, noxious 28
Marshall, Barry 120
McTaggart, Lynne 185
Meade, Michelle 92–3
media
  hyperpartisan 54
  social 4, 27–8, 36, 51, 123, 139, 178
memory
  generalized 17
  transactive 91–2
mental health 70–1, 73–4, 76
  care 78
  clinical encounters 77
  crisis 77–8
  elite performance, challenges of 97
  emergency support for 78
  expertise by experience in 76–81
  healthcare services 77
  research 5, 71, 80–1
Merton, Robert 40, 153
methodology 80, 114–15, 127
Mill, John Stuart 34–5
mind-body problem 86–7
misinformation 36–7, 123, 139, 143–4, 176–7
Moberger, Victor 15, 24–7
Moore, G.E. 39–40
moral
  AI-assisted enhancement 233–41
  expertise 7, 197, 232–46
  information 243
  judgment 179–80, 197, 236–7
  virtue 189–204
Morone, Joseph 35
Moses, Robert 195–6
Mothersill, Mary 219–20
Murray, Charles 179–80

Nagel, Thomas 72
Nehamas, Alexander 220
network theory 224, 226–9
neutrality 34–5
  issue of 192
  Principle 34–5
  problem of 31
  science, of 31–2
  technical 195
New York Times 176
Newton, Isaac 21
Nguyen, Thi 217–18, 228, 261

No marginalization 267-8
"Nobel effect" 23-4
Nobel Prize 16, 23-4
"no-passing-the-buck principle" 123
normative
    aesthetic reasons 214
    ignorance 7, 277-8, 282-3
    question 215, 218-21, 223-4, 226, 228-9
nostalgia 38, 41-2
noxious markets 28

objectivity 72-4
"On Bullshit" (essay) 24-5; see also Frankfurt, Harry.
Open Dialogue model 68-9
open-loop
    communities 46
    professions 5, 53-4, 56-7, 59-60
Oreskes, Naomi 161-2, 172-4
Owens, David 122

Palermos, S. Orestis 88-9, 91-2
pandemic; see COVID-19 pandemic.
partiality 79
pathos 6-7, 202
peer reviews 14, 18-19, 35, 51
performance
    collaborative 86, 89-90, 93
    ecology 94-5
    effective 94
    expert 45-6, 85, 87, 95, 239-40
    group 89-91, 93, 95-7, 99-100
    practice, in 224-9
    skilful 135-41, 143-4
    team 96-7
Periodic Table of Expertises 149, 151, 160
Persson, Ingmar 232-4
Pew Research Center 53
phenomenology 75
phronesis 203
Plato 17, 44-5, 47, 59-60
    Charmides, The 14
    Republic 176
    Socratic dialogues 14
pleasure, society, and the self 221-4
political
    decision-making 6
    dimension to ignorance 282
    ignorance 282
    polarization 44-5, 55, 147, 172-3
    pollsters 55-6
    scientific claims, implications of 112
    theorists 55-6

Popper, Karl 127
populism and populists
    democracy 44-5
    epistemic 37-43
    expertise, suspicion regarding 5
    leaders 158
    technological 156
practitioners, mercenary 55-6
prejudice 119-20, 222, 254
probabilities
    objective 117-20
    subjective 113, 117-18
problem
    equality, of 33-4
    expertise on technical and scientific matters, of 14
    Extension of Expertise (PoEE), of 136-9
    legitimacy, of 152-3
    priors, of 119-21
    pseudoscience, of 13-28
    reflexivity, of 79
professional
    associations 51
    competition 50
    disagreements 56
    entrepreneur, as 51-2
    malpractice and fraud 55
    networks 51
professionalization, evolving 49-52
professions
    definition and description of 47-9
    institutionalized forms of expertise, as 49
    open-loop 5, 53-4, 56-7, 59-60
Protevi, John 96
pseudo-authority 36-7
pseudophilosophy 25-7
pseudoscience 26-8, 127, 179-80
psychology
    moral 232-3
    narrative 75
    organizational 91-2, 97
    psychiatry and 73
    sport 86-7
public expertise 3, 7, 274-83
public-information campaigns 278-80

qualitative research 75-6
quality-assurance devices 48-9
Quast, Christian 241, 262

racialization 270-1
racism 27, 54, 179-80
Raoult, Didier 41-2

INDEX

Rapinoe, Megan 96
Raz, Joseph 171–2
Reagan administration 172–3
realism, perspectival 132–3, 136, 139–40, 143–4
Rehabilitation Act 190–2
"replication crisis" 115
rights
　cultural 126–7, 134, 136, 144
　human 6, 126, 131
　participatory 193
　Right to Enjoy the Benefits of Scientific Progress (REBSP) 126–8, 130–44
　Right to Expertise (RtE) 143–4
　Right to Science 6, 126–44
　inductive 109–10, 120–1
Roberts, Robert C. 181–2
Robinson, Brian 201–2
Rolin, Kristina 24
Ross, Philip 18
Rudner, Richard 109–17, 123–4

Sangomas 253–71
Savulescu, Julian 234–8
scaffolding 79–80
scepticism 31–2, 35, 40, 54–5, 120, 129, 147, 153, 159, 173–4, 176–7, 242
science
　advancement and progress 128–36
　anti-science 23
　climate-change 71, 147, 161–2
　computer 16, 50
　democracy, role in 147–64
　denialism 24
　formative aspirations of 154
　ideology of domination, as 31–2
　norms of 153
　philosophy of 4, 71–2
　Right to Science 6, 126–44
　role and status of 147
　"scientization of lay expertise" 140
　self-evaluation, self-criticism, and self-correction, capacity for 22–3
　society, in 156–60
　Studies 148–54
　technical and political phases 154–6
　technology studies (STS), and 147
　Truth and Democracy 31–2; *see also* Kitcher, Philip.
　"value freedom" 107–8
scientific
　elite 6
　evidence 108–9
　expertise, studies of 18–19

knowledge 6, 33–5, 40, 50–1, 73, 127, 130–9, 141, 143–4, 147, 149
method 32–3
perspective 132, 140
self-regulation 47–8, 51
sexism 27
Shanteau, James 18
shared mental models 90–1
Simmons, Joseph 175
"sincerity/care condition" 256–7
Singer, Peter 198
skill accounts 260–1
skills, collaborative embodied 85–100
Smith, Jacqui 16–17
social
　experts, indicators of trust in 36–7
　ignorance, feature of 278
　information 5, 36–7, 42–3
　justice 160, 192
　media 4, 27–8, 36, 51, 123, 139, 178
　networking 36, 41–2, 54–6
　"parity principle" 99
　psychologists 55–6
　science 55–6, 65, 71, 148–9
　socialization 6, 147–8, 150–3, 159, 164
societies, high-tech 3
society, fractal model of 157
sociologists 4, 55–6
sociology 4, 32–3, 55–6, 148, 152–3
Socrates 13–15, 20, 27, 197
Socratic AI 7, 232–46
Socratic dialogue 7, 14, 244–5
sophistry 20, 27
Southern Africa 7, 253–4
specialist expertise 149–52, 156–7, 164
specialization 3, 36–7, 44, 50–2, 58, 94, 99–100, 224–5
　institutionalization of expertise, and 44
　professional 50–1
Sperber, Murray 17–18
spiritualism 21
sport 3, 5, 45–6, 85–8, 90–2, 94–5, 97–9, 157
Statesman 47; *see also* Plato.
Steele, Katie 112
stereotyping 60
Stichter, Matt 66–7, 239–40, 243–4
Studies of Expertise and Experience (SEE) 137–8, 149, 152–60, 162–4
subjectivity and disagreement 71–6

Tassella, Alberto 235–6
teamwork 90, 94, 97
technocracy 39–41

technological populism 156
Tekin, Şerife 72–4
Thagard Equilibrium 193
Thorndike, Edward 18
Thunberg, Greta 161–2
Time magazine 185
Tollefsen, Deborah 88–9, 91–2
transactive memory systems 91–2
transparency 6–7, 31–4, 59–60, 191–3, 201
Trump, Donald 117, 147, 161–2, 180
trust
   epistemic 33–4, 171–2, 178
   experts, in 172–4
   fidelity to 6–7, 201
   public 163, 172–3
   trustworthiness 36, 175–6
truth-centred accounts 258–60, 268–9
Tufekci, Zeynep 176
Turner, Stephen 33–4
Twitter 36

ufology 25–6
United Nations (UN) Declaration of Human Rights (UNDHR) 6, 126, 128
United Nations Educational, Scientific and Cultural Organization (UNESCO) 126
universalism 40, 154, 220

vaccination 147–8, 160–4
value judgements 109–10

value, -s
   aesthetic 7, 213–29
   epistemic 36, 228–9
"Van Gogh fallacy" 22
vanity 181
veracity 6–7, 159, 176–7, 201–3
vices 6, 25, 60, 181–2, 185, 277
Video Assistant Referee (VAR) 3
virtue, -s 180–5, 200
   epistemic 45, 60, 194
   intellectual 182–5, 189–204
   moral 189–204
virtuosos 45–6, 264

Wakefield, Andrew 160
Wallace, Alfred Russel 21
Warren, Robin 120
Weber, Max 40
West, John 23
Whitcomb group 181–4
Wilholt, Torsten 112
Williams, Bernard 219
Winch, Peter 148
Winsberg, Eric 112
wisdom, practical 6–7, 59, 184–5, 203
Wittgenstein, Ludwig 26, 148
Wood, Jay 181–2
Woodhouse, Edward 35

YouTube 41–2